About Island Press

Island Press is the only nonprofit organization in the United States whose principal purpose is the publication of books on environmental issues and natural resource management. We provide solutions-oriented information to professionals, public officials, business and community leaders, and concerned citizens who are shaping responses to environmental problems.

In 1994, Island Press celebrated its tenth anniversary as the leading provider of timely and practical books that take a multidisciplinary approach to critical environmental concerns. Our growing list of titles reflects our commitment to bringing the best of an expanding body of literature to the environmental community throughout North America and the world.

Support for Island Press is provided by Apple Computer, Inc., The Bullitt Foundation, The Geraldine R. Dodge Foundation, The Energy Foundation, The Ford Foundation, The W. Alton Jones Foundation, The Lyndhurst Foundation, The John D. and Catherine T. MacArthur Foundation, The Andrew W. Mellon Foundation, The Joyce Mertz-Gilmore Foundation, The National Fish and Wildlife Foundation, The Pew Charitable Trusts, The Pew Global Stewardship Initiative, The Rockefeller Philanthropic Collaborative, Inc., and individual donors.

About Ecotrust / Interrain Pacific

Founded in 1991, Ecotrust is a nonprofit organization based in Portland, Oregon, and designed to foster conservation-based development in the coastal temperate rain forests of North America. Ecotrust works in Alaska, British Columbia, Washington, and Oregon in places whose global ecological importance is complemented by local commitment to the interdependent goals of environmental conservation and economic opportunity. Ecotrust aims to promote a more effective and enduring conservation strategy in the Pacific Northwest and around the world by stimulating and supporting community-level examples of conservation-based development.

Interrain Pacific is a nonprofit information access affiliate of Ecotrust based in Portland, Oregon, that promotes conservation-based development by using geographic information systems (GIS), other information technologies, and internet connectivity to enhance understanding of social and ecological patterns of change in the coastal temperate rain forest region and beyond.

The Rain Forests of Home

□ □ □ □ □ □ □ □ □ □ □ □ □

The Rain Forests of Home

Profile of a North American Bioregion

Edited by

Peter K. Schoonmaker, Bettina von Hagen, and Edward C. Wolf

Foreword by

M. Patricia Marchak and Jerry F. Franklin

Ecotrust / Interrain Pacific

ISLAND PRESS
Washington, D.C./ Covelo, California

Grateful acknowledgment is expressed for permission to publish the following previously copyrighted material: On pages 215–216, "From the Tower," by Brad Matsen, in *Reaching Home: Pacific Salmon, Pacific People,* by Natalie Fobes, Tom Jay, and Brad Matsen. Copyright © 1994. Reprinted by permission of Alaska Northwest Books. On page 308, from "Mixing Business With Pleasure Keeps This Banker in Touch with the Community," by JoAnne Booth, from *Ruralite,* Tillamook County edition, March 1995.

The rain forests of home : profile of a North American bioregion/
edited by Peter K. Schoonmaker, Bettina von Hagen, and Edward C.
Wolf : foreword by Patricia Marchak and Jerry F. Franklin.
p. cm.
Developed from a conference held at the Whistler Resort near
Vancouver, B.C. in late Aug. 1994.
Includes bibliographical references and index.
ISBN 1-55963-479-0 (cloth). — ISBN 1-55963-480-4 (pbk.)
1. Rain forest ecology—Northwest Coast of North America—
Congresses. 2. Rain forests—Northwest Coast of North America—
Congresses. I. Schoonmaker, Peter K. II. Von Hagen, Bettina.
III. Wolf, Edward C.
QH104.5.P32R35 1997
574.5′2642′09795—dc20 96-32773
CIP

Printed on recycled, acid-free paper

Manufactured in the United States of America
10 9 8 7 6 5 4 3 2 1

Contents

Foreword □ M. Patricia Marchak and Jerry F. Franklin □ ix

Preface □ Spencer B. Beebe □ xiii

Acknowledgments □ xv

Introduction □ Peter K. Schoonmaker, Bettina von Hagen, and Edward C. Wolf □ 1

1. Oceanography of the Eastern North Pacific □ David K. Salmon □ 7
2. Climate of the Coastal Temperate Rain Forest □ Kelly Redmond and George Taylor □ 25
3. The Influence of Geological Processes on Ecological Systems □ David R. Montgomery □ 43
4. Vegetation from Ridgetop to Seashore □ Paul Alaback and Jim Pojar □ 69

CONCEPTS IN ACTION **Restoring and Managing Ecosystems**

□ Bears in the Greater Kitlope Ecosystem □ Ken Margolis □ 89

□ Restoration as Practice in the Mattole □ Seth Zuckerman □ 91

□ Crossing the Land of Chalayuck □ Kurt Russo □ 94

□ Stopping Spartina's Hostile Takeover of Willapa Bay □ Wendy Sue Lebovitz and Dan'l Markham □ 96

□ Restoring Knowles Creek □ Thomas C. Dewberry □ 99

5. Terrestrial Vertebrates □ Fred L. Bunnell and Ann C. Chan-McLeod □ 103
6. Streams and Rivers: Their Physical and Biological Variability □ Robert J. Naiman and Eric C. Anderson □ 131
7. The Terrestrial/Marine Ecotone □ Charles A. Simenstad, Megan Dethier, Colin Levings, and Douglas Hay □ 149

CONCEPTS IN ACTION **Forestry and Fisheries Initiatives**

□ The Kennedy Lake Technical Working Group □ Ian Gill, Peter K. Schoonmaker, and Kim Hyatt □ 189

□ Institute for Sustainable Forestry □ Seth Zuckerman □ 192

□ Clatsop EDC's Salmon Program □ Jim Hill □ 195

□ The Pacific Forest Trust □ Constance Best □ 197

□ The Skeena Watershed Committee □ Evelyn Pinkerton □ 201

□ Ecoforestry: An Approach to Ecologically Responsible Forest Use □ Mike Barnes and Twila Jacobsen □ 204

□ Heritage Stocks: A "Good News" Strategy to Save Salmon □ Guido R. Rahr III □ 207

□ The Prince William Sound Science Center □ Nancy Bird □ 209

8. Pacific Salmon: Life Histories, Diversity, Productivity □ Willa Nehlsen and James A. Lichatowich □ 213

9. Environmental History □ Richard J. Hebda and Cathy Whitlock □ 227

10. Pre-European History □ Wayne Suttles and Kenneth Ames □ 255

11. Traditional Ecological Knowledge □ Nancy J. Turner □ 275

CONCEPTS IN ACTION **New Approaches to Learning and Decision Making**

- □ Validating Vernacular Knowledge: The Ahousaht GIS Project □ Edward Backus □ 299
- □ Local Science in Willapa Bay, Washington □ Kathleen Sayce □ 301
- □ Oregon's Watershed Health Program □ Mary Lou Soscia □ 304
- □ The Tillamook Bay National Estuary Project □ Marilyn Sigman □ 306
- □ The Coastal Studies and Technology Center □ Mike Brown and Neal Maine □ 309

12. "The Great Raincoast": The Legacy of European Settlement □ William G. Robbins □ 313

13. Economic and Demographic Transition on the Oregon Coast □ Hans D. Radtke, Shannon W. Davis, Rebecca L. Johnson, and Kreg Lindberg □ 329

CONCEPTS IN ACTION **Conservation-Based Development**

- □ WoodNet □ Todd Thomas □ 349
- □ Financing Small Business on the Olympic Peninsula □ Patty Grossman □ 352
- □ ShoreTrust Trading Group □ Nancy Hauth □ 354
- □ Rural Development Initiatives □ Jennifer Pratt □ 357

14. From Ecosystem Dynamics to Ecosystem Management □ Ken Lertzman, Tom Spies, and Fred Swanson □ 361

15. A Vision for Conservation-Based Development in the Rain Forests of Home □ Peter K. Schoonmaker, Bettina von Hagen, and Erin L. Kellogg □ 383

Contributors □ 407

Key Organizations □ 409

Index □ 411

Foreword

We do not think of the coastal temperate rain forest as virginal. Among other things, its sex is ambiguous, and anything that has endured millennia cannot be expected to be chaste. Where it has never been cut (or not for a long time), the forest has a multilayered canopy. In many spots, its trees shelter the earth from the sun so effectively that no tangle of brush grows on the forest floor; cedar groves are examples of the peace thereby achieved. In other spots, sun rays peep and shimmer between layers of fir and hemlock, and understories promise a green future. Where fires or logging have denuded the ground, bumptious broom, salal, and brash weeds of all glorious varieties jostle for a spot in the sun; eventually, alder and cottonwoods and then the softwoods will grow among the weeds and, gradually, shove them aside.

Biologists, ecologists, and foresters enumerate the many species of flora and fauna in the coastal temperate rain forest. The story they tell is one of diversity and regeneration, rather than a history of destruction, and that is appropriate if we are to move on. But the history of human impact on the forest must be understood too. The beginnings of that history, the stories of the peoples who crossed the Bering Sea land bridge, are lost in the mists of time. We know from the evidence of their descendants that the first peoples of the coast, in the course of millennia, created advanced societies that influenced but did not diminish their environment. The history of serious incursions into the coastal rain forests by North Americans of European descent begins just over a century ago as forest companies and settlers moved in. Forest products companies were a more significant agent of deforestation in this region than anywhere else in Europe or North America. By the turn of the century, industrial forestry had already become big business.

Unlike many other forested regions, farming never caused significant deforestation along the rain forest coast. In fact, many loggers in the early days were dismayed to discover that having cut the trees they could not sell the land to farmers: the soil holds trees but is not much good for cabbages. So the coastal forests of western North America were used and abused more for industrial forestry than for farming or urban development. In the decades after World War II, with the full flowering of a mass production and mass consumption society in North America and in Europe, market demand for construction lumber, pulp, and newsprint increased at a quickening pace. The northwestern United States and British Columbia provided a large portion of these products.

Mass production of forest products requires massive production capacities.

Over the past fifty years huge, integrated industrial operations—lumber mills with multiple lines; huge pulpmills; newsprint and other paper mills—have been established throughout the forest regions of northern California, Oregon, Washington, and British Columbia. From time to time these are modernized or new facilities are built, each time eliminating a tier of unskilled jobs no longer needed in the new operation. Modern mills are automated; "operators" rather than old-fashioned workers scan the printouts, supervise the sorting, test the output. The mills produce ever more, but they employ fewer workers, and they have done so steadily since the mid-1960s.

How are communities to cope with such patterns of change? Some small communities have the internal and environmental resources to create new rural sources of income. Other towns are adapting to the new higher-value but lower-labor input sectors of the industry. Not all towns are likely to become tourist attractions or centers for higher-value wood industries. Some might tap urban skills and become service-industry locations for geographically "footloose" industries. Some towns, however, will simply cease to exist, except for a shrinking contingent of elderly residents unwilling to relocate. Those who seek solutions to the dislocation of coastal communities once based on industrial forestry must accept the possibility that not all communities can, or perhaps even should, be saved.

Paradoxically some new communities may be created as well. Some will be science-based research towns centered on biological or marine stations or perhaps recreation and retirement centers. But there will also be new forestry towns based on plantations. Those who care about the temperate rain forests know that plantations of hybrid cottonwood or acres of genetically uniform Douglas-fir do not a forest make. Like them or not, we may be obliged to recognize that such unnatural forests (if not grown on good forestland and not displacing natural forests or their regeneration) may become significant sources of employment and regional income.

Perhaps the greatest challenges of all lie in the realm of the social contract. Old understandings seem to be null and void: who, today, would champion the traditional role of a professional "priesthood" entrusted with resource management? Who would turn to science and technology as the ultimate sources of wisdom? The future of the coastal temperate rain forest requires professionals, citizens, and communities to become more deeply involved in management of natural resources, and we have not figured out how to make that happen without rancor. The interdisciplinary approach exemplified by this book, in which natural and social science perspectives are brought to bear on common challenges, offers a place to begin.

How are local communities and individuals to be fully engaged in this process? How can their legitimate claims to stewardship be reconciled with larger national interests and with the global context within which we all operate? Nothing has yet annulled the adage that "knowledge is power." Today, *shared* knowledge is essential. Technologies and data can easily be shared, but educational programs that put them in citizens' hands—and call on the local knowledge that residents possess—are also essential.

The social contract can be renegotiated in promising ways. The Scientific Panel for Sustainable Forest Practices in Clayoquot Sound, for example, successfully integrated the knowledge of four First Nations leaders with the understanding of an interdisciplinary group of resource scientists. One reason for that success was the full sharing of knowledge—and full acknowledgment of its limitations—by all participants.

Local efforts to develop new approaches to social consensus and decision making have arisen spontaneously, among them the Applegate Partnership in southern Oregon, the Quincy Library Group in northern California, and the Willapa Alliance in coastal southwestern Washington. At their best, formal government efforts can take note of these initiatives and seek to create new arenas and opportunities for similar social innovation.

Our current search for solutions still takes the affluent society as its context. We inherited a model of society in which all is subordinate to economic growth. Our culture, like our paper and wood, is mass produced; our children learn that individualism is the greatest good, that personal achievement and private accumulation are the highest goals toward which a human being can strive. These things will change in time simply because we cannot continue to consume: we will either learn to work together to support both society and ecosystem, or we will die out. The economic model that has guided western society for several centuries offers no lasting basis for sustainability.

In coming to grips with the social and environmental dilemmas on this rain forest coast, and in the boreal, tropical, and subtropical forests as well, we need to think in longer time spans. The communities of the coastal temperate rain forest have much in their favor; it may be here that we first achieve what community activist Lynn Jungwirth of the Watershed Institute in Trinity County, California, has called "affluent subsistence." The world would do well to watch closely the stirrings on this rain forest coast.

M. Patricia Marchak
Vancouver, British Columbia
Jerry F. Franklin
Seattle, Washington

Preface

Healthy human communities and healthy environments are fundamentally interdependent. Productive and diverse ecosystems maintain options for people. Environmental and economic deterioration, in contrast, reinforce one another as they erode options in a downward spiral.

Ecotrust was founded on the belief that ecological restoration and community revival are mutually reinforcing as well. This hopeful idea is born not of untempered optimism but of experiences which have taught us that the release of human creativity and opportunity lies at the heart of sustainability. We call the approach founded on this idea "conservation-based development." It is the search for an enduring conservation strategy, one that systematically integrates the interests of both people and place. It recognizes that functioning life support systems are the foundation of human development.

We work in the coastal temperate rain forests of North America because rain forests are important in themselves and because this is where we live. We also begin here because rain forests have become a metaphor of our time, a reflection of the current course of civilization. If it is important to save the rain forests of the tropics, why not in the United States and Canada as well?

To navigate the path of conservation-based development in the rain forests of home, we use what we call the "New Bearings" compass with which Ecotrust organizes its strategies and resources. True north is understanding—not research or data per se, but a synthesis of science and vernacular knowledge, that is, the wisdom of elders and the knowledge of people who live in a place. We are attempting to map large-scale, long-term patterns of cultural, social, economic, and ecological change through time and space and to explore the implications of these patterns for communities and for the region as a whole. Our aim is to help residents of the coastal temperate rain forest harness the forces of change to their advantage.

To this end, in August 1994 we helped to assemble scientists and thinkers from a wide array of disciplines and experience in the first conference on the temperate rain forests of North America. This book, one result of that meeting, for the first time offers an integrated description of the characteristics, history, culture, economy, and ecology of this well-defined region—the boundaries of home. We learned from this meeting that there is indeed a place where coastal forest and Native cultures developed together, each shaped by the dynamics of change and adaptation over a relatively short period of time and tied by a distinctive set of physical and biological conditions that distinguish it from other regions in the world. We call this a "bioregion." It is no accident that the extraordinary richness of the coastal interface between mountains and sea sustained a uniquely prosperous and cultured First Nations population, and today continues to be the best place in the world to produce an extraordinary array of fish, shellfish, forest, and farm products.

Many before us have described the devastating effects of industrial colonization and development on the distinctive cultures, magnificent forests, and abundant fish and wildlife of this bioregion. Ecotrust prefers to explore the comparative advantages created by a unique combination of people and place. Does this bioregion, like every other in the world, have distinctive opportunities for the people who call it home? Are there ways in which to improve farming, fishing, and forestry to produce healthy, high-quality goods and services in an environmentally restorative way while respecting the rights of the people who have known the coast as home for millennia? What fundamental synergies tie the environmental quality of rural communities of the Pacific Coast to the growing urban centers of San Francisco, Portland, Seattle, Vancouver, and Juneau? Only by unleashing the creativity and resourcefulness of the people who live here can we begin to find out, and together navigate a better course into the future.

Spencer B. Beebe

Acknowledgments

We are deeply grateful to the elected and hereditary leaders of the Haisla Nation who shared with us and others their home, the Kitlope Valley, which provided the initial inspiration for a conference that would raise the profile of the coastal temperate rain forest bioregion. We are also indebted to the Squamish Nation, particularly the Wolf Clan of the Cheakamus Valley, and to Chief and Councilor Randall Lewis for welcoming us to the traditional territory of the Squamish people where we hosted the conference that launched this book. Our cosponsors—the British Columbia Ministries of Aboriginal Affairs and Forests, the Center for Streamside Studies at the University of Washington, the Department of Forest Science at Oregon State University, the Environmental Studies Program of the University of Victoria, the School of Resource and Environmental Management at Simon Fraser University, and the Pacific Northwest Research Station of the USDA Forest Service—joined Ecotrust in making the conference possible. Interrain Pacific (formerly Pacific GIS), particularly its founder Edward Backus, cofounder Randall Hagenstein, and Andy Mitchell (now with ESRI), created the GIS maps of the coastal temperate rain forest that provided the focus for the conference and subsequent work. We are especially grateful to the Forest Service, for its financial support of the conference, and to the staff of the Whistler Conference Center for its gracious and efficient assistance.

Space does not permit us to thank everyone whose assistance, work, or thought made this book possible. We do wish to thank all the contributors, though, for taking time from other commitments to write and then revise their chapters and profiles. The authors of each chapter have acknowledged those whose substantive reviews sharpened their final work.

Neither the conference nor this book could have been completed without our Ecotrust colleagues Elizabeth Coleman and Nancy Hauth; their contributions to the project were substantial, and this book is a testament to their hard work and commitment. Marko Muellner of Interrain Pacific devoted painstaking hours to the preparation of a number of illustrations and color plates. Bob Moore, who returned to Oregon after more than two decades in New Zealand's temperate rain forests, gave us invaluable editorial assistance just when we needed it. We also wish to thank our other Ecotrust colleagues and our Board of Directors, chaired by Jack Hood Vaughn, for providing the collegial setting and organizational support for a project of this magnitude.

Barbara Dean of Island Press encouraged us at every turn since we first proposed this book to her in 1994. Barbara Youngblood and Bill LaDue, also of Island Press, steered us calmly through the many shoals of book production.

We have made every effort to assemble a complete and accurate portrait of the coastal temperate rain forests of North America—the rain forests of home—and hope it is a work that will prove useful for many years to the people of this extraordinary bioregion. We are indebted to all who have lent inspiration and substance to this work; we alone are responsible for any errors that remain.

Peter K. Schoonmaker
Bettina von Hagen
Edward C. Wolf

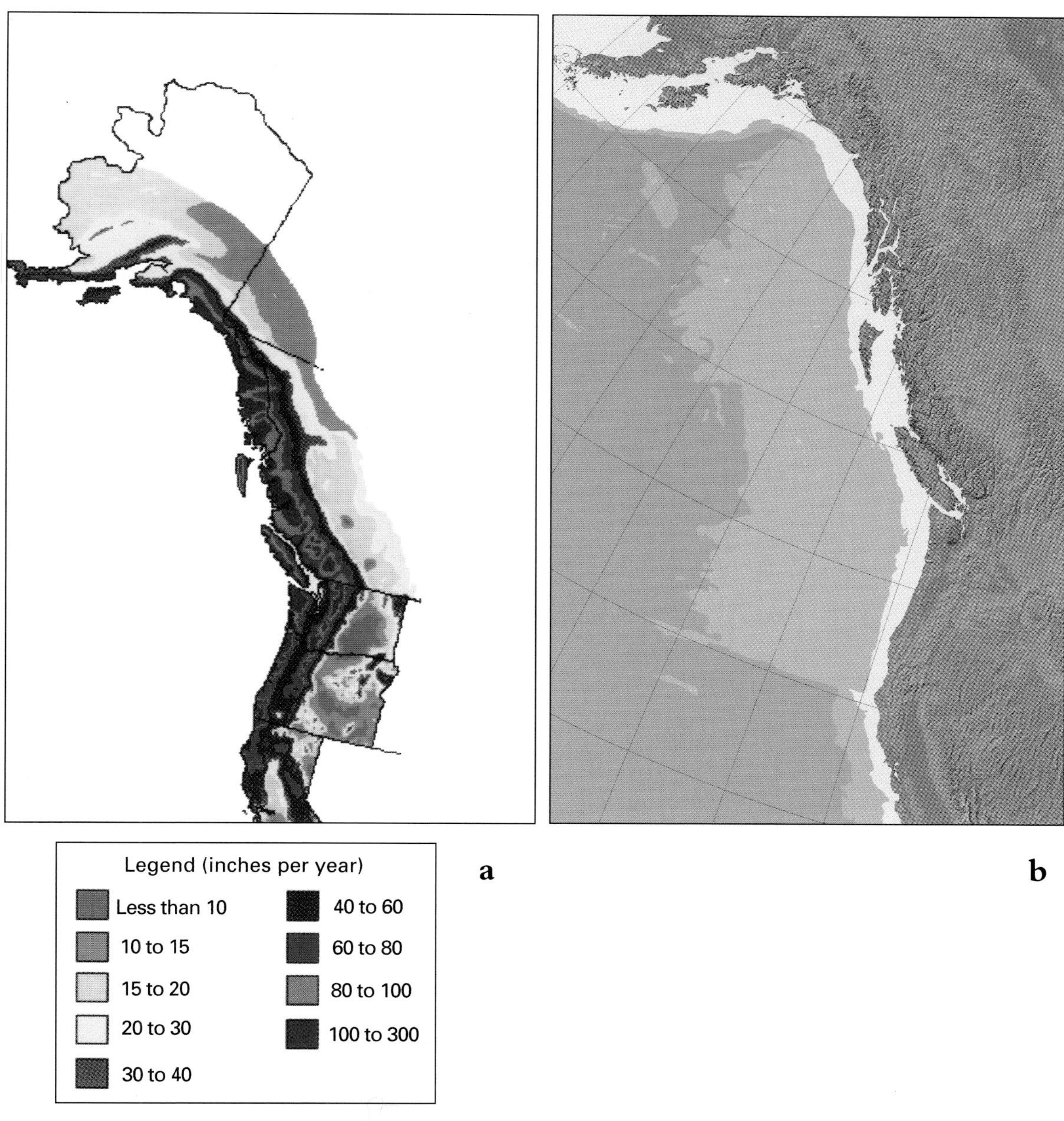

Plate 1. (*a*) Average annual precipitation (1961–1990) and (*b*) physiography of North America's coastal temperate rain forest bioregion. Coastal mountains comb 2 meters or more of precipitation each year from oceanic air masses; moderate maritime temperatures ensure that most of this falls as rain. *Sources:* (*a*) U.S. data are from Taylor and Daly (1995); Canadian data are from Bryson and Hare (1974). Greater detail is presented for California, Oregon, and Washington. See Redmond and Taylor in this volume. (*b*) Ecotrust, Pacific GIS, and Conservation International (1995).

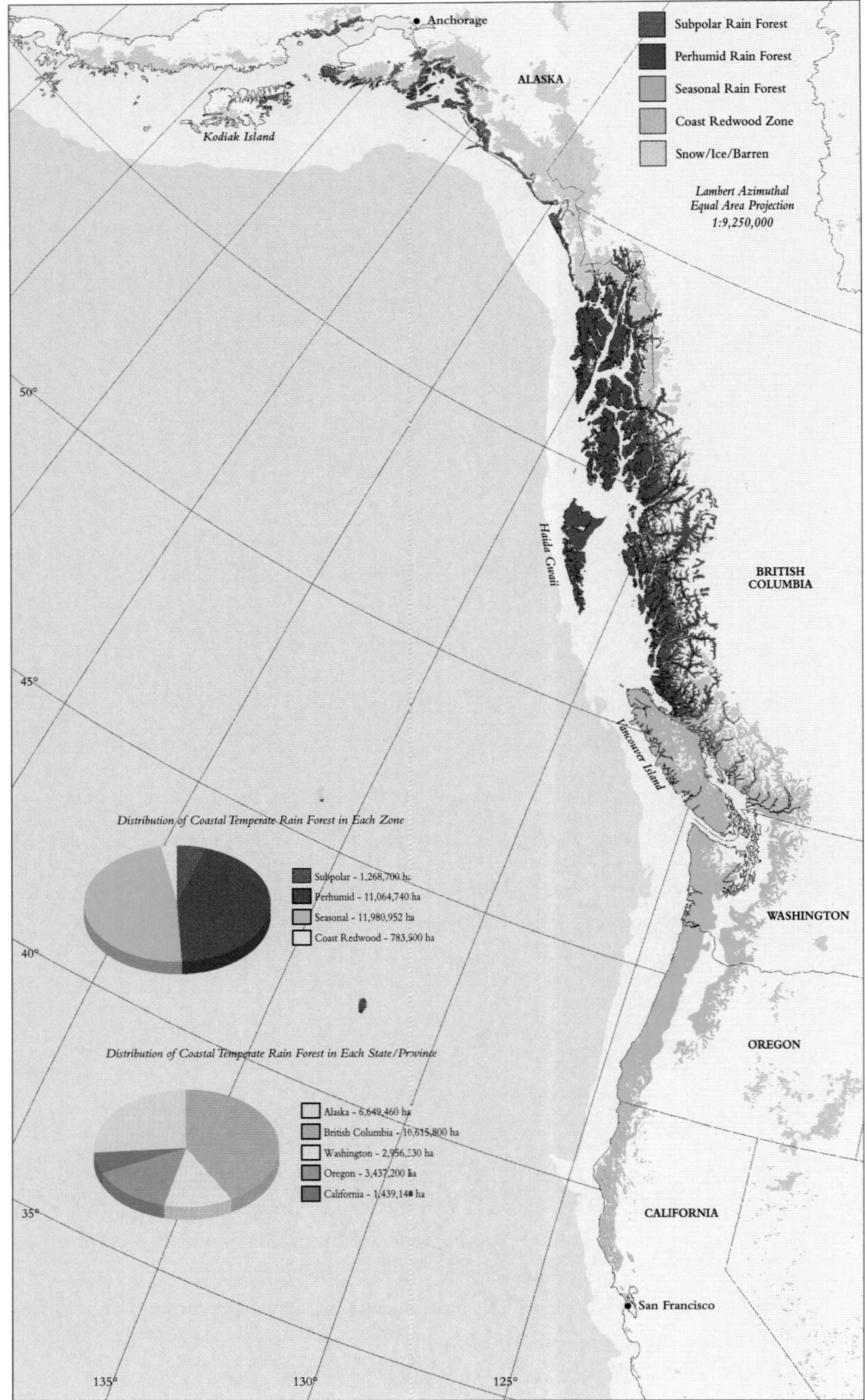

Plate 2. Original distribution of the North American coastal temperate rain forest. Covering some 25 million hectares, coastal forests between Alaska and northern California once occupied slightly more land than the combined total of all other areas of coastal temperate rain forest worldwide. *Source:* Ecotrust, Pacific GIS, and Conservation International (1995).

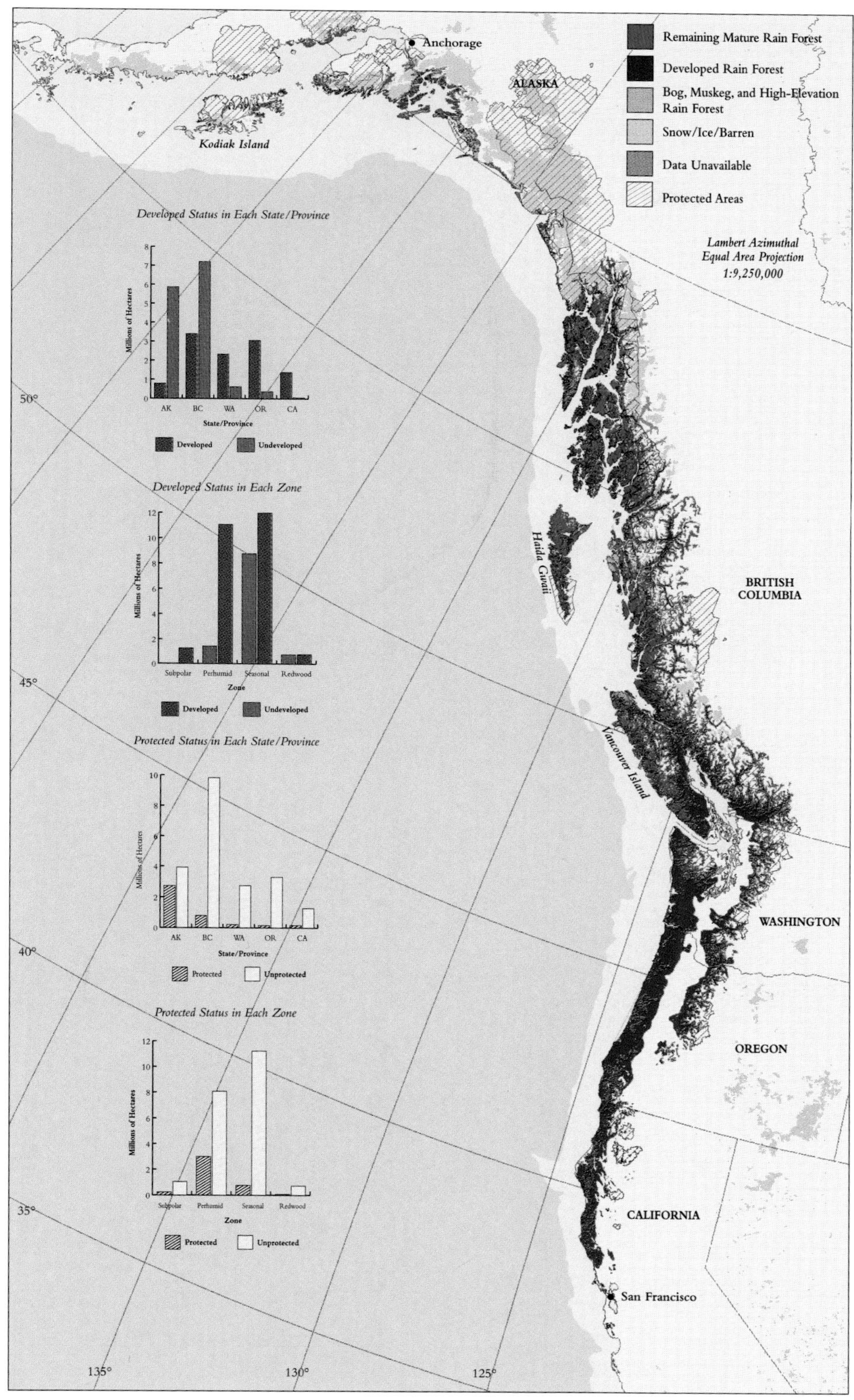

Plate 3. Current status of the North American coastal temperate rain forest. Overall 44 percent of this coastal rain forest has been altered by logging, farming, or urban development; the main impacts of "development" are visible from Vancouver Island south. *Source:* Ecotrust, Pacific GIS, and Conservation International (1995).

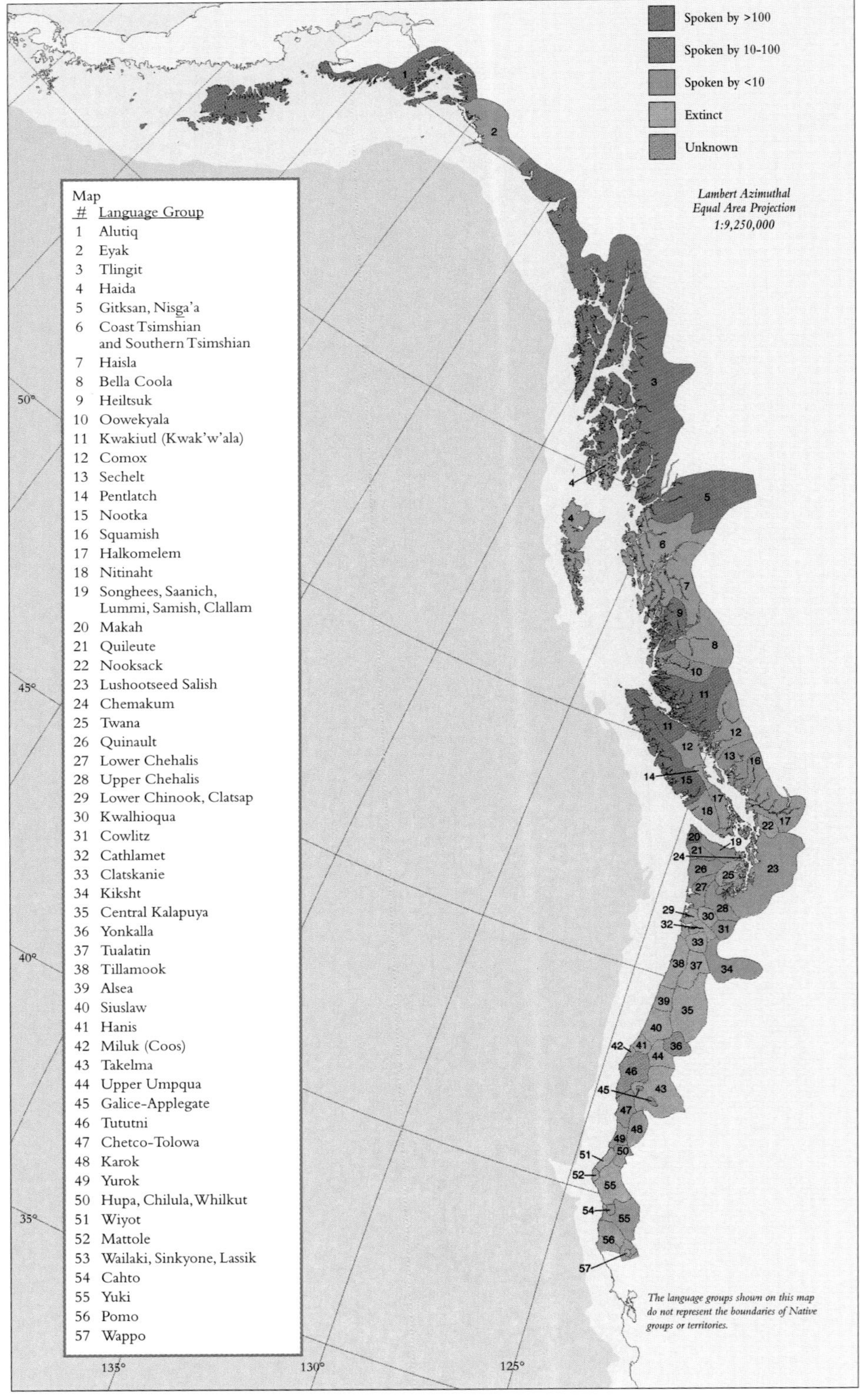

Plate 4. Current status of First Nations language groups of the coastal temperate rain forest. More than sixty distinct languages were once spoken along the rain forest coast. Today, two-thirds of these are believed to be extinct or are spoken by fewer than ten individuals. *Source:* Ecotrust, Pacific GIS, and Conservation International (1995).

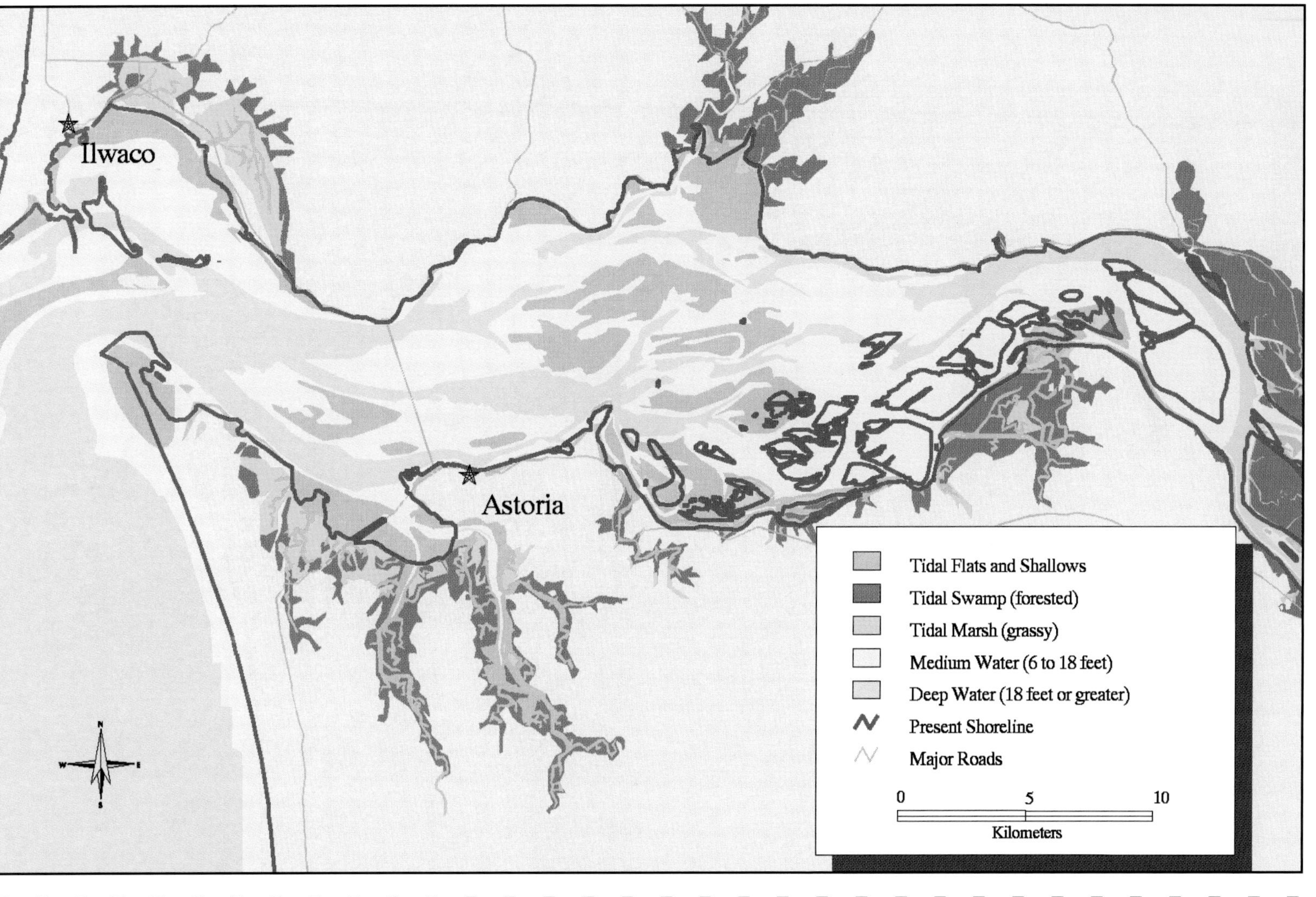

Plate 5. Habitat change along the Columbia River estuary: Map depicts the condition of the estuary and shoreline position in 1868–1875; present shoreline is shown in red. *Source:* Lower Columbia River Bi-State Water Quality program (1995).

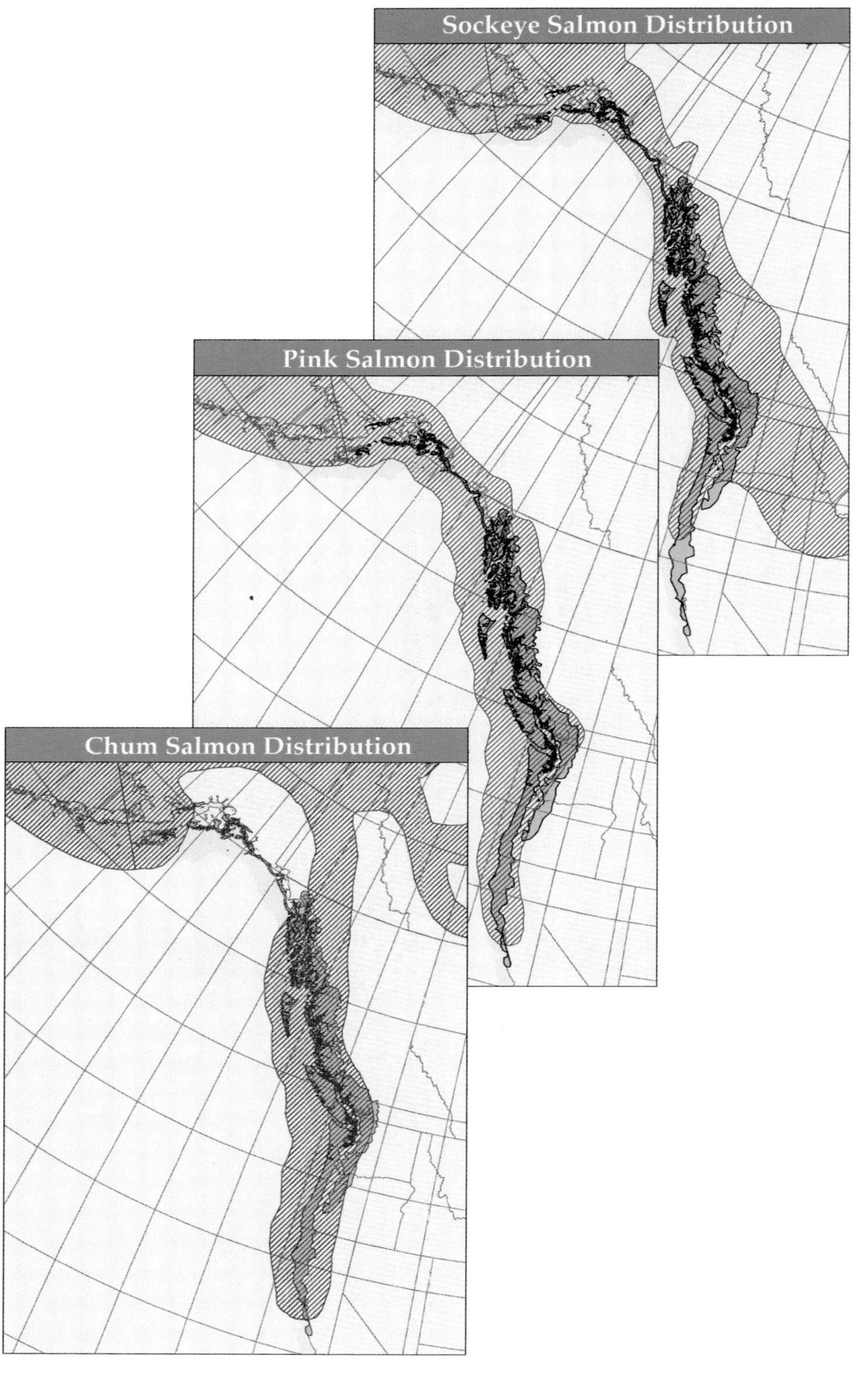

Plate 6. Distribution of anadromous salmonids of the coastal temperate rain forest: chum, pink, and sockeye salmon. Nearly all of the streams, rivers, and estuaries in the bioregion are (or were) part of the salmon's extended ecosystem. *Source:* Oregon State University.

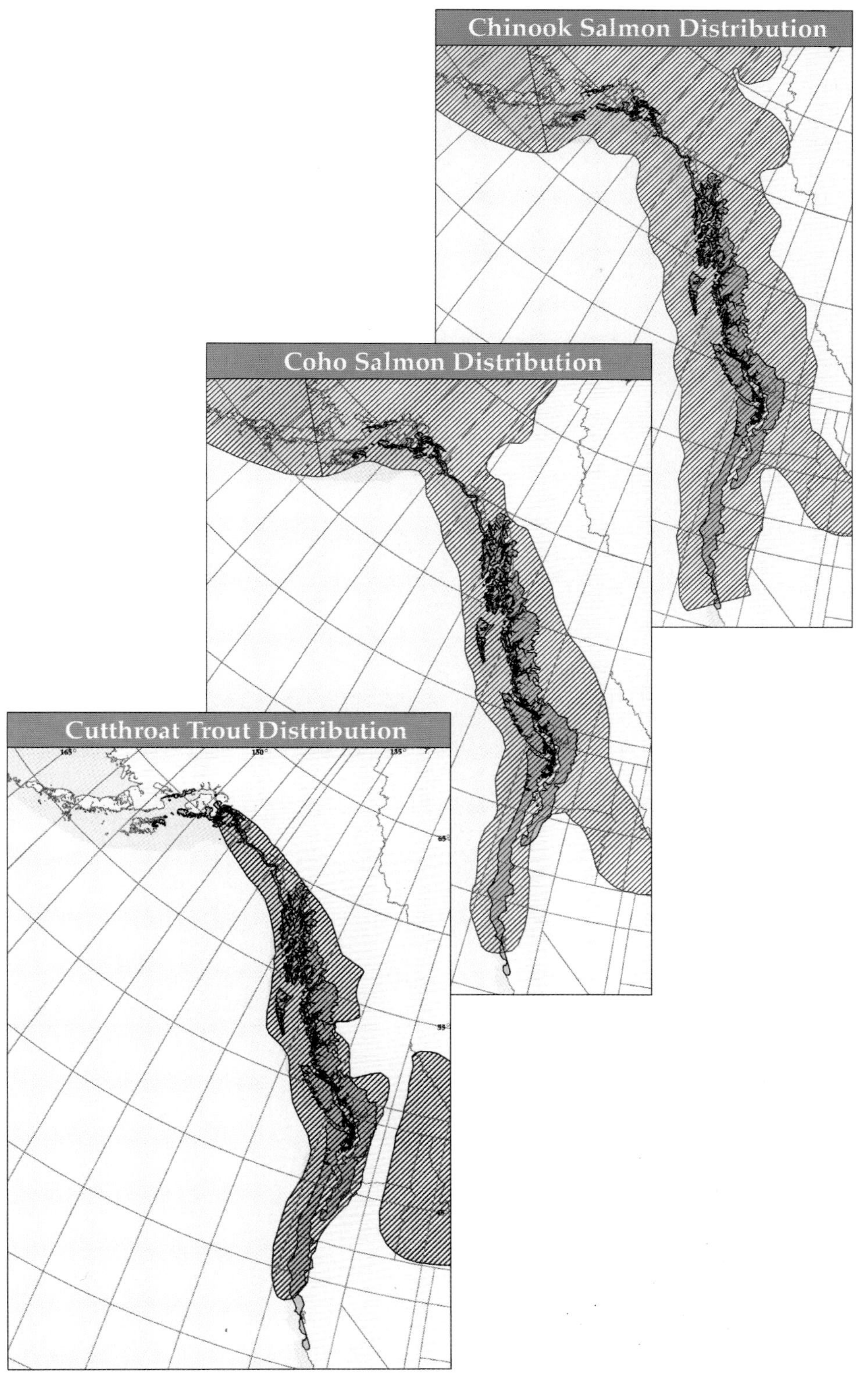

Plate 7. Distribution of anadromous salmonids of the coastal temperate rain forest: sea-run cutthroat trout, coho salmon, and chinook salmon. *Source:* Oregon State University.

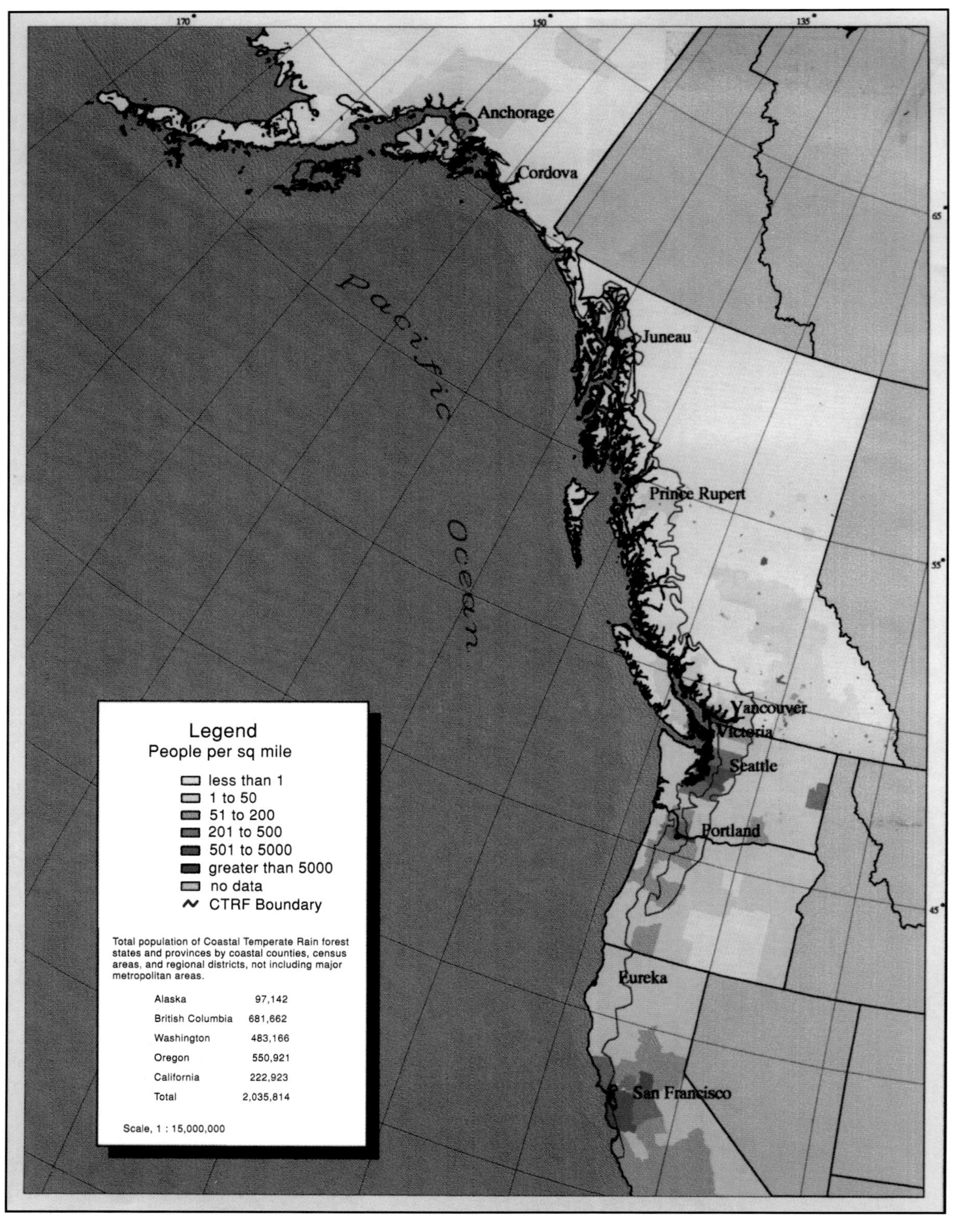

Plate 8. Human population density along North America's rain forest coast. Outside the urban areas, many stretches of the coast support a density comparable to the First Nations populations that preceded European settlement. *Source:* Compiled by Ecotrust and Interrain Pacific from public sources, 1996.

Introduction

Peter K. Schoonmaker, Bettina von Hagen, and Edward C. Wolf

□ □

Stretching from the redwoods of California to the vast stands of spruce and hemlock on Kodiak Island, Alaska, the coastal temperate rain forests of North America are characterized by an unparalleled interaction between land and sea. The marine, estuarine, and terrestrial components combine to create some of the most diverse and productive ecosystems in the temperate zone. Originally found on almost every continent (Figure 1), only half of the world's coastal temperate rain forests still stand. Half of those that remain are in North America (Plate 2).

For thousands of years the coastal temperate rain forests of North America supported one of the highest densities of nonagricultural human settlements on the continent. Given the rich forestland, abundant salmon runs, fertile floodplains, and magnificent scenery, coastal watersheds from northern California to the Gulf of Alaska should continue to sustain robust economies and vibrant communities. Although this region supports many innovative sustainability initiatives, the cultural, economic, and environmental well-being of many coastal communities has been stressed. The region is faced with diminishing natural capital, declining employment in traditional resource-based industries, and outward migration of its young people to the cities. What is the future for the region if these trends continue? And what are the opportunities for a brighter future given the coastal temperate rain forest's ecological, economic, and cultural characteristics? This book describes these characteristics while offering practical examples of local action designed to restore and maintain the ecological, economic, and cultural health of North America's rain forest coast.

The origins of this book go back to a small group who gathered in the Kitlope watershed in north coastal British Columbia in the summer of 1993. By day we collected information about the condition of this ecosystem, home to the Haisla Nation and the largest pristine temperate rain forest watershed in North America. At night our small interdisciplinary team—botanists, zoologists, ecologists, anthropologists, and Haisla people—shared the day's findings around the fire. As we made connections from one discipline to another, a portrait of the watershed emerged. And we began to envision a portrait of the bioregion that encompasses the Kitlope: the coastal temperate rain forest of North America, a bioregion that many in the group knew intimately yet incompletely. We drafted a preliminary sketch of this rain forest coast, home to most of us, that encompassed people and place.

From this sketch we developed the agenda for a conference that was held in late August 1994. Scientists, First Nations leaders, and sustainable development practitioners gathered at the Whistler Resort near Vancouver, B.C., to explore the environment and people of the coastal temperate rain forest. Over three days, the forty presenters and sixty other participants discussed the bioregion from a variety of perspectives ranging from academic research to community organizing. From this exploration arose a common understanding of the biological character and cultural history of the coastal temperate rain forest—and the socioeconomic challenges facing residents in the bioregion.

The book joins people and place by weaving chapters on environment and culture with case studies that describe what coastal residents are doing to integrate environmental conservation with socioeconomic development. Chapters 1, 2, and 3 describe the major environmental forces that shape North America's rain forest: the ocean/atmosphere system and landforms. Chapters 4 and 5 discuss the region's characteristic flora and vertebrate fauna, with an emphasis on species diversity and the structure of biological communities. Chapters 6 and 7 are integrative: moving from flowing fresh water and the streamside zone to a focus on nearshore dynamics, they complete, in a sense, a hydrological cycle begun in the first chapter. Chapters 8 and 9 stretch our scales of observation: Pacific salmon tie the bioregion together from headwaters to open ocean, and the history of the coastal rain forest since the Late Pleistocene extends our perception of environmental patterns and processes back to the last glacial era.

We continue with a review of human occupation of the coastal temperate rain forest bioregion since that glaciation, emphasizing in particular the influence of environment on culture. Chapter 10 surveys pre-European history; Chapter 11 examines the traditions of local and indigenous knowledge that continue today in the practices of First Nations people. The focus on history continues in Chapter 12 with an examination of the environmental and cultural upheaval brought by European settlers, as well as the advent of industrial

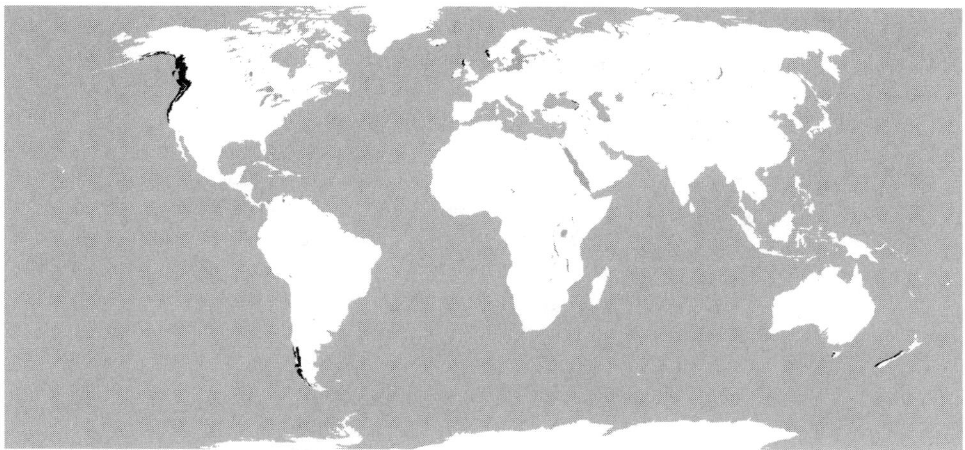

Figure 1. Original distribution of coastal temperate rain forests.
Source: Interrain Pacific.

resource extraction in the twentieth century. Recent economic and cultural trends in the communities of the Oregon coast are the subject of Chapter 13, bringing this history up to the present and exploring the implications of these trends for the future of the bioregion. Chapter 14 explores the science- and policy-based approach to resource stewardship known as ecosystem management, an approach with roots in the patterns of natural disturbance that have shaped the coastal temperate rain forest. The book concludes with Chapter 15: a vision for integrating local and regional interests and initiatives around information, conservation, economic development, and policy reform.

Interspersed among the chapters are four sections titled "Concepts in Action," profiling community-level organizations or sustainability initiatives selected from throughout the coastal temperate rain forest region (Figure 2). Focusing on specific locations and describing common themes that bind the region together, these case studies illustrate by example some of the principles developed in the main chapters. The cases reflect a growing sense that the principles of environmental stewardship will emerge largely through practice. Each section highlights some of the most innovative socioenvironmental experiments unfolding in North America:

- Restoring and managing ecosystems
- Sustainable forestry and fisheries initiatives
- New approaches to learning and decision making
- Conservation-based development

This book is more than a tour guide. It represents current thinking about the status of the bioregion. It is the first systematic attempt to describe the coastal temperate rain forest of North America, to detail its history both before

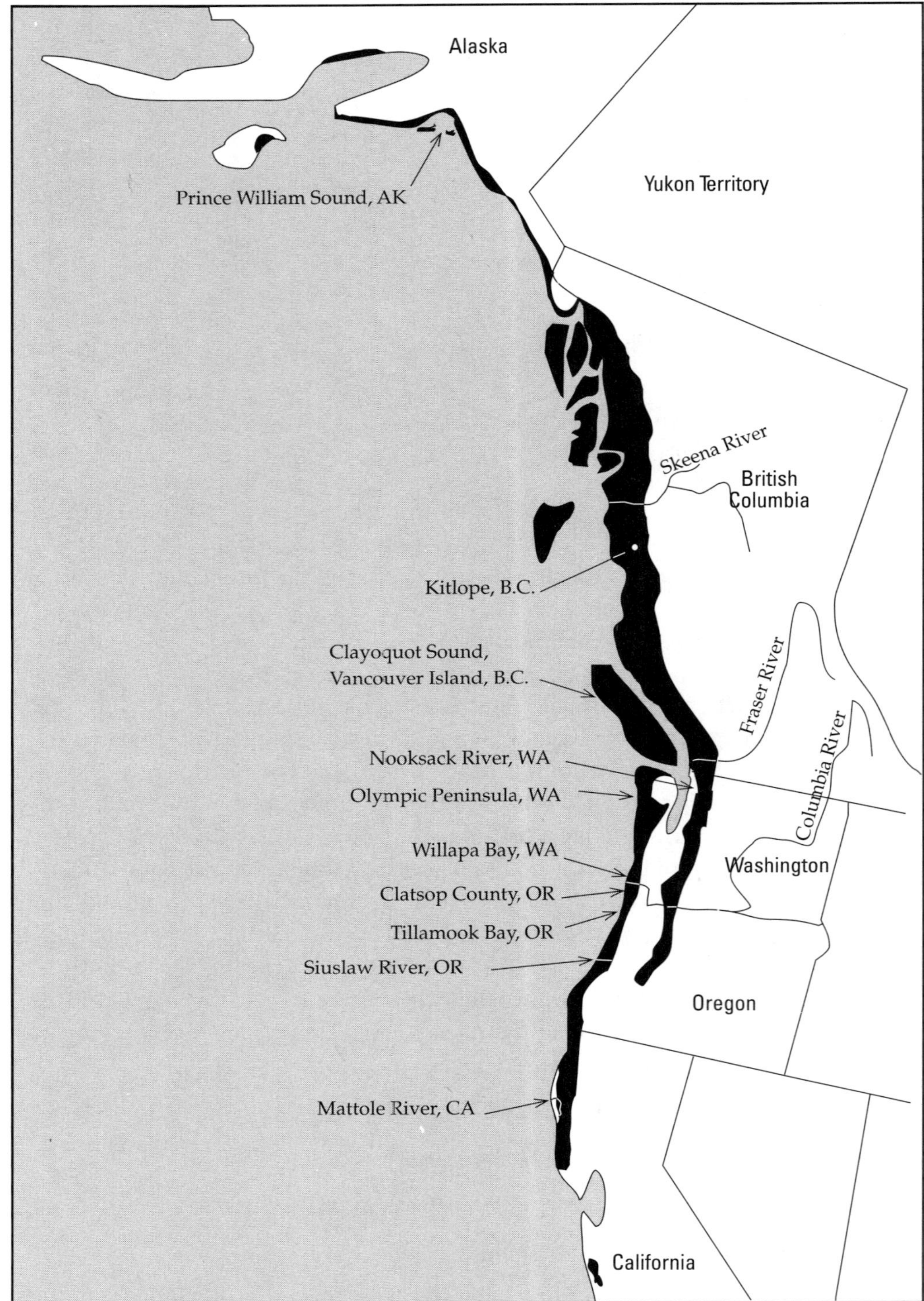

Figure 2. Selected locations of community initatives profiled in "Concepts in Action" sections.

and after European contact, and to explore the implications of its natural and human history for the future course of development and conservation in the region. In addition, this book uses the context of the coastal temperate rain forest to explore the principles and practice of environmental and cultural stewardship, which previously have been explored mainly in the context of developing tropical regions.

We hope this book will provide the general reader and specialist alike with a compelling portrait of the region and with resources for exploring this rain forest coast in more depth. We have created the book with a broad readership in mind: the fisheries biologist who wants some background on geomorphology, the county planner who needs a broader view of the bioregion, the activist who wishes to learn about community initiatives up and down the coast. Chapters are amply referenced from the latest professional literature and from standard and seminal works about the bioregion. Several themes are common to many chapters. That these themes arose independently from various disciplines is a testament to the cohesion of the coastal temperate rain forest bioregion.

Theme One: The rain forest coast is a dynamic region along a range of scales in time and space. While environments and cultures share similarities throughout the bioregion, they also change markedly with altitude, latitude, and distance from the shore. From predator/prey relationships to ocean circulation, from biotic responses to windstorms, fire, or glaciation to human migration patterns, the components of the bioregion are almost never in equilibrium.

Theme Two: The bioregion is both productive and diverse biologically. These attributes are expressed differently, however, than in tropical rain forests. The terrestrial component of the bioregion is less species-rich than in the tropics, but it exhibits great genetic and structural diversity and accumulates several times more biomass per hectare. The productivity of the marine and estuarine components of this bioregion dwarfs that of most other marine and aquatic ecosystems.

Theme Three: The region's fecundity and diversity have evolved together with indigenous cultures since the last ice age. Coupled with current human populations that are comparable to those in pre-Columbian times (excluding the few urban centers within the region), this natural capital offers some of the most favorable conditions for sustaining environmental, economic, and cultural health.

Theme Four: Resource management was no less complex and perplexing for First Nations people than it is for us today. They had the advantage of experience and a sense of place, while Euro-Americans are descendants of people who were more recently prone to wander; those who arrived here only in the last two centuries have simplified and degraded parts of the coastal rain forest, but many have also developed a sense of place and stewardship.

Theme Five: Global and extraregional effects are not limited to Euro-American colonization. The bioregion trades energy and matter along a spectrum of time and space scales. While we focus here on local and bioregional characteristics, we must remember that the bioregion's boundaries are porous—ecologically (El Niño effects, exotic species invasions), socially (recent immigrants, electronic communications), and economically (global markets for timber and fish, effects of tourism and retirement).

Theme Six: The links between environment, economy, and culture are especially evident along the rain forest coast; the corollary of this is that damage to one component of this trio degrades the other two. While such degradation is usually self-evident, sometimes it is not: the decline of salmon south of Vancouver Island has crippled coastal fisheries in Oregon and Washington, yet booming fisheries in northern British Columbia and Alaska may mask a long-term cyclical decline that we cannot yet detect.

Theme Seven: Natural events unfold at scales in time and space that human beings are ill-equipped to perceive. The importance of disturbance—from chronic, cyclical oceanic/atmospheric change to acute, catastrophic events like earthquakes, landslides, fires, and windstorms—is a theme that runs through many chapters. The consensus is that an understanding of disturbance—natural and anthropogenic—is a key to sound ecosystem management. Furthermore, we must acknowledge our ignorance of many natural processes. We must act conservatively using the best science available, treat all actions as experiments, and be ready to adapt accordingly.

And Theme Eight: Many of the case studies and chapters maintain that the integration of local and global, environment and people, is essential to restoring and maintaining the natural capital upon which people along the coast depend. We conclude the book by examining barriers to this integration and discussing strategies for overcoming them: we call the sum of these strategies conservation-based development. Despite our increased understanding, despite all the practical examples, clearly more work must be done to bring about this synthesis. Our next step is to fuse theory, practice, and thoughtful empirical testing.

This book is rich with examples of how individuals, organizations, and communities throughout the coastal temperate rain forest can reconceive "progress" and reconcile their aspirations with the character of the bioregion. We hope the book will encourage many to articulate and follow local visions and to "learn by doing."

A note on units: The binational scope and multidisciplinary nature of this book make it especially challenging to choose consistent units of measure. For the most part, the chapters use metric measures of weight, volume, distance, and area. References to measurement in the case studies are either English or metric, depending on the convention of the place being profiled.

1. Oceanography of the Eastern North Pacific

DAVID K. SALMON

□ □

The eastern North Pacific Ocean is part of a vast and continuous expanse of water that forms the oceans of the world. The ocean and atmosphere act as a strongly coupled system—transferring heat, moisture, and momentum back and forth and interacting with landmasses to create the weather and hence the climate of the earth. In this respect, the eastern North Pacific does not act in isolation. Its behavior is strongly influenced by events in tropical, subtropical, and subarctic regions.

The coastal rain forest owes its physical and ecological characteristics to the interactive processes that occur between the ocean, atmosphere, and landmasses. Physical energy is transferred back and forth extensively between them: radiative energy, winds, precipitation, ocean waves. Energy is also transferred ecologically via processes such as migrations of salmon and other anadromous species into freshwater systems.

The ocean tends to oscillate between two climatic extremes for periods that last between ten and twenty years. These oscillations manifest themselves as changes in surface ocean and air temperatures, the strength of the wind, amounts of precipitation, and numerous other processes related to climate including floods and crop failures. Fluctuations in the abundance of marine phytoplankton, zooplankton, and many fish species often closely parallel these climatic changes.

Overview of the Region

Circulation in the upper layers of the North Pacific Ocean (about the upper 1500 meters) is driven principally by the wind. The ocean surface and near-surface atmosphere transfer heat, moisture, and momentum to one another. Therefore, the principal large-scale patterns of the North Pacific Ocean's circulation have much in common with the major components of North Pacific winds (Figure 1.1). The northeast trade winds, a band of westward-flowing air, dominate the equatorial and tropical regions of the North Pacific. These trade winds extend across the entire Pacific Basin. In the subtropics the surface winds are dominated by the eastward-flowing westerlies, which extend in a band across the North Pacific. Together the trade winds and the westerlies drive a large-scale oceanic feature called the North Pacific subtropical anticyclone, which consists of a clockwise-flowing system of currents made up of the westward-flowing North Equatorial Current, the northward- and northeastward-flowing Kuroshio Current off Japan, the eastward-flowing North Pacific Current (sometimes called the West Wind Drift), and the southward-flowing California Current (Figure 1.2).

Many of the storms that traverse the North Pacific are generated in the Kuroshio Current region, where tremendous amounts of heat are lost to the atmosphere from the warm waters of tropical origin that make up the Kuroshio. Other areas where storms develop include the central Aleutian Archipelago and the eastern Gulf of Alaska, particularly in the vicinity of large glaciers that front the coastal region. These storm systems are important to the ecology of the coastal rain forest because of their direct influence on the physical environment of the eastern North Pacific. These storms affect streamflow, cloud cover, fog, ambient air, stream and upper ocean temperatures, and the amount of stored precipitation available as spring runoff. Storms are also important in the moderation of continental temperature fluctuations (and associated processes) in the coastal temperate rain forest. The region is temperate largely because of the moderating influence of the marine component of the climate system.

Between about 45 and 60 degrees north latitude, extending across the Pacific Basin is an atmospheric feature called the Aleutian Low pressure system (Figure 1.3), a region of statistically low barometric pressure, enhanced by a nearly continuous procession of storms (except during summer) that move generally northeastward after forming in the Kuroshio region, the Bering Sea, or the Gulf of Alaska. The Gulf of Alaska is one of the most active meteorological regions on earth (Wilson and Overland 1987), especially during winter, when it is not uncommon for storms to generate 15-meter-high seas and 100-knot winds. A storm moves through this region on the average of every four or

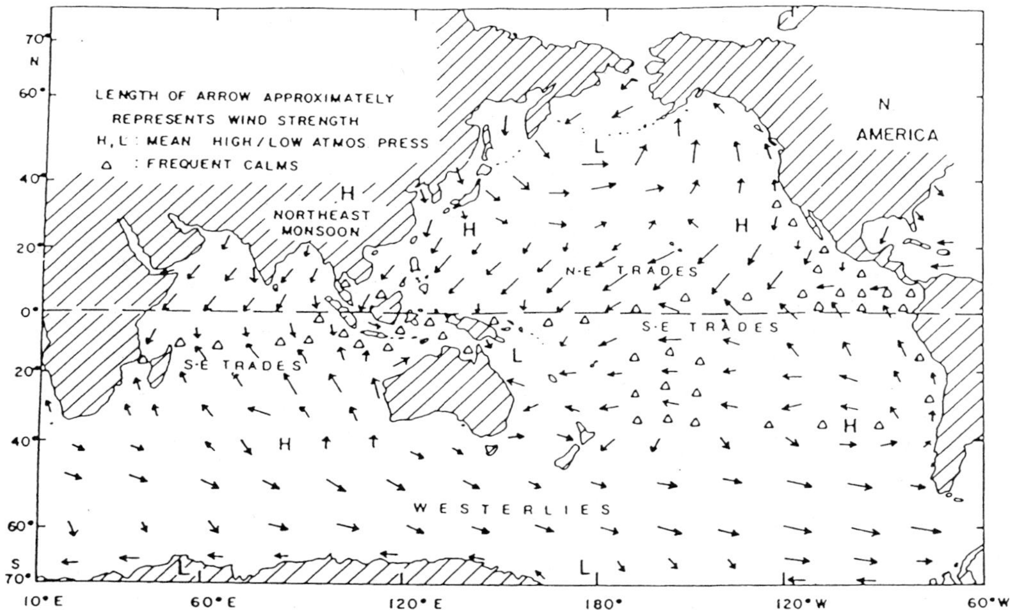

a. Winds in February

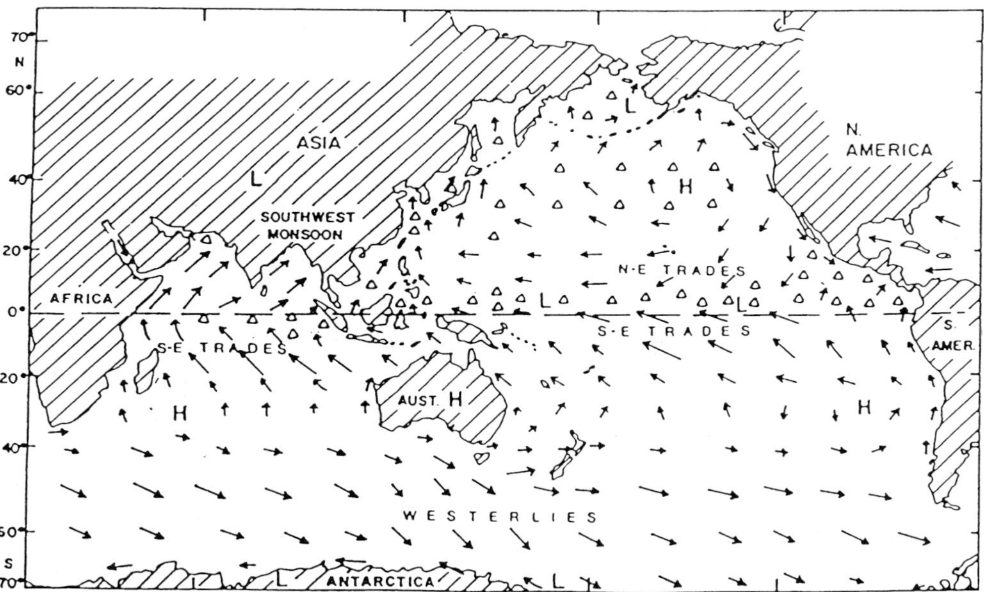

b. Winds in August

Figure 1.1. Winds and mean atmospheric pressure highs and lows in the Indian and Pacific oceans: (*a*) February (northeast monsoon); (*b*) August (southwest monsoon).

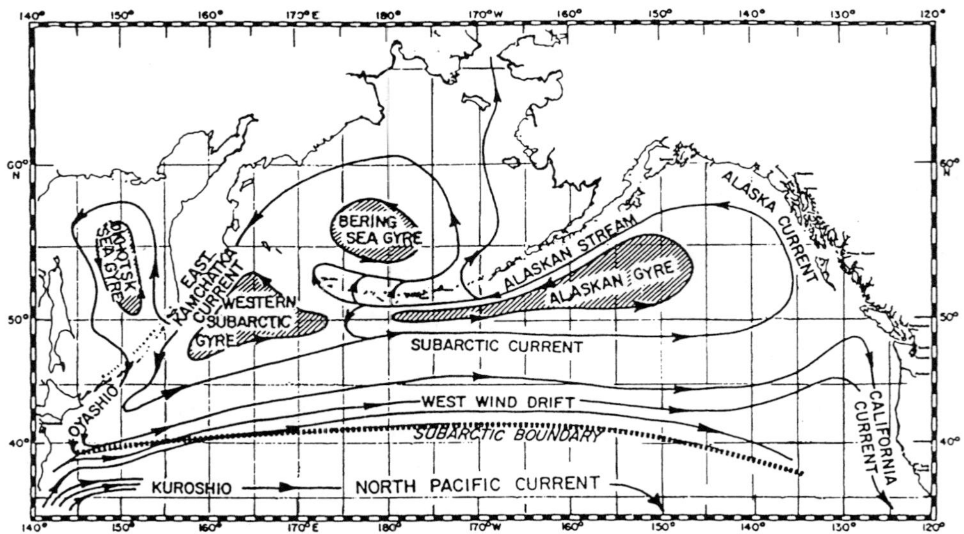

Figure 1.2. Circulation in the subarctic Pacific Ocean south of the Aleutian Islands, showing the Alaskan and Western Subarctic gyres. The Kuroshio, California, and Subarctic currents (also known as the north branch of the North Pacific Current) are also shown. *Source:* Pearcy (1992), based on Dodimead et al. (1963). Reprinted by permission of the Washington Sea Grant Program.

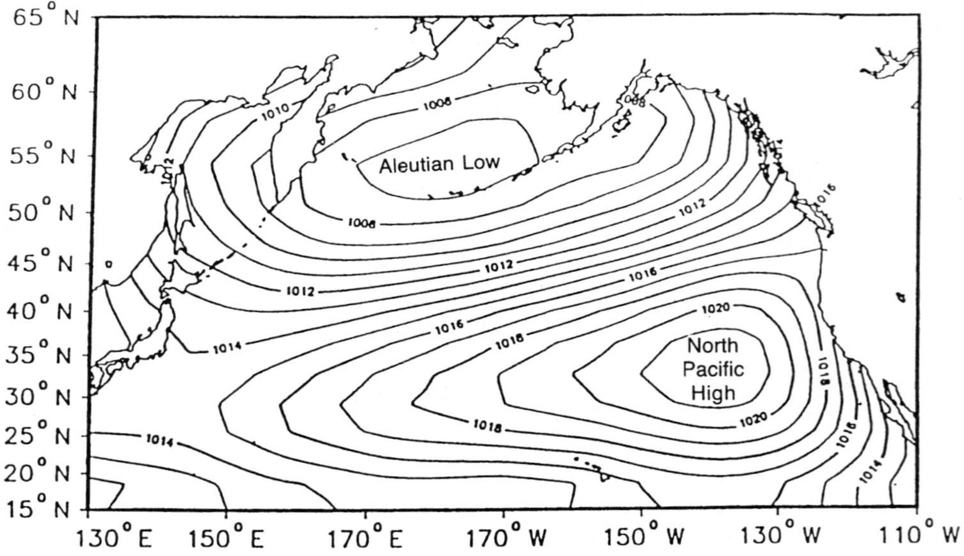

Figure 1.3. Mean North Pacific sea level pressure: 1946–1988.

five days throughout the winter months (Hartman 1974; Wilson and Overland 1987). The coastal rain forest owes many of its physical characteristics to the effects of atmospheric and oceanic processes associated with the Aleutian Low: frequent and intense coastal storms, strong and persistent winds, fog, rain, and snow. The physical morphology of the coastal region is literally shaped by the actions of wind and waves on the continental landmass.

The Aleutian Low dominates the atmospheric circulation between 45 and 60 degrees north latitude during fall, winter, and early spring. During summer the storm activity abates and the low-pressure cells are displaced (they often move over land) by a high-pressure system called the North Pacific High. The North Pacific High is statistically present year round in the subtropical Pacific region between about 20 and 45 degrees north. It is the dominant factor influencing the weather on the west coast of North America during summer, when it intensifies and expands northward over the subarctic region of the Pacific Ocean.

The winds in the Aleutian Low drive two large ocean gyres, one in the Bering Sea and the other in the Gulf of Alaska (Figure 1.2). The predominant ocean circulation in both regions is counterclockwise, as are the wind systems that drive the currents. These wind systems are directly associated with the storms that frequently form in the regions or move there after forming off eastern Asia in warm waters of tropical Pacific origin. The Alaska gyre consists of the eastward-flowing North Pacific Current (which forms the southern boundary of the gyre) and the northwestward- and westward-flowing Alaska Current. The Bering Sea and the Gulf of Alaska are extremely important to the ecology of the temperate rain forest region because they play a major role in supporting the oceanic phases of the life histories of Pacific salmon. Pacific salmon characteristically spend more than half of their lives (depending on the species) feeding on zooplankton produced in these oceanic regions. In addition these regions are economically valuable to the coastal rain forest area because they support large stocks of commercially important fishes including halibut, sablefish, walleye pollock, Dungeness, king, and tanner crab, albacore, scallops, shrimp, sea urchins, and sea cucumbers, to name just a few.

The Alaska Coastal Current flows within about 50 kilometers of the coast of the entire Gulf of Alaska. It is a westward-flowing current driven by fresh water and wind. It carries the entire load of runoff from the rivers bordering the Gulf of Alaska (Royer 1979, 1981; Schumacher and Reed 1980). The strength of the coastal current is strongly influenced by seasonal variations in both surface winds and freshwater discharge from rivers. The ecology of the nearshore regions in the subtropical and subarctic North Pacific is quite distinct from the waters of the offshore ocean gyres largely because of continental effects associated with differential precipitation and changes in the winds and ocean currents associated with the continental margins. This is particularly true in the Gulf of Alaska region, where the coastal mountain ranges ringing the gulf act as a barrier to the inland passage of storms, resulting in large amounts of precipitation (2 to 3 meters per year) in the coastal region as the storm systems are forced against the high coastal mountains in the nearshore region of the gulf. The storm systems transport tremendous amounts of heat from the ocean, as well, affecting the temperature extremes in the coastal regions. The coastal rain forest probably extends so far poleward because of these moderating effects.

Coastal upwelling and downwelling are oceanic conditions that represent the interaction between wind stress, the rotation of the earth, and frictional effects between the ocean and the continental boundary. In the Northern Hemisphere, these interactions move water to the right of the prevailing winds. Thus upwelling is associated with equatorward winds along the coast and offshore movement of surface waters, while downwelling is associated with poleward winds along the coast and onshore motion of surface waters. During fall, winter, and spring, the coastal regions of northern British Columbia and Alaska are characterized by downwelling conditions that occur in response to prevailing southeasterly and easterly winds. The coastal region of the Gulf of Alaska is characterized as a coastal temperate environment because the prevailing winds moderate the cold subarctic climate through the importation of warm moist subtropical air masses and warm subtropically derived ocean currents. The continental shelf regions of the subarctic North Pacific support high densities of economically and ecologically important marine populations.

During spring and summer, the prevailing winds blow from the north along the west coast of North America, creating an upwelling along the west coast of the United States and southern British Columbia. This upwelling greatly moderates the climate and ecology of the coastal region by bringing relatively cold nutrient-rich subsurface waters to the surface. The nutrient-rich waters support high levels of primary production, which in turn supports tremendous zooplankton and fish biomass. Coastal upwelling regions are among the most productive marine ecosystems on the planet. The ecology of coastal upwelling systems is strongly modulated regionally by seasonal and interannual variability in the winds and by more distant variability in both atmospheric and oceanic conditions in the tropical Pacific region.

The southward-flowing California Current and the northward-flowing Alaska Current relate most directly to variability in the coastal rain forest region. As the North Pacific Current nears the North American continent, it splits and becomes the Alaska Current and the California Current. The relative amounts of water that flow poleward and equatorward in these systems are thought to be closely related to the strength and size of the Aleutian Low.

When the Aleutian Low is intensified, it also tends to expand its region of influence further south. It is thought that under these conditions more warm water of subtropical origin flows poleward into the Alaskan Current than during periods of a weakened Aleutian Low. A weakened and contracted Aleutian Low would result in less water moving into the Alaska Current and more cold waters with subarctic physical properties moving south into the California Current. These processes have implications for fisheries. Shifts in the position of the Aleutian Low cause shifts in the positions of the large-scale North Pacific ocean frontal systems—the boundary regions of the large current systems in terms of their physical and ecological characteristics. On time scales of several years, large-scale displacements (to the north or south) of these current systems and associated frontal systems result in the expansion or contraction of the ranges of various phytoplankton, zooplankton, and fish populations.

El Niño, La Niña, and the Southern Oscillation

The terms El Niño, La Niña, and the Southern Oscillation are often used by oceanographers and atmospheric scientists to describe processes that have their origins in the tropical Pacific and Indian oceans with profound consequences for the northeastern Pacific and the coastal rain forest zone. The Southern Oscillation is an atmospheric indicator of a process in which El Niño and La Niña events represent extremes in conditions in the tropics, including ocean temperature, strength and persistence of the trade winds, strength of the coastal upwelling off South America, and precipitation across the entire tropical Pacific. For excellent detailed treatments of the large-scale climatological, ecological, and social effects of the El Niño–Southern Oscillation, see Philander (1990) and Glantz et al. (1991).

The physical and biological characteristics of the El Niño–Southern Oscillation provide a framework for discussing climate and ecology in the North Pacific. Indeed, the cycle plays a dominant role in forcing environmental change in the ocean, in the atmosphere, and on land in the North Pacific on time scales of a few years to decades. El Niño affects the coastal rain forest in a number of ways. Variability of North Pacific marine phytoplankton, zooplankton, and fish populations, for example, can often be linked to the El Niño cycle.

El Niño Phase

How does an El Niño event occur in the tropical Pacific? Before the onset of an El Niño event, oceanic conditions are characterized by warmer surface waters in the western equatorial Pacific and colder surface waters in the eastern tropical Pacific where strong coastal upwelling brings relatively cold, nutrient-rich, deep waters to the surface. The western tropical Pacific is characterized by frequent convective activity and intense rains; the eastern tropical Pacific is much drier. The coastal upwelling marine ecosystem along the west coast of South America is highly productive—a result of nutrient-rich deep waters being brought to the surface and fostering the growth of phytoplankton. In turn, large zooplankton populations thrive on the phytoplankton and hence support higher levels of fish, seabirds, and marine mammals.

A typical El Niño warm event occurs when the northeast and southeast trade winds weaken, causing a diminished westward circulation in the equatorial ocean current system. The trade winds sometimes weaken to the extent that there are occasional westerly wind bursts along the equator. The surface waters of the upwelling regions off South America warm partly in response to the eastward flow of heat from warmer western tropical Pacific waters. As tropical atmospheric convection zones change their positions, the moisture-producing areas move eastward toward the coast of South America. The coastal desert regions of the eastern tropical Pacific receive abundant rain and the terrestrial ecosystem flourishes. The coastal upwelling is suppressed and the marine ecosystem in the eastern tropical Pacific suffers because of the absence of nutrient-rich waters in the upper layers of the ocean. These changes drastically reduce all levels of productivity.

In the eastern North Pacific, the effects of El Niño events often manifest themselves in an intensified and expanded Aleutian Low, as well as higher air and surface ocean temperatures. Animal and plant populations are affected by the changes in their environment that can be brought about by the El Niño cycle. They can also be affected by the presence of species that move poleward in conjunction with warming conditions along the west coast of North America. Numerous examples of these phenomena can be found in Wooster and Fluharty (1985). Pacific mackerel, for instance, are known to move north into British Columbia waters in response to El Niño warm events. Since mackerel can prey heavily on juvenile salmon inhabiting the coastal region, these poleward excursions of mackerel can significantly affect regional fisheries.

La Niña Phase

In the tropical Pacific, La Niña conditions are characterized by strong trade winds, by strong convective activity and precipitation in the western tropical Pacific, by low surface ocean temperatures in the eastern equatorial and coastal upwelling regions, by dry conditions in the eastern tropical atmosphere and adjacent arid continental regions, and by high productivity in the coastal upwelling ecosystems along the west coast of South America.

La Niña conditions can often be related to low surface ocean and air temperatures in the eastern subtropical and subarctic North Pacific, as well as high surface ocean temperatures far away from landmasses in the central North Pacific. Storm activity in the Aleutian Low weakens and high-pressure systems known as blocking ridges tend to prevail over the subarctic North Pacific during the winter months. These ridges are important because they deflect storm systems away from the Gulf of Alaska—either northward into the Bering Sea or southeastward toward the west coast of the continental United States. The occurrence of these ridges during the winter months declined greatly from 1976 through 1988, relative to the previous three decades, apparently because of an intensification of the Aleutian Low during that period (Salmon 1992).

Temporal Variability

Changes in the ocean occur on a variety of time scales—from fractions of seconds to thousands of years. The annual cycle of the seasons dominates many records of North Pacific atmospheric and oceanic variables including air and upper ocean temperatures, precipitation, and winds. Periods of fluctuation greater than one year are of great ecological importance. Time scales of two to three, five to seven, and ten to twenty years tend to dominate the range of year-to-year variations in both physical and biological systems (Table 1.1). The two to three and five to seven year time scales are closely related to Southern Oscillation phenomena. The ten to twenty year variability probably involves two signals—one at about fourteen years that is related to the Southern Oscillation (Quinn et al. 1987; D. Ware, pers. comm.) and one at about twenty years, which may be related to tides forced by lunar variations (Royer 1989). Variability on all of these time scales has important implications for fish populations including salmon, herring, mackerel, hake, sardines (Mysak et al. 1982; Ware 1990; Beamish and Bouillon 1993), bluefin tuna (Mysak 1986), and Pacific halibut (Parker et al. 1995).

The ten to twenty year time scales apply to areas larger than the entire North Pacific and significantly alter the biological, chemical, and physical environment. The most recent warm climatic regime began in about 1976 and lasted until at least 1988. Researchers remain uncertain whether that regime has actually ended. After an apparent shift in 1989 (the first La Niña cold event to occur since about 1974), an El Niño warm event of nearly unprecedented duration dominated the Pacific region at least through 1994. Preceding the 1976–1988 period was a cold climatic phase that lasted from about the early 1960s until 1976. In many ways, the characteristics of these climatic regimes significantly oppose each other. Indeed, climatic conditions in the North Pacific during these periods can be largely characterized as being in one or the other of these two phases.

Table 1.1. Time Scales in the North Pacific

Parameter	Period of Oscillation (Years) 2.5–4	5–7	14–17	18–20
Eastern North Pacific sea surface temperature	X	X		
Central North Pacific sea surface temperature	X	X		
Gulf of Alaska air temperature	X	X		X
Gulf of Alaska ocean currents (line P) and chemical properties	X	X		
Gulf of Alaska wind field	X	X		
Sea level height on west coast of North America		X		
Precipitation (snow accumulation on Mt. Logan)				X
Pacific mackerel year-class strength			X	
Pacific herring year-class strength	X		X	
Pacific halibut recruitment				X
Sardine year-class strength		X	X	
Southern Oscillation Index	X	X	X	

In the North Pacific, the period from 1976 through 1988 can most simply be described as a decade of El Niño-like conditions. The Southern Oscillation Index (an index of conditions in the tropical Pacific that is negative during El Niño events and positive during La Niña events) was consistently negative. Coastal upwelling regions along the Pacific coasts of North and South America were associated with lower productivity and warm ocean temperatures. The Aleutian Low intensified during the same period (Trenberth 1990; Salmon 1992), sending numerous intense storms and warm moist air masses into the eastern North Pacific. As a result, eastern North Pacific surface ocean and air temperatures were warm but central and western North Pacific Ocean surface temperatures were cold. Precipitation in the coastal mountain ranges ringing the Gulf of Alaska increased significantly compared to the previous cold regime.

Increases in greenhouse gases in the atmosphere, and corresponding temperature changes over large regions of the planet, may also play a role in the recently observed atmospheric and oceanic variability in the Pacific. The magnitude of their influence relative to other climatic factors is still largely unknown. The most recent decadal climate regime has involved numerous extremes of many physical processes, such as high and low regional temperatures and unusually high and low precipitation. This pattern suggests that increases in atmospheric greenhouse gases have had an effect on the amplitude of these physical processes in the latter portion of this century. Nevertheless, large areas of the North Pacific cooled from 1969 to 1988, suggesting that decade-scale variability overshadows global warming. Although many regions of the Northern Hemisphere experienced a warming trend from 1969 through 1988, the dominant pattern of temperatures and surface winds in the North Pacific appears to be associated with El Niño (Weare et al. 1976; Salmon 1992). Past records of atmospheric and oceanic variables suggest that a return to La Niña-type conditions (cold events and increased upwelling) is highly likely within a few years.

Implications for Fisheries

Some fish populations have fluctuated dramatically throughout history, even in the absence of human fishing pressures (Soutar and Isaacs 1974; Baumgartner et al. 1992). Although the reasons for these fluctuations are uncertain or difficult to prove, hypotheses point to changes in the physical environment, catastrophic events caused by disease or some other agent (such as pollution), competition, and predation. Discussions of the links between fish abundance and climate can be found in Shepard et al. (1984), Wooster and Fluharty (1985), Beamish and McFarlane (1989), Kawasaki et al. (1990), Beamish and Bouillon (1993), and Beamish (1995).

The work of Venrick et al. (1987) has shown that a significant summer increase in chlorophyll (indicating higher phytoplankton production) in the central North Pacific might be related to long-term changes in both winter winds and ocean surface temperatures. In particular, they postulate that an increase in the strength of winter winds, accompanied by a drop in ocean surface temperature, has changed the carrying capacity of the near-surface ecosystem in the central North Pacific. Conversely, others have recently suggested that El Niño warm conditions have drastically reduced phytoplankton and zooplankton production in the upwelling region off southern California (Roemmich and McGowan 1995).

Brodeur and Ware (1992) have shown that the distribution and abundance of zooplankton in the Gulf of Alaska are strongly related to wind-driven ocean conditions. In particular, years of low zooplankton abundance in the Gulf of Alaska correspond to years of weak winter winds and lower than normal surface ocean temperatures in the Aleutian Low, while high zooplankton abundance tends to occur during years of intensified winter winds in the Gulf of Alaska and correspondingly higher than normal ocean surface temperatures. The Gulf of Alaska had twice the summer zooplankton biomass in the period 1980 to 1989 (and an increase in pelagic fish and squid abundance) than it had in the years 1956 to 1962. These two time periods were characterized by an intensified and weakened Aleutian Low, respectively. Various mechanisms might account for these observations:

- Stronger winter winds increase the circulation of the Alaska gyre, causing greater upwelling of nutrient-rich waters in the Gulf of Alaska. (This central gyre upwelling and coastal upwelling are driven by different processes.) If primary production is limited by nutrient availability, then the increased upwelling in the central Gulf of Alaska will support greater primary—and hence secondary—production.
- The intensified winter winds increase the mixed-layer depth (to about 100 meters), which might slow primary production in the spring and allow the zooplankton to graze more efficiently.
- Changes in the wind field produce changes in the relative amount of subtropical water that flows northward at the branch point of the California and Alaska currents.

Fulton and LeBrasseur (1985) have shown that zooplankton community structure (and therefore prey size distributions for juvenile salmonids) is significantly altered in conjunction with movement of the subarctic boundary in the North Pacific, particularly in response to El Niño events. These shifts of the subarctic boundary and associated changes in flow alter zooplankton communities in both the Gulf of Alaska and the California Current area.

Beamish and Bouillon (1993) have observed a long-term relationship between the intensity of the Aleutian Low and the production of chum, pink, and sockeye salmon over the entire North Pacific Ocean. They note that periods of high salmon production coincide with an intensified Aleutian Low, a warm climatic regime, and high copepod (a crustacean constituent of zooplankton) abundance. Salmon feed heavily on copepods and other zooplankton during their stay in the ocean. Over the years, the intensity of the Aleutian Low fluctuates in concert with the Southern Oscillation, indicating that year-to-year changes in the Aleutian Low are closely related to the Southern Oscillation cycle in the tropics (Emery and Hamilton 1985; Salmon 1992). These correlations do not necessarily establish cause and effect, but they do demonstrate that climatic and oceanographic factors are strongly linked to the abundance of marine zooplankton and fish populations.

Mysak et al. (1982) have shown that recruitment of Pacific herring and sockeye salmon fluctuates coherently with the five- to six-year period of variability that occurs off British Columbia and southeastern Alaska. Parker et al. (1995) have shown that trends in the abundance of Pacific halibut in the Gulf of Alaska over the last sixty years correspond to the 18.6-year lunar nodal tidal cycle (Figure 1.4). The mechanisms that might link these variables are tidally modulated mixing and changes in flow that affect nutrient availability, as well as changes in primary and secondary production dynamics. These examples characterize a common theme found in abundance records of marine popula-

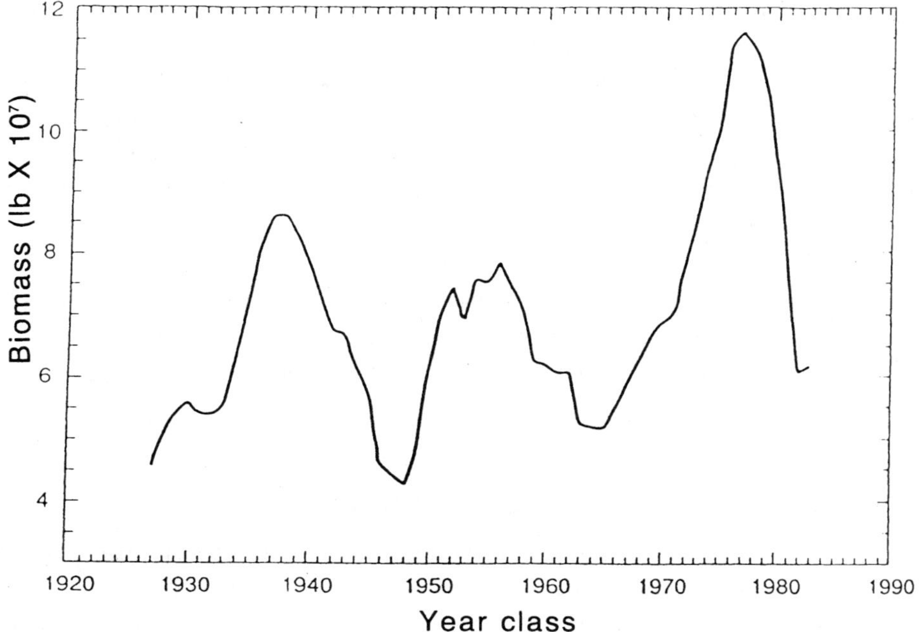

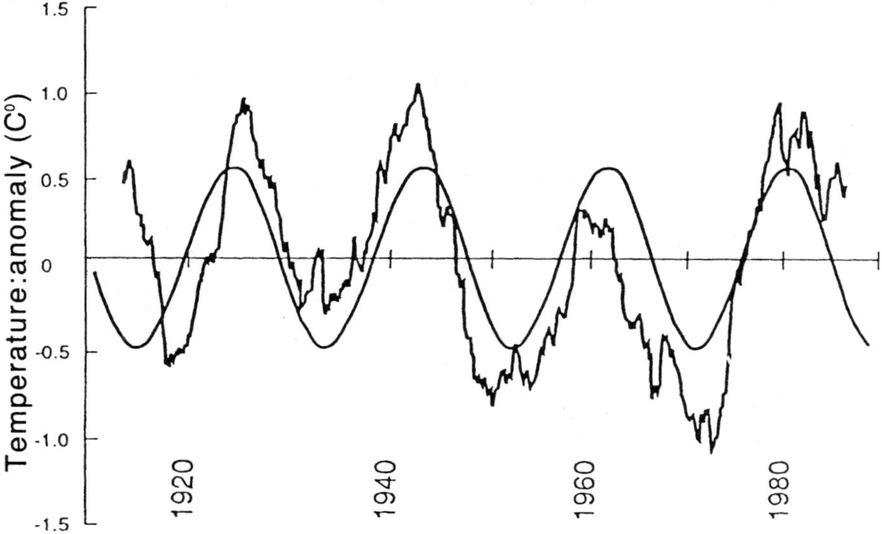

Figure 1.4. (*a*) Pacific halibut recruitment biomass: 1927–1983. The series has been smoothed with a three-year moving average. (*b*) Air temperature anomalies at Sitka, Alaska (five-year average), low-pass filtered, with "best fit" 18.6-year lunar tidal signal superimposed. Halibut is among the marine species whose abundance seems to vary with a predictable climate and tidal cycle.

tions—namely, that periodic or cyclical climatic processes play a major role in constraining the abundance of marine plant and animal populations.

In general, it appears that zooplankton and many ecologically and economically important fish populations in the subarctic North Pacific respond favorably to warming (Brodeur and Ware 1992; Thomas and Mathisen 1993). In the subtropical upwelling regions of the California Current, El Niño warm events reduce phytoplankton and zooplankton abundance by affecting the upwelling regime (McGowan 1984). In the tropical Pacific, El Niño warm events and longer-time-scale warming effects have disastrous implications for populations of phytoplankton, zooplankton (McGowan 1984; Roemmich and McGowan 1995), and certain pelagic fish and bird species including anchovies and seabirds that prey on them (Glantz 1984).

Links to the Terrestrial Environment

Weather systems directly connect the marine and terrestrial environments. Interactions between continental and marine high- and low-pressure systems generate the weather patterns that we observe. The Gulf of Alaska acts as a graveyard for storms because it is ringed by high coastal mountains. Storms that enter the gulf usually dissipate their energy there because they cannot move over the coastal mountains. This coastal mountain effect results in high precipitation in the coastal region and modifies the winds that approach the North American continent from the ocean. The warm gulf waters that flow eastward modify the coastal air temperatures as well. The high coastal mountains play a similar role along the west coast upwelling region, blocking the passage of storms and modifying the wind and precipitation regimes.

One of the direct biological links between the marine and terrestrial environment in the coastal rain forest region is the introduction of marine-derived nutrients into the terrestrial system by migrating salmon. Significant amounts of marine-derived nitrogen and other organic materials enter freshwater systems as adult salmon return to rivers and streams to spawn and die (Kline et al. 1990, 1993). The nutrients from decaying carcasses nourish bacteria, plankton, young salmon, and other juvenile fish. Plants too use these marine-derived nutrients. In fact, the nitrogen content of the leaves of some trees near Bristol Bay, Alaska, was found to be essentially 100 percent marine-derived nitrogen (T. Kline, pers. comm.).

The flow of materials moves in both directions, for the land also supplies organic and inorganic compounds to the sea. The main vehicles for the transport of materials from land to sea are river systems and the winds. Much inorganic material is the result of continental weathering and erosion. Large riverine inputs of silicates and other minerals and substantial wind-driven inputs of continental iron reach the sea. Both of these nutrients are vitally important to the growth of certain marine plants (diatoms) and other species. It is also thought that iron might be a limiting nutrient in plant production in the offshore North Pacific region (Martin 1991).

The atmospheric and oceanic phenomena associated with the Southern Oscillation play a direct role in modifying the climate of the terrestrial environment of the North Pacific. El Niño–Southern Oscillation events are associated with floods and droughts over large regions of the continents. For example, El Niño events have been implicated in flooding in California, increases in the number of intense storms in the South and North Pacific, and droughts in Africa, South America, and the South Pacific (Glantz 1984).

The climate plays a dominant role in constraining the distribution and abundance of both marine and terrestrial plant and animal populations in the North Pacific. A better understanding of how changes in the North Pacific alter the structure and function of coastal ecosystems is a step toward better management of natural resources within these populous and intensively managed regions.

References

Baumgartner, T., A. Soutar, and V. F. Bartina. 1992. "Reconstruction of the history of Pacific sardine and northern anchovy populations over the past two millennia from sediments of the Santa Barbara Basin, California." *CalCofi Report.*

Beamish, R. J., ed. 1995. "Climate change and northern fisheries populations." *Can. J. Fish. Aqua. Sci.* (special issue) 121:449–459.

Beamish, R. J., and D. R. Bouillon. 1993. "Pacific salmon production trends in relation to climate." *Can. J. Fish. Aqua. Sci.* 50:1002–1016.

Beamish, R. J., and G. A. McFarlane, eds. 1989. *Effects of Ocean Variability on Recruitment and Evaluation of Parameters Used in Stock Assessment Models.* Vancouver, B.C.: Department of Fisheries and Oceans.

Brodeur, R. D., and D. M. Ware. 1992. "Long-term variability in zooplankton biomass in the subarctic North Pacific Ocean." *Fish. Oceano.* 1:32–38.

Dodimead, A. J., F. Favorite, and T. Hirano. 1963. *Salmon of the North Pacific Ocean. Part 2: Review of Oceanography of the Subartic Pacific Region.* International North Pacific Fisheries Commission Bulletin 13.

Emery, W. J., and K. Hamilton. 1985. "Atmospheric forcing of interannual variability in the northeast Pacific Ocean: Connections with El Niño." *J. Geophys. Res.* 90:857–868.

Fulton, J. D., and R. J. LeBrasseur. 1985. "Interannual shifting of the Subarctic boundary and some of the biotic effects on juvenile salmonids." In W. Wooster and D. Fluharty, eds., *El Niño North: Niño Effects in the Eastern Subarctic Pacific Ocean.* Seattle: Washington Sea Grant.

Glantz, M. H. 1984. "Floods, fires and famine: Is El Niño to blame?" *Oceanus* 27:14–19.

Glantz, M. H., R. W. Katz, and N. Nicholls. 1991. *Teleconnections Linking Worldwide Climate Anomalies.* Cambridge: Cambridge University Press.

Hartmann, D. L. 1974. "Time spectral analysis of mid latitude disturbances." *Month. Weath. Rev.* 102:348–362.

Hood, D. W., and S. T. Zimmerman, eds. 1987. *The Gulf of Alaska: Physical Environment and Biological Resources.* Washington, D.C.: Minerals Management Service, U.S. Department of the Interior.

Kawasaki, T., S. Tanaka, Y. Toba, and A. Taniguchi, eds. 1990. *Long-Term Variability of Pelagic Fish Populations and Their Environment.* New York: Pergamon Press.

Kline T. C., J. J. Goering, O. A. Mathisen, P. H. Poe, and P. L. Parker. 1990. "Recycling of elements transported upstream by runs of Pacific salmon I: $d^{15}N$ and $d^{13}C$ evidence in Sashin Creek, southeastern Alaska." *Can. J. Fish. Aqua. Sci.* 47:136–144.

Kline T. C., J. J. Goering, O. A. Mathisen, P. H. Poe, P. L. Parker, and R. S. Scanlan. 1993. "Recycling of elements transported upstream by runs of Pacific salmon II: $d^{15}N$ and $d^{13}C$ evidence in the Kvichak River watershed, Bristol Bay, southwestern Alaska." *Can. J. Fish. Aqua. Sci.* 50:2350–2365.

Klyashtorin, L. 1993. "Can we predict U.S. salmon production?" Unpublished manuscript.

Martin, J. H. 1991. "Iron, Liebig's law and the greenhouse." *Oceanography* 4:52–55.

McGowan, J. A. 1984. "The California El Niño, 1983." *Oceanus* 27:48–51.

Mysak, L. A. 1986. "El Niño, interannual variability and fisheries in the northeast Pacific Ocean." *Can. J. Fish. Aqua. Sci.* 43:464–497.

Mysak, L. A., W. W. Hsieh, and T. R. Parsons. 1982. "On the relationship between interannual baroclinic waves and fish populations in the northeast Pacific." *Biol. Oceano.* 2:63–102.

Parker, K. S., T. C. Royer, and R. B. DeRiso. 1995. "High latitude climate forcing by the 18.6 year lunar nodal cycle and low frequency recruitment trends in Pacific halibut." *Can. J. Fish. Aqua. Sci.* (special issue) 121:449–459.

Pearcy, W. G. 1992. *Ocean Ecology of North Pacific Salmonids.* Seattle: University of Washington Press.

Peterson, D. H., ed. 1989. "Aspects of climate variability in the Pacific and the western Americas." *Geophys. Mono.* 55. Washington, D.C.: American Geophysical Union.

Philander, S.G.H. 1990. *El Niño, La Niña, and the Southern Oscillation.* San Diego: Academic Press.

Quinn, W. H., V. T. Neal, and S. E. Antunez de Maylo. 1987. "El Niño occurrences over the past four and a half centuries." *J. Geophys. Res.* 92:663–678.

Roemmich, D., and J. McGowan. 1995. "Climatic warming and the decline of zooplankton in the California Current." *Science* 267:1324–1326.

Royer, T. C. 1979. "On the effect of precipitation and runoff on coastal circulation in the Gulf of Alaska." *J. Phys. Oceanog.* 7:92–99.

———. 1981. "Baroclinic transport in the Gulf of Alaska, Part II: A freshwater driven coastal current." *J. Mar. Res.* 39:251–266.

———. 1989. "Upper ocean warming in the North Pacific: Is it an indicator of global warming?" *J. Geophys. Res.* 94:18175–18183.

Salmon, D. K. 1992. "On interannual variability and climate change in the North Pacific." Ph.D. diss., University of Alaska, Fairbanks.

Schumacher, J. D., and R. K. Reed. 1980. "Coastal flow in the northwest Gulf of Alaska: The Kenai Current." *J. Geophys. Res.* 85:6680–6688.

Shepard, J. G., J. G. Pope, and R. D. Cousens. 1984. "Variations in fish stocks and hypotheses concerning their links with climate." *Rapp. P.-v. Reun. Cons. Int. Explor. Mer.* 185:255–267.

Soutar, A., and J. D. Isaacs. 1974. "Abundance of pelagic fish during the 19th and 20th centuries as recorded in anaerobic sediment off the Californias." *Fish. Bull.* 72:257–273.

Thomas, G. L., and O. A. Mathisen. 1993. "Biological interactions of natural and enhanced stocks of salmon in Alaska." *Fish. Res.* 18:1–17.

Trenberth, K. 1990. "Recent observed interdecadal climate changes in the northern hemisphere." *Bull. Am. Met. Soc.* 71:988–993.

Venrick, E. L., J. A. McGowan, D. R. Cayan, and T. L. Hayward. 1987. "Climate and chlorophyll *a*: Long-term trends in the central North Pacific Ocean." *Science* 238:70–72.

Ware, D. M. 1990. "Climate, Predators and Prey: Behavior of a link oscillating system." In T. Kawasaki, S. Tanaka, Y. Toba, and A. Taniguchi, eds., *Long-Term Variability of Pelagic Fish Populations and Their Environment.* New York: Pergamon Press.

Weare, B., A. Novato, and R. Newell. 1976. "Empirical orthogonal function analysis of Pacific sea surface temperatures." *J. Phys. Oceanog.* 6:671–678.

White, W. B., and N. E. Clark. 1975. "On the development of blocking ridge activity over the North Pacific." *J. Atmos. Sci.* 32:489–502.

Wilson, J. G., and J. E. Overland. 1987. "Meteorology (of the Gulf of Alaska)." In *The Gulf of Alaska: Physical Environment and Biological Resources.* Report 86-0095. Washington, D.C.: National Oceanographic and Atmospheric Administration.

Wooster, W. S., and D. L. Fluharty, eds. 1985. *El Niño North: Niño Effects in the Eastern Subarctic Pacific Ocean.* Seattle: Washington Sea Grant.

2. Climate of the Coastal Temperate Rain Forest

KELLY REDMOND AND GEORGE TAYLOR

□ □

The area of the coastal temperate rain forest is largely defined by climatic factors. In this chapter we describe the climatic characteristics that distinguish this region. If the natural and human inhabitants of the region are to survive as healthy individuals and communities, the management of their interaction must take account of climate and its variability. These interactions take place in the face of a largely unpredictable and constantly fluctuating physical environment. Indeed, it is impossible to arrive at an accurate evaluation of past management practices if these natural fluctuations—which occur on time scales from seconds to centuries—have not been factored in.

The coastal mid-latitude rain forests occupy a narrow coastal strip near the Pacific Ocean (Figure 2.1). The marine influence attenuates considerably at the crest of the first major range parallel to the coast, whether the Sierra Nevada, the Cascades, the British Columbia Coast Range, or the Chugach Range. In Oregon, for example, the annual precipitation decreases from approximately 95 inches at the Santiam summit to 14 inches at Sisters (18 miles east) to 8 inches near Redmond (37 miles east) (Plate 1).

The location of the coastal temperate region is intimately connected with a particular combination of climatic properties: a close proximity to cool and moist oceanic air, a subdued range of temperature extremes, and a high frequency of clouds, fog, and precipitation. With the exception of wind, these forests are exposed to fewer climatic hazards than North American forests to their east and south.

The north–south orientation of the coastline is almost perpendicular to the

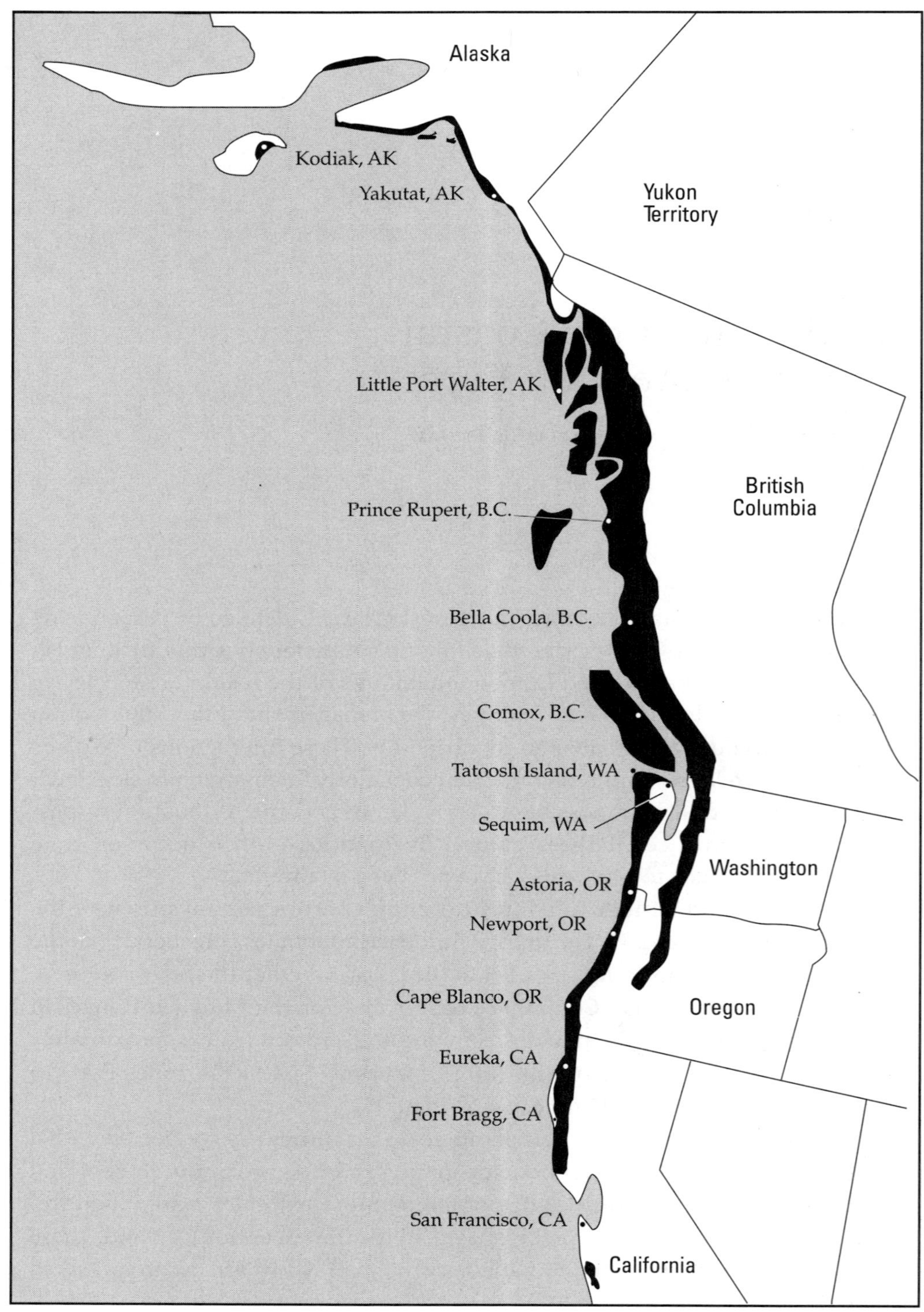

Figure 2.1. Selected locations mentioned in Chapter 2.

prevailing atmospheric flow at upper levels. The steep topography leaps from the sea nearly everywhere. The effect of this sudden rise in elevation on moisture-laden air is to produce copious precipitation. On both local and regional scales, the interaction between ocean, land, and topography creates considerable complexity of climate elements.

In summer, the equator-to-pole temperature difference—which drives the general circulation of the atmosphere—reaches an annual minimum in the Northern Hemisphere. This causes the prevailing west-to-east flow to shift northward and decline in intensity. The core of this flow impinges on the Canadian and Alaska Panhandle coasts. High pressure forms at the surface over the eastern Pacific west of the continental United States. Associated large-scale subsidence (a sinking of the air mass) inhibits the formation of precipitation, particularly along the southern half of the coastal forests.

All major ocean surface currents are wind-driven. The clockwise airflow around the oceanic summer high-pressure cell creates a southward-moving ocean current along the West Coast. This current, combined with the effect of the earth's rotation, produces ocean upwelling along the West Coast. This southward-flowing current and upwelling usually begins its summer existence abruptly in an event known as the "spring transition." The cool water brought to the surface by upwelling greatly modifies the coastal summer climate.

As autumn approaches, the hemispheric circulation expands and strengthens. October brings the wettest conditions for the northern coastal forests. In autumn, too, the subtropical high-pressure cells contract in strength and size and move southward. The south-flowing coastal ocean current slackens, eventually stops, and is replaced by a northward winter flow. The winter current does not produce upwelling and its associated cool band of water.

As winter arrives, the storm track becomes more active and moves southward, reaching its most southerly position in January and February. The month of greatest precipitation similarly advances from October at high latitudes to February at low latitudes. By early November, the main storm season is in full force from Vancouver Island to southern Oregon.

As the season further progresses to spring, the hemispheric circulation slowly begins to contract back toward high latitudes and declines in intensity. This is a gradual process, completed over several months, during which the frequency of cyclonic storms striking the coast slowly diminishes.

Precipitation and Humidity

The mid-latitude rain forests of western North America are among the wettest areas of the world apart from the tropics. Coastal areas in the Pacific Northwest receive 60 to 80 inches of precipitation a year—and more than 200 inches typically falls in the higher coastal mountains immediately to the east. Farther north, annual precipitation exceeds 200 inches at some sea level sites and over 250 inches at higher elevations. Henderson Lake, in northern British Columbia, has recorded 320 inches in one year. Even greater amounts are likely in ungauged locations away from sea level.

Unlike subtropical regions, where the wet season tends to occur during summer, mid-latitude coastal areas receive the bulk of their annual precipitation during winter. The wettest months are in early autumn (September and October) to the north; in late autumn/early winter (October–December) midway through the coastal rain forest range; and later in winter (January–February) to the south (Figure 2.2). Although significant precipitation can occur during the warm season, average totals during those months are generally lower than in winter. Summer precipitation varies strongly with latitude (Figure 2.3).

Another prominent feature of this region, perhaps more noted by inhabitants than the amount, is the frequent and persistent nature of precipitation. In much of the region, rain or snow falls on more than half the days of the year. The number of days per year with measurable precipitation increases from 50 to 100 in southern California to two to three times that much in northern British Columbia and southeastern Alaska (Figure 2.4). At their southern end, the coastal forests do not seem to be found where measurable precipitation

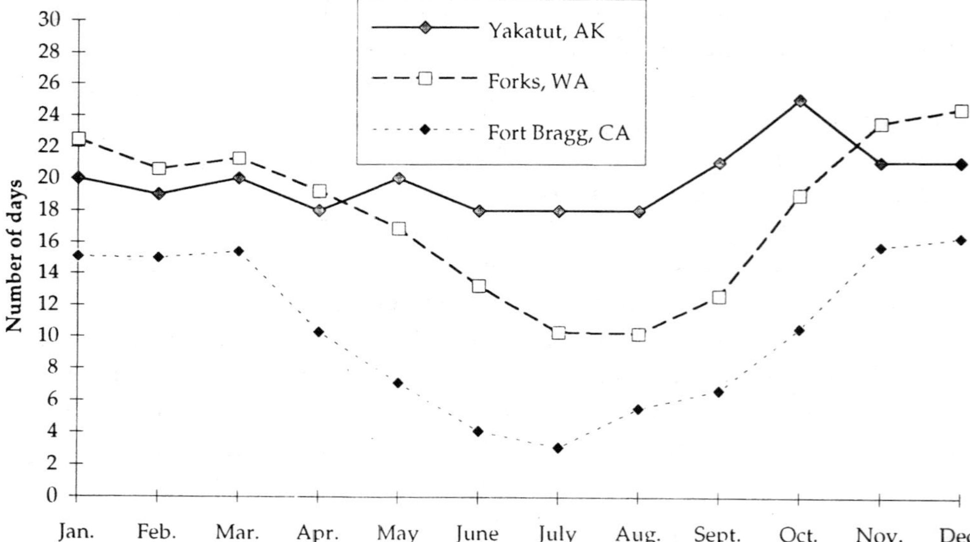

Figure 2.2. Average number of days with measurable precipitation near the south (Fort Bragg, California), middle (Forks, Washington), and north (Yakutat, Alaska) portion of the coastal temperate rain forest. *Sources:* Western Regional Climate Center, Oregon Climate Service, and Environment Canada.

(0.01″ or more) falls on fewer than 100 days. Some very extreme cases of precipitation frequency have been reported, especially in the northern half of the region. Little Port Walter in the Alaska Panhandle has seen rain fall on 285 days of a 365-day period. Most coastal stations have noted periods where measurable precipitation has fallen every day for six consecutive weeks or more. Yakutat, Alaska, has seen rain on 45 consecutive days; Prince Rupert, British Columbia, on 57 days; and Little Port Walter, Alaska, has recorded rain on 69 consecutive days.

Station	Jan.	Feb.	Mar.	Apr.	May	June	July	Aug.	Sept.	Oct.	Nov.	Dec.	Annual Total (inches)
Kodiak Island, AK	10	8	7	7	8	7	5	8	11	10	9	10	63.25
Seward, AK	9	8	6	6	6	4	3	8	15	16	9	10	67.74
Yakatut, AK	8	7	7	7	6	5	5	8	12	15	10	10	151.41
Juneau, AK	8	7	6	5	6	6	8	10	12	14	9	8	54.26
Little Port Walter, AK	10	8	8	6	5	4	4	6	10	16	12	11	225.13
Annette, AK	10	9	8	8	6	5	4	6	9	15	11	11	103.18
Prince Rupert, BC	10	8	7	7	6	5	4	6	10	15	11	11	100.71
Kitimat, BC	12	10	8	6	4	3	3	4	8	15	14	12	107.50
Bella Coola, BC	15	10	7	5	4	3	3	4	8	15	14	13	66.04
Port Hardy, BC	13	9	8	6	4	4	3	3	7	14	14	14	73.53
Estevan Point, BC	13	11	10	8	5	4	3	3	5	12	14	13	125.23
Comox, BC	14	10	10	5	4	3	3	3	4	11	17	17	46.77
Vancouver Airport, BC	13	11	9	6	5	4	3	3	6	10	15	15	45.96
Neah Bay, WA	14	11	10	7	4	3	2	2	4	10	15	17	104.42
Astoria Airport, OR	15	11	11	7	5	4	2	2	4	9	15	16	66.42
North Bend, OR	16	12	12	7	5	3	1	2	3	7	16	17	63.78
Gold Beach, OR	15	13	13	7	5	2	1	1	3	7	15	17	79.40
Eureka, CA	16	13	14	8	4	1	0	1	2	7	17	16	37.53
Monterey, CA	19	14	17	9	2	1	0	1	2	5	15	15	18.70
Long Beach, CA	21	21	17	6	1	0	0	1	2	2	14	14	11.80
San Diego, CA	18	15	18	8	2	1	0	1	2	4	15	16	9.90

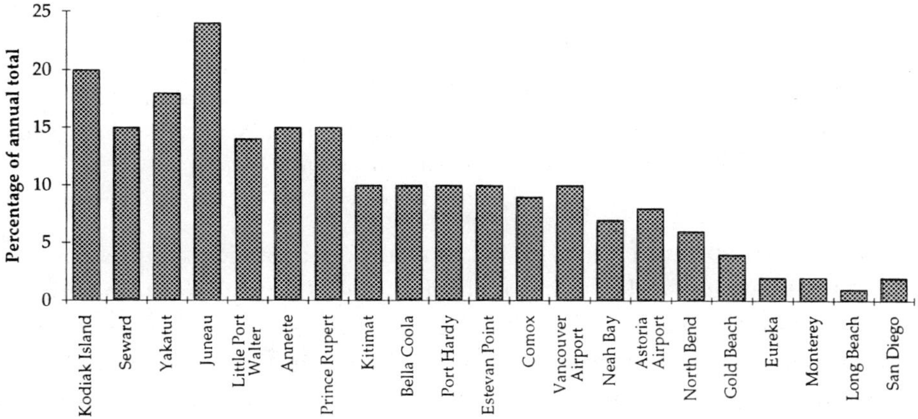

Figure 2.3. Percentage of annual precipitation that falls each month along a coastal transect from Kodiak, Alaska, to San Diego, California, and percentage of total observed from June to August.

Station	Jan.	Feb.	Mar.	Apr.	May	June	July	Aug.	Sept.	Oct.	Nov.	Dec.	Annual Total
Kodiak Island, AK	19	17	16	16	19	16	14	15	18	18	17	18	203
Seward, AK	14	13	13	12	15	13	12	15	18	17	13	15	170
Yakatut, AK	20	19	20	18	20	18	18	18	21	25	21	21	239
Juneau, AK	18	17	18	17	17	16	16	18	20	24	19	20	220
Little Port Walter, AK	22	19	21	19	18	15	14	15	19	26	23	23	234
Annette, AK	20	19	20	19	17	15	14	15	18	24	22	22	225
Prince Rupert, BC	21	19	21	19	18	17	16	17	19	25	23	22	237
Kitimat, BC	19	17	19	17	17	14	14	13	18	23	22	19	212
Bella Coola, BC	19	16	17	16	15	16	13	13	16	21	20	19	201
Port Hardy, BC	22	19	20	19	16	14	11	12	15	21	23	22	214
Estevan Point, BC	23	20	21	19	15	13	10	11	13	20	23	23	211
Comox, BC	19	16	16	13	11	10	8	8	10	16	19	20	166
Vancouver Airport, BC	19	16	16	13	12	10	7	7	9	15	19	21	164
Neah Bay, WA	20	18	19	17	14	11	8	9	12	18	22	22	190
Astoria Airport, OR	21	19	21	18	15	12	8	8	10	15	21	22	190
North Bend, OR	19	17	20	16	12	9	4	6	7	12	20	19	161
Gold Beach, OR	15	14	16	11	7	5	2	3	5	9	15	15	117
Eureka, CA	14	14	16	11	8	5	2	3	4	9	15	15	116
Monterey, CA	10	9	11	7	4	3	2	3	3	4	8	9	73
Long Beach, CA	5	5	6	3	1	0	0	1	1	2	4	5	33
San Diego, CA	6	5	7	4	2	1	0	1	2	2	5	6	41

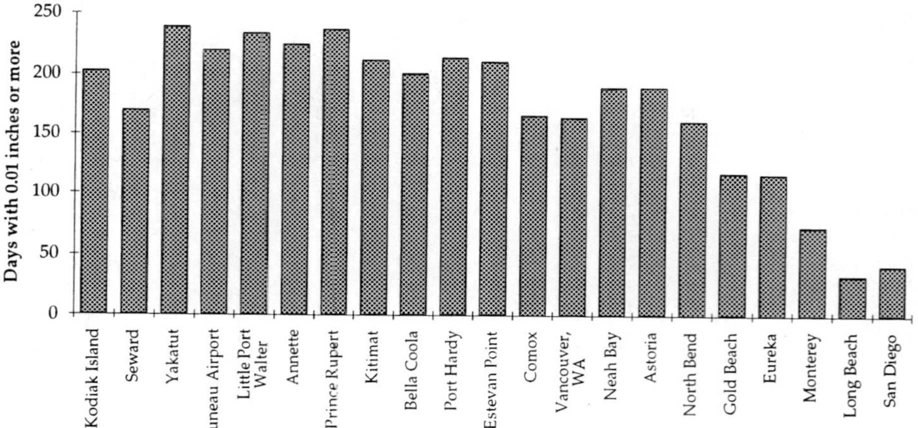

Figure 2.4. Average number of days with measurable precipitation along a coastal transect from Kodiak, Alaska, to San Diego, California. *Sources:* Western Regional Climate Center, Oregon Climate Service, and Environment Canada.

The most important factors influencing annual average precipitation in the region are elevation and distance from the coast. Locally, elevation is the critical factor; but on a regional basis, distance from the coast becomes increasingly important. Stations a short distance inland in Oregon and Washington receive precipitation on about 35 percent of all the hours of a typical January (and about 20 percent of all the hours in a year). Coastal plains and inland valleys in northern Oregon see rain falling during about 20 percent of the hours in January (and about 11 percent of the hours in a year).

Although air masses in the Northwest have sufficient amounts of water to allow precipitation to occur at sea level and over flat terrain, orographic (terrain-induced) influences on precipitation are very significant. The primary orographic effect is precipitation increasing with elevation; in general, the higher the elevation, the greater the precipitation. There may be an elevation at which precipitation reaches maximum values, but if it exists it is likely above the tree line, which is rather low through much of the coastal rain forest region.

There are a number of "rain shadows" where locations downwind from mountains are shielded from precipitation, sometimes to a considerable extent. Along the north coast of the Olympic Peninsula, for example, precipitation decreases from over 100 inches near the coast to about 15 inches annually at Sequim. Parts of Puget Sound exhibit the same effect. A similar pattern is noted on the east side of Vancouver Island.

The presence of land greatly increases precipitation because of increased friction and steepness. It has been estimated that only about 30 inches of precipitation falls over the open ocean just a few miles west of the Oregon beaches where 60 or more inches falls annually (Reed and Elliott 1973; Elliott and Reed 1973). Accurately mapping precipitation in mountainous territory has proved to be difficult. There are many methods of interpolating precipitation from monitoring stations to grid points. Some methods provide estimates of acceptable accuracy in flat terrain, but none has been able to explain the extreme and complex variations in precipitation that occur in mountainous regions. Significant progress has recently been achieved, however, through the development of an analytical model called PRISM that generates gridded estimates of monthly and annual precipitation. PRISM is uniquely suited to regions with mountainous terrain because it incorporates a conceptual framework that allows the spatial scale and pattern of orographic precipitation to be quantified and generalized (Daly et al. 1994). This modeling method is being used to produce precipitation maps for every state in the continental United States; detailed maps for British Columbia and Alaska are not yet under way.

Analysis of long-term time series of precipitation in the coastal temperate region reveals considerable year-to-year variation. Typically the change from one year to the next is 15 to 18 percent of the annual mean, and the standard deviation is 13 to 16 percent of the annual mean. This is actually less variation than is typically observed farther away from the coast, where the change from one year to the next is closer to 18 to 22 percent of the mean. These figures apply to the northern United States and Canada; further south, they are several percentage points higher. Over a 100-year period, a typical range observed in annual precipitation is from about 60 to 140 percent of the long-term mean. In more variable climates away from the strong oceanic influence, this range is more nearly 50 to 200 percent (Figure 2.5).

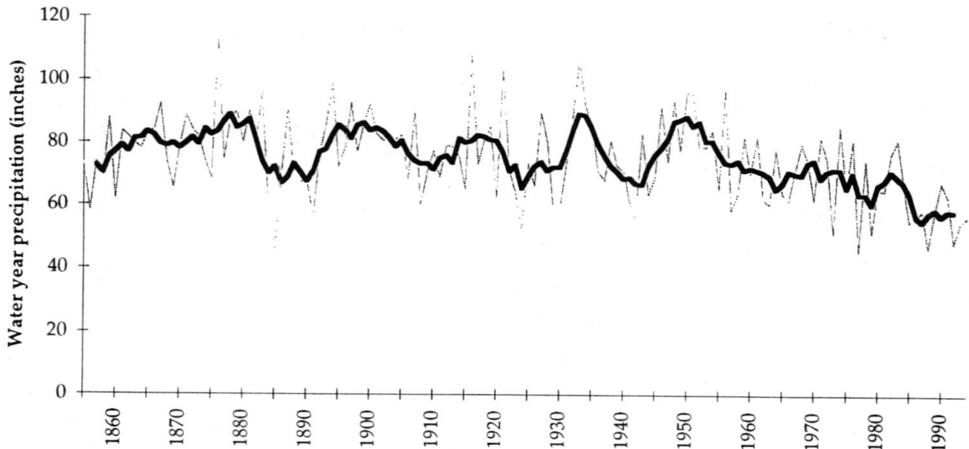

Figure 2.5. Water year (October–September) precipitation at Astoria, Oregon: 1855–1994. Tick mark is on ending year. Thin line: annual totals; thick line: five-year running average. *Source:* Oregon Climate Service archives.

Precipitation records in the vicinity of the coast do not show much correspondence with records from the Great Basin and its northward extension into Canada. The Dust Bowl years, for example, do not appear to have been remarkably dry along the Canadian coast. Despite repeated attempts to find cyclical patterns, there is no convincing evidence that precipitation follows such patterns. Cycles can always be identified for short periods, but they seldom persist for very long after their "discovery."

Where precipitation varies so greatly in space, a seemingly minor station relocation can cause considerable changes in *reported* average precipitation. Where spatial gradients of 5 to 10, and even 20 to 30, inches per mile can occur, a gauge relocation of just a few hundred yards in the horizontal or vertical can result in significant changes in long-term averages. These spurious effects, masquerading as apparent changes in climate, are notably difficult to detect in a region where long-term gauges are relatively few and widely separated.

The presence of the Pacific, combined with generally mild temperatures, causes average relative humidity throughout the rain forest region to be rather high. Table 2.1 shows long-term average relative humidity at 4 P.M. (in general, the minimum values for the day) at several locations within the coastal rain forest area.

Streamflow

When it falls as rain, the abundant precipitation soon appears in the many rivers and streams in the region. In the higher and more northerly portions, an

Table 2.1. Mean 4 P.M. Relative Humidity (percent)

Site	Jan.	Feb.	Mar.	Apr.	May	June	July	Aug.	Sept.	Oct.	Nov.	Dec.
Yakutat, AK	78	78	74	72	74	76	79	80	79	81	82	83
Annette, AK	74	73	68	66	64	68	69	71	72	77	76	78
Prince Rupert, BC	78	73	71	69	71	76	78	80	77	78	78	80
Tofino, BC	83	79	76	73	72	73	73	76	75	80	82	85
Hoquiam, WA	89	83	71	71	67	71	72	77	74	75	90	89
Astoria, OR	78	76	69	70	69	71	69	71	70	74	78	81
North Bend, OR	80	79	74	73	73	74	73	74	75	79	80	82
San Francisco, CA	72	66	68	64	60	58	61	63	55	65	64	63

increasing percentage falls as snow and does not appear until the spring melt. In either case, precipitation and streamflows correlate extremely well on an annual basis.

During this century, many of the rivers and streams that originate in or flow through the coastal rain forest have been dammed for hydroelectric and flood control purposes. This has made streamflows more uniform by mitigating the highest flows in winter and spring and maintaining higher minimum flows in summer and autumn. Over the past fifty years, increases in reservoir storage capacity on the Columbia River system have intercepted the dominant spring snowmelt peak, reducing the large late-spring pulse of fresh water emptied into the sea near Astoria. Because of this delay the amount of fresh water sent to the sea in autumn and early winter has increased. The freshwater plume that used to travel south on summer ocean currents has been reduced, and the northward-flowing plume in winter has increased. Declines in salinity along the coast have consequently been noted as far as 1000 kilometers north of the mouth of the Columbia River (Ebbesmeyer and Tangborn 1992). The implications for estuaries and the species that depend on them are unknown.

The Columbia River drains regions well east of the coastal temperate rain forest, and the effects of its freshwater plume are felt far up and down the coastline by the marine life that inhabits the nearshore environment. The estimated natural flow of the Columbia has been monitored since 1879, providing residents and policymakers a detailed record of the natural variability of this vast hydrological system (Figure 2.6). The severely dry period from 1928 to 1932, for example, is widely used as the "design drought" for the hydroelectric system. The dry period beginning in 1985 has historical significance—one of many factors associated with the decline of the salmon stocks in the Columbia Basin.

Temperature

The pervasive marine influence in the coastal temperate region strongly moderates the annual and daily cycles of temperature. This influence begins to

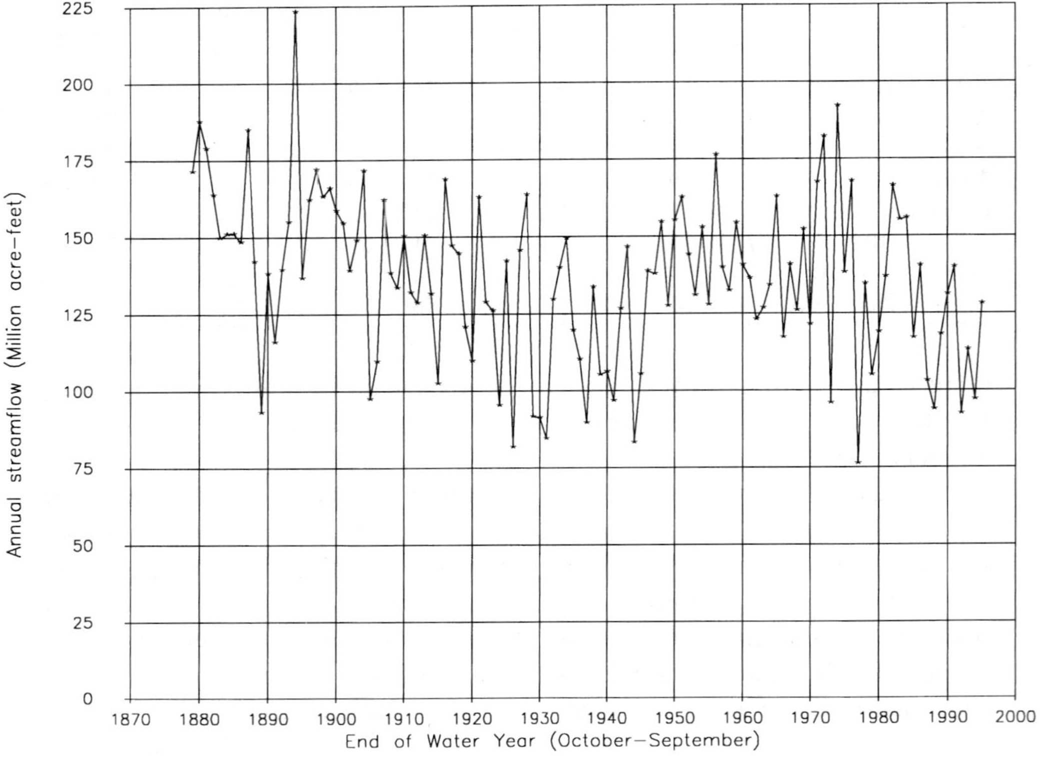

Figure 2.6. Annual water year (October–September) streamflow of the Columbia River at The Dalles, Oregon. Reconstructed natural flow after adjusting for storage, evaporation, and diversion. *Source:* Northwest Power Planning Council.

diminish just inland of the beach at many locations, but it persists inland wherever there are gaps from streams traversing the coastal mountains. For example, temperature is much more variable as one moves inland from Newport on the Oregon coast to Corvallis in the Willamette Valley (elevation 230 feet) to Madras just east of the Cascades (elevation 2230 feet) (Table 2.2). The Coast Range provides some insulation from the moderating effects of the ocean so noticeable at Newport, and the additional barrier of the Cascade Mountains causes temperatures at Madras to be considerably more extreme than at Corvallis.

A distinguishing feature of the temperate region is the lack of extended periods of high temperature. It is most remarkable, for example, that in over 100 years of observation the temperature at Eureka, California, has never exceeded 86°F. Probably no other station in the United States (alpine sites excepted) can claim this distinction. At some coastal locations the temperature can climb to 90 or even 100°F, but only for a day or two. At this point atmospheric feedback processes set in motion a landward push of cool marine air.

Table 2.2. Temperatures at Three Oregon Sites

Parameter	Newport	Corvallis	Madras
Mean maximum in warmest month	65.1°	81.2°	87.2°
Mean minimum in coldest month	50.0°	33.0°	21.1°
Days with maximum 90°F or more	0.5	13.5	33.3
Days with maximum 32°F or less	0.6	2.7	13.8
Record high temperature	100°	108°	112°
Record low temperature	1°	–14°	–40°
Annual heating degree days at 65°F	5132	4818	6444
Annual cooling degree days at 65°F	0	203	277

Even very warm locations, such as Medford, Oregon, which averages about 95°F for a maximum temperature in late July, has never spent more than eleven consecutive days above 100°F. The heat stress that extended warm temperatures impose on forest growth in other regions is thus seldom experienced by the coastal rain forest.

At the other end of the scale, extremely cold temperatures are not experienced nearly so frequently as they are just to the east of the first mountain ranges parallel to the coast. It is not uncommon for Yakutat, Alaska, to be at +40°F while nearby Whitehorse, 180 miles east, is at –40°F. When cold air does visit the coastal temperate rain forest, the usual cause is a buildup of high pressure over the continental interior; cold air has piled up to a depth sufficient to spill over gaps in the mountains and rush down the river corridors as a strong and gusty wind. The Fraser River and the Columbia Gorge are notorious for these dramatic events. One such event in December 1983 resulted in several days with easterly gusts over 80 mph in the Columbia Gorge during "fair" weather. As in the case of precipitation, temperature trends are characterized by significant year-to-year variations as well as noticeable long-term trends.

Winds

Local winds in the coastal temperate region are dominated by large-scale pressure patterns over the North Pacific and onshore. During winter and, to a lesser extent, autumn and spring, frequent cyclonic storms reach the area from the west, greatly influencing winds and other weather elements. Summer months see fewer strong storms and are more typically characterized by sea/land breeze regimes.

When the area is under the influence of cyclonic storms, large-scale winds tend to follow a particular pattern, although local terrain, the location, and the intensity of storms can alter this pattern significantly. A typical storm behaves as follows:

- As the storm approaches, winds from the south or southeast begin to blow. Wind speeds usually increase as the associated front nears, with the strongest winds just preceding the passage of the front. The strongest wind speeds of the year typically occur in this situation. Precipitation tends to be steady.
- As the cold front passes, pressure increases. Winds decrease and become more gusty, and the direction shifts to the northwest quadrant. Precipitation begins to vary rapidly in intensity.
- As the frontal system continues moving eastward, high pressure builds onshore, causing steady decreases in wind speeds and gustiness. If storms are closely bunched together, this step may be skipped entirely.
- As another storm approaches, south winds herald the repetition of the process.

During summer, the North Pacific High, a quasi-stationary area of high pressure off the coast, exerts a significant influence on western United States weather. The high moves northward in summer as the jet stream weakens and moves poleward. As a result, summer Pacific storms, already less vigorous than their winter counterparts, tend to be diverted to the north. Southern Alaska and northern British Columbia may receive substantial summer rainfall from these Pacific storms.

Wind is probably the single most important climate-related agent of disturbance in the rain forest: it exerts a profound effect on the vegetation and wildlife as well as the economy of the region. Several times each year, very strong winds hit the coast. Wantz and Sinclair (1981) have published estimates of extreme winds in the northwestern United States. They calculate that two-year recurrence values for one-minute average speeds along the coast are as high as 90 km/h (56 mph), while 50-year events produce winds of approximately 120 km/h (74 mph). Peak gusts may be 40 percent higher. More recent coastal measurements have shown that peak gusts above 100 mph occur nearly every year on exposed ridges. Cape Blanco, Oregon, a particularly windy location, averages more than 250 hours per year with a mean hourly wind speed in excess of 50 mph.

The most damaging storm in recent decades was the Columbus Day storm of October 11–13, 1962, which affected the entire coastal rain forest region. During that storm, Naselle, Washington, reported a peak gust of 160 mph, while Mount Hebo, Oregon, reported 131 mph and Portland (Morrison St. Bridge) 116 mph. Winds at Cape Blanco were estimated at 175 to 195 mph. The storm caused some $250 million in damage and took 50 lives. Fortunately the Columbus Day storm was a rare event. Lesser storms, however, can cause major damage along the coast: windthrow (downing of large trees), property damage, and marine-related damage. Although some areas near the coast are sheltered by the terrain, most of the coastal vicinity is quite exposed to wind damage during strong storms.

The strong canyon winds noted earlier for the Columbia and Fraser rivers are seen along the entire length of the coast where inlets and fjords channelize the flow. These sustained winds can last for days during periods of high pressure inland.

Cloud Cover and Solar Radiation

Another aspect of the pervasive marine influence is that the area tends to be very cloudy throughout the year. Along the coast, average annual sky cover (over a period of twenty years) ranges from 82 percent at Yakutat, Alaska, to 80 percent at Annette, Alaska, to 76 percent at Forks, Washington, and Astoria, Oregon, to 69 percent at Eureka, California, to 49 percent slightly inland at the San Francisco airport. At the northern end of the coastal temperate rain forest, only forty to sixty clear days occur in a typical year.

Tatoosh Island, Washington, just off the northwestern tip of the Olympic Peninsula, is representative of most areas along the immediate coastline with 60 to 80 percent average monthly cloud cover. Cloudiness often drops significantly a few miles inland, however, especially during the warm season. The best proxy indicator of cloud cover is probably air temperature: maximum daily air temperature is inversely related to cloudiness. The effects of cloud cover on temperature are generally much more pronounced in summer, when the incoming solar radiation is at its peak. During winter, when the area is dominated by large-scale storm systems, cloud cover tends to be much more uniform throughout the area.

Snow

Snow is relatively rare along the immediate coastline in the southern part of the temperate rain forest; in the northern half, snowfall is fairly heavy. The percentage of precipitation falling in the form of snow remains below 2 percent northward through Vancouver Island. On the northern British Columbia coast (Prince Rupert), the snow contribution reaches about 5 percent. At the northern end of the region, Yakutat receives an average of about 13 percent of its annual water from the sky as snow. Although these percentages sound small, Yakutat's snowfall is 200 inches (equivalent to nearly 20 inches of water). Some Alaska coastal sites report averages up to 350 inches per year.

As one moves inland or upward, the amount of snowfall reported per year increases steadily. Laurel Mountain, Oregon, for example, in the Coast Range at 3590 feet above sea level, averages 110 inches of snow per year. Assuming a ratio of snow to water of 10:1, this represents about 10 percent of Laurel Mountain's average annual precipitation of 116 inches.

The snowpack that can develop in the higher elevations and more northerly latitudes produces a protective covering for sensitive young seedlings that might otherwise be harmed by temperatures below freezing. In this way snow can extend the range of the rain forest into areas that would otherwise be limited by severely cold conditions.

Teleconnections

Like the biological components of coastal forests themselves, the entire global climate system is highly interconnected. Although it is extremely complicated, the climate system functions as a single entity. The term "teleconnection" has been used in climatology for the past half century to describe apparently connected climatic events separated by long distances—sometimes simultaneous and sometimes with time lags. In recent years it has become apparent that behavior of the tropical Pacific Ocean has widespread atmospheric effects around and beyond the entire Pacific Basin. The ocean surface between South America and the date line, on the equator, occasionally experiences anomalous warm El Niño or cool La Niña conditions.

Of particular relevance to the coastal forests, El Niño events—taken over the past sixty years—tend to be associated with wet winters in the southwestern United States, dry winters in the Pacific Northwest, and somewhat wetter than average winters further north in the Yakutat–Kodiak portion of coastal Alaska; temperatures tend to be above average along the entire west coast of the continental United States (Redmond and Koch 1991). La Niña events are associated with dry winters in southern California and the southwestern deserts, with wet winters in the Pacific Northwest, and with somewhat drier than average winters in southern and southeastern Alaska; temperatures are likely to be below average. The effects are not negligible in magnitude. Indeed, they amount to 30–40 percent differences in October–March precipitation. The combined effects of temperature and precipitation magnify the effects on snowfall, which later express themselves in streamflow in a similar sense.

Furthermore, these relationships are predictive. Pressure patterns in the Southern Hemisphere—themselves closely associated with tropical sea surface temperatures—during (Northern Hemisphere) summer and autumn provide considerable information about likely precipitation and temperature for the following winter in the Pacific Northwest. The Southern Oscillation Index is often used to characterize the tropical Pacific (Figure 2.7). Spring and summer streamflow on the Columbia River is high following La Niña events

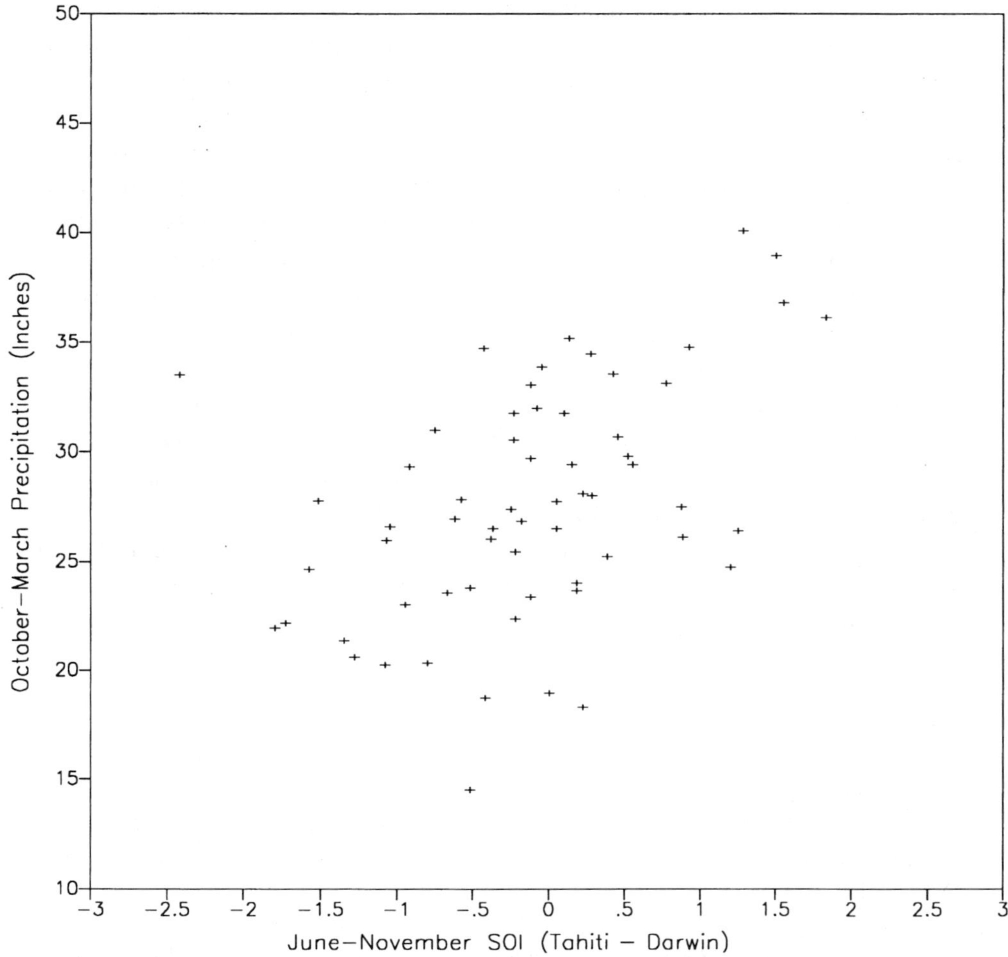

Figure 2.7. Relationship between June–November Southern Oscillation Index (SOI) and subsequent October–March precipitation averaged by area for the state of Washington based on 61 winters beginning 1933–1934. When SOI > 0 (La Niña or cold tropical sea temperatures), the state typically experiences a wet winter. When SOI < 0 (El Niño or warm tropical sea temperatures), the state typically experiences a drier than normal winter. *Source:* Western Regional Climate Center.

and low following El Niño events. These relations are not perfect—there are exceptions—but they apply to the vast majority of cases.

Weather, Climate, and Decision Making

Weather and climate, both local and regional, are the predominant reasons for the existence of the coastal temperate rain forest. An understanding of the distribution, seasonality, and history of weather and climate parameters will permit improved decision making on such critical issues as harvest and regrowth of trees, animal habitat, the effects of industrial and residential development, surface and groundwater supply, and preservation of native ecosystems. Effective policies relating to human interactions with the coastal temperate rain forest must be consistent with, and allow for, the influence of a constantly fluctuating physical environment. In many cases, quantifiable information from past climatic records is available to inform policy decisions. Moreover, sound management of natural resources requires continuous monitoring of the physical environment as well as methods to place current conditions into proper historical context. Climate archives exist to serve that purpose.

This chapter has highlighted the most salient characteristics of the climate of the coastal temperate rain forest. And there is much more information in both paper and electronic form—narrative and descriptive text, as well as extensive sets of data. Much of this information, moreover, can now be accessed via computer. Such information can be obtained from the state climate offices in Alaska, Washington, Oregon, and California; from the Western Regional Climate Center in Reno, Nevada; and from Environment Canada in Vancouver, British Columbia.

References

Bryson, R. A., and F. K. Hare, eds. 1974. *Climates of North America. Vol. II. World Survey of Climatology.* Amsterdam: Elsevier.

Daly, C., R. P. Neilson, and D. L. Phillips. 1994. "A statistical-topographic model for mapping climatological precipitation over mountainous terrain." *J. App. Met.* 33(2):140–158.

Ebbesmeyer, C. C., and W. Tangborn. 1992. "Linkage of reservoir, coast and strait dynamics, 1936–1990: Columbia River Basin, Washington Coast, and Juan de Fuca Strait." In M. E. Jones and A. Laenen, eds., *Interdisciplinary Approaches in Hydrology and Hydrogeology.* Minneapolis, Minn.: American Institute of Hydrology.

Elliott, W. P., and R. K. Reed. 1973. "Oceanic rainfall off the Pacific Northwest coast." *J. Geophys. Res.* 78(6):941–948.

Redmond, K. T. 1985. "An inventory of climate data for the state of Oregon." Report SCP-3. Corvallis: Office of the State Climatologist, Oregon State University.

———. 1992. "Climate." In S. D. Hobbs et al., eds., *Reforestation Practices in Southwestern Oregon and Northern California.* Corvallis: Forest Research Laboratory, Oregon State University.

Redmond, K. T., and R. W. Koch. 1991. "Surface climate and streamflow variability in the western United States and their relationship to large-scale circulation indices." *Water Res. Res.* 77(9):2381–2399.

Reed, R. K., and W. P. Elliott. 1973. "Precipitation at ocean weather stations in the North Pacific." *J. Geophys. Res.* 78(30):7087–7091.

Taylor, G. H., and A. Bartlett. 1993. "The climate of Oregon—coastal area." Special Report 913. Corvallis: Oregon Climate Service and Oregon Agricultural Experiment Station.

Taylor, G. H., and C. Daly. 1994. "Development of isohyetal analyses using the PRISM model." Unpublished technical report. Corvallis: Oregon Climate Service.

———. 1995. "Development of a model for use in estimating the spatial distribution of precipitation." Paper presented at the Ninth Conference on Applied Climatology, Dallas, Texas.

Wantz, J. W., and R. E. Sinclair. 1981. "Distribution of extreme winds in the Bonneville Power Administration service area." *J. App. Met.* 20:1400–1411.

3. The Influence of Geological Processes on Ecological Systems

David R. Montgomery

□ □

The geology and landforms of coastal western North America define a key component of the environmental template governing development of the region's ecosystems. Full consideration of the consequences of human use of landscapes requires a framework within which the effects of geological processes across spatial scales can be examined. At the regional scale, the tectonic setting governs topography and the pattern of large-scale disturbances, such as the the distribution of volcanoes and the frequency of earthquakes. Below the regional scale, broad differences in topography, tectonic setting, geology, and glacial history define geomorphic provinces. Within these provinces, local differences in topography and bedrock lithology distinguish "lithotopo units" associated with different extents, styles, or rates of geomorphic processes. Within lithotopo units, the type and distribution of hillslopes, hollows, channels, and floodplains govern processes that control habitat characteristics at even finer scales.

Each level of this conceptual hierarchy provides a context for understanding how geological and geomorphological processes influence ecosystem structure and dynamics. Given the variability of natural landscapes in time and space, such stratification is necessary if we are to understand how ecosystems are likely to respond to human activity. In the coastal temperate rain forest, large differences in landscape form and processes over a variety of scales should inform efforts to assess, and address, the ecological impacts of human decisions.

The physiographic variability of the coastal temperate rain forest region of western North America makes it an excellent place to explore such hierarchical approaches to linking geology and ecology. This region extends from northern California to southeastern Alaska and from the Pacific coast inland to encompass the coastal mountain ranges, the Cascade Range, and the intervening valleys (Plate 1*b*). Generally mild winters allow development of the forests that define the region. Differences in geology and tectonic setting lead to diverse landforms and geomorphological processes. Human activities that relate directly to these land-shaping processes in the region include the construction of large coastal and inland cities and their associated suburbs, the clearing of valleys for agriculture, and the practice of industrial forestry in the mountains. The increasing intensity of resource use in the region reflects dramatic increases in human population, along with the adoption of technological innovations such as the "steam donkey" in the late ninteenth century and the chainsaw and bulldozer after World War II.

The variety of geological settings and processes in this region is, of course, also responsible for its splendid and varied landscapes: the glaciated mountains of British Columbia and Alaska, islands of the Alexander and Haida Gwaii (Queen Charlotte) archipelagoes, forested valleys of the Olympic Peninsula, volcanoes of the Cascade Range, and the marine terraces of Oregon and California. Each of these distinctive landscapes reflects fundamental differences in geological processes, materials, and history. The region's tectonic setting further shapes the geomorphological influences on coastal temperate rain forest ecosystems.

Geology, Geomorphology, and Ecology

Landforms reflect the interactions among tectonic processes that govern the development of relief, the processes of erosion, the structure and lithology of the underlying rocks, and the influences of climate. Competition between the tectonic forces that drive rock uplift or subsidence and the geomorphological processes that erode rock and fill basins shapes regional physiography. The structure and lithology of underlying rocks, together with the influences of climate and climate history, modulate the intensity of erosion. Tectonic processes drive the uplift of rocks that become sculpted into a landscape; they also define the physical boundaries that control landscape dissection. Mountain ranges grow only when rock uplift exceeds erosion, a condition that may be satisfied for only a fraction of the time that a range exists. In the absence of rock uplift, landscapes erode to subdued topography (Davis 1899). Where rock uplift, lithology, and climate are constant, erosion may keep pace and a landform may persist for long periods (Hack 1960).

Lithology and structure determine a landscape's ability to resist erosion: weak or highly fractured rocks erode easily, compared to stronger or massive rocks. Regional climate determines the amount of precipitation available to drive the erosive processes that sculpt landscapes. Precipitation and temperature also influence the weathering processes that break down rocks and produce sediment. Physiography, in turn, influences the regional climate. Climate and topography appear to govern the distribution, frequency, and intensity of many geomorphological processes. Thus there is a feedback cycle between erosion and landform: erosion shapes topography, which itself controls such processes as runoff generation and shallow landsliding. Other processes, such as deep-seated bedrock landsliding and bedrock river incision, are more strongly influenced by geological structure and the strength of bedrock. These interactions illustrate how tectonics, geology, and climate shape the regional landforms that constrain biological and geomorphological processes.

Ecological systems, including patterns of human use and occupation, reflect the spatial and temporal variability of environmental conditions. A complex interplay of tectonic, geochemical, hydrological, and geomorphological processes governs the distribution of habitat types and natural disturbances in a landscape. This physical template strongly influences ecological processes. Geological processes influence ecological systems over a wide range of time and space (see Likens and Bormann 1974; Southwood 1977; Swanson et al. 1988; and Turner 1989). The converse is also true, as biological processes influence or alter the physical environment.

Human actions, in particular, dramatically alter the distribution, frequency, and magnitude of geochemical, hydrological, geomorphological, and biological processes in most regions of the world (Turner et al. 1990). Understanding human influence on environmental systems involves assessing the influence of geological and geomorphological processes on landscapes and ecosystems. Methods of sustainable resource management at the landscape scale require the application of such understanding in a spatial context in order to determine the likely distribution, frequency, and magnitude of landscape-scale environmental change. A simple hierarchical framework (Figure 3.1) offers one way to address geological influences on ecosystem structure and dynamics.

Plate Margins and Regional Tectonic Setting

Earth's surface is broken into plates consisting of dense oceanic crust, lighter continental crust, or both. As these plates move relative to one another, they either spread apart, collide across a convergent margin, or slide past one another along a transverse margin. Slow circulation of material in earth's mantle drives the motion of the plates, with new crust created where upwelling forms

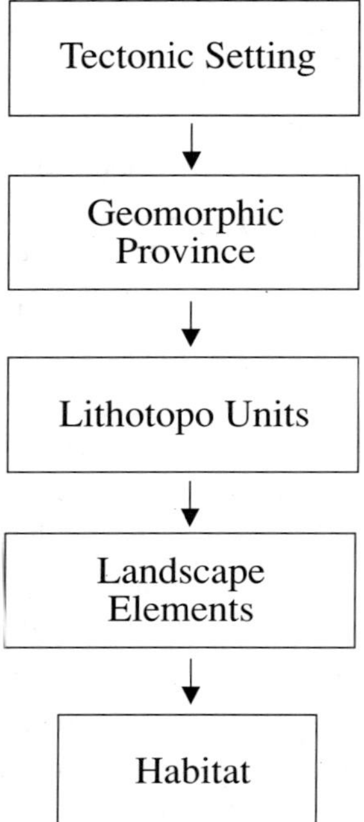

Figure 3.1. Hierarchical framework for understanding the influence of geological processes on ecological systems.

a spreading center or rift zone. Oceanic and continental crust collide in several distinct ways at convergent margins. In the coastal temperate rain forest region, the crust of the Pacific and Juan de Fuca plates either subducts beneath the North American plate or moves northward along the strike-slip faults of transverse margins.

These two types of plate margin—convergent and transverse—create distinct topographic settings, lithologies, and large-scale environmental disturbance regimes. The western coast of North America, for example, is defined by the San Andreas Fault Zone in California, the Cascadia subduction zone extending from northern California to southern British Columbia, and the transverse Queen Charlotte Fault Zone from northern British Columbia to southeastern Alaska (Figure 3.2). These tectonic boundaries shape the basic form of the region, determining the location of mountains with respect to the coast, the distribution of volcanoes, the structure and lithology of bedrock, and the size and frequency of earthquakes. The nature of the plate margins in the coastal temperate rain forest region changes as plates move past and collide with one another (Atwater 1970). The size and relief of the mountain ranges in the forest region reflect both the conditions at tectonic boundaries and their history. Hence large-scale tectonic processes influence local climates,

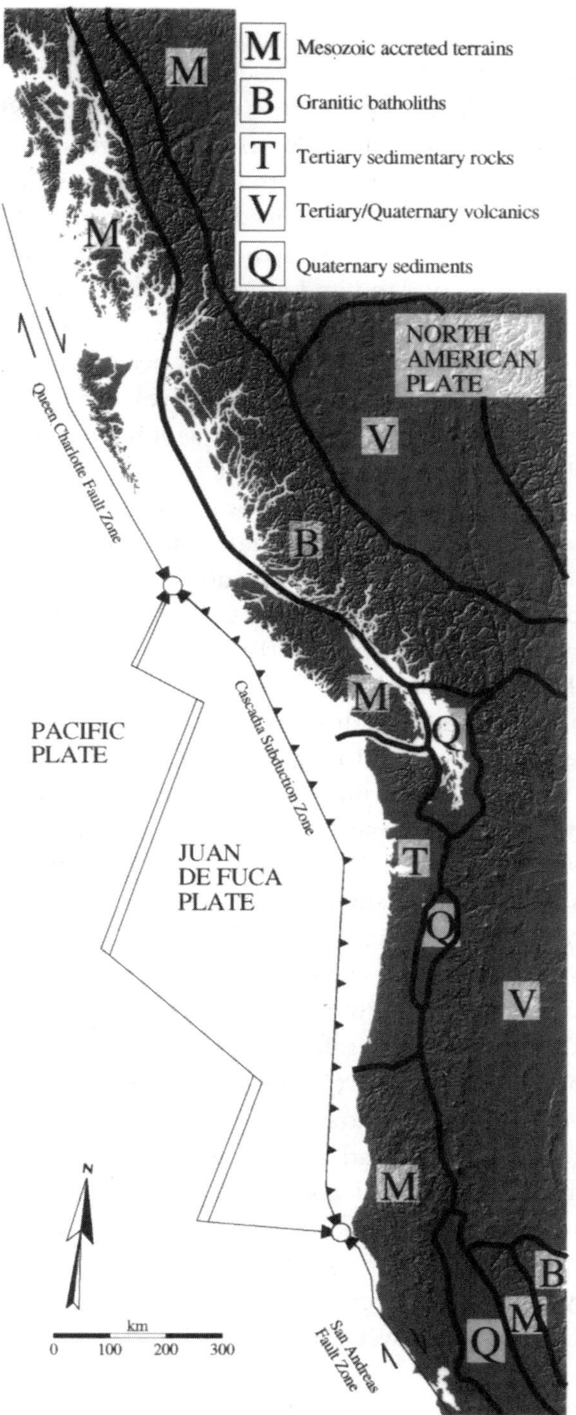

Figure 3.2. Tectonic setting and generalized geology of the West Coast of North America.

types of geomorphological processes, upland disturbance processes, and even stream habitat characteristics throughout the region.

In the typical subduction zone, one converging plate sinks under a less dense plate. Along the west coast of North America, this process involves collision of denser oceanic lithosphere with lighter continental crust. In the Cascadia margin, sedimentary rocks derived from continental debris pile up on the edge of the continent, creating a sedimentary wedge that forms low-elevation coastal mountains on land and a prism of sediments on the continental shelf. Partial melting of the sinking slab feeds a system of volcanoes that rise inland from the plate margin. The collision of oceanic and continental plates along the west coast of North America thus gives rise to a coupled system of two mountain ranges: a lower-lying coastal sedimentary wedge and a higher-elevation active volcanic arc further inland.

In contrast to the convergent nature of subduction zones, transverse margins involve lateral movement of two plates along a transform fault. Oblique convergence across a transform fault results in slip parallel to the fault and compression across the fault, the latter leading to uplift in the vicinity of the fault. Transform margins can transport rocks hundreds to thousands of kilometers from their point of origin (Umhoefer 1987). The movement of crustal materials along the complex fault geometries of transverse margins can result in local mass convergence and topographic uplift. Uplift of the coastal ranges of central and northern California, for example, reflects a combination of local fault geometry, compression across the plate margin, and migration of the Mendocino "triple junction," which seperates three plates in the region. Transverse margins may create a single belt of potentially complex uplift, rather than the two-range pattern typical of subduction zone margins.

Large-scale tectonic processes also control the complex geology of western North America. Distinct suites of rock types tend to occur in tectonically similar settings. The coastal temperate rain forest region has several such lithological assemblages (Figure 3.2): (1) Mesozoic (Cretaceous and older) accreted terrains consisting of partially to highly metamorphosed sedimentary and igneous rocks; (2) Cretaceous and Tertiary granitic batholiths comprising the uplifted and eroded roots of volcanoes; (3) Tertiary sedimentary wedges consisting of marine sedimentary and volcanic rocks; (4) Tertiary and Quaternary intrusive and extrusive volcanic rocks associated with active volcanic arcs; and (5) Quaternary alluvium (material deposited by rivers) and glacial deposits filling structural troughs between uplifted sedimentary wedges and the active volcanic arc.

Long-term geological processes govern the distribution of these lithological assemblages. Tectonic convergence associated with subduction zones scrapes off ocean floor sediments, creating coastal mountains of compressed and uplifted sedimentary rocks. Convergence also results in the accretion of seamounts and islands to western North America over millions of years. Much of the western United States accreted to the North American continent through such processes over the last 250 million years (Jones et al. 1982). Movement along transform faults subsequently shuffled and rearranged many of these accreted terrains. The distribution of granitic and volcanic rocks also reflects interaction of long-term erosional and tectonic processes. A change from a subduction boundary to a transform fault, for example, shuts off the source feeding magma to stratovolcanoes, which eventually erode, exhuming the cooled granitic roots of the volcanic arc. The association of large granitic bodies with present transform margins reflects this history of changing tectonic conditions along the plate margin.

Geological processes and disturbance regimes controlled by the tectonic setting affect ecological systems at the regional level over spans of time long enough to influence the origin of distinct plant and animal species. Disturbances associated with volcanoes and earthquakes, for example, are components of large-scale disturbance regimes affecting large areas (hundreds to thousands of square kilometers) over long periods or at infrequent intervals (Figure 3.3); thus they may be a significant regional influence on population dynamics (see Holt 1993). Populations of plants and animals isolated by

Figure 3.3. Toutle River (Washington), downslope of Mount St. Helens, showing the influence of large-scale volcanism on riverine habitat. (Photo courtesy of D. Montgomery.)

mountain uplift, glacial advances, or rising and falling sea levels can gradually diverge and form new species. Mountain uplift in the coastal temperate rain forest region coincided with speciation of the various Pacific salmon species (Stearly 1992). The present diversity and distribution of salamanders in coastal California reflect the influence of tectonic motion along the boundary of the Pacific and North American plates (Yanev 1980). Tectonic processes have shaped the evolutionary dynamics of ecological systems in North America over long periods of time. The regional tectonic setting, however, also influences geological and geomorphological processes that affect ecosystems over shorter time spans and smaller areas.

Geomorphic Provinces

The coastal temperate rain forest region encompasses distinct geomorphic provinces, conceptually similar to the "ecoregions" proposed by Omernik (1987), that differ in the large-scale geological and climatic processes controlling the distribution of relief, slopes, stream profiles, and geomorphological processes that influence habitat characteristics and dynamics. Although there are many ways to define them, geomorphic provinces reflect tectonic setting, bedrock types and structures, climate, and glacial history. These criteria distinguish areas where different geomorphological processes, each with characteristic links to ecological systems, dominate land form.

Bedrock Lithology and Structure

A variety of tectonic settings control the general suites of rock types and structures found within the coastal temperate rain forest region, as well as regional physiography. Each tectonic setting in this region possesses a distinct set of rock types with differing characteristics. The suite of rock types and structures associated with each tectonic setting, in turn, determines the mechanical properties that influence geomorphological processes and landforms. Marine sandstones, for example, produce coarse soils conducive to shallow landslides and debris flows. Expansive siltstones and micaceous sandstones typically disintegrate when wetted and then dried, producing fine, easily moved stream sediment. Gorges incised deeply through thick accumulations of glacial sediments promote landsliding. Volcanic rocks generally are well drained and also weather to clay-rich soils prone to slow-moving earth flows. Fresh volcanic rock can support deep valleys with steep slopes. Weathered volcanic and metavolcanic rocks and glacial sediments, by contrast, generally are mechanically weak and prone to landslides. Weathered granite produces coarse soils prone to debris flow, as well as sediment that rapidly breaks down into sand in stream channels. Fresh granite persists as boulders and cobbles in streambeds.

Alluvium is quite erodible, but it is deposited on gentle slopes where streams migrate laterally rather than cutting deeply into underlying materials. Each of these rock assemblages weathers and interacts with topography differently, leading to characteristic styles of landsliding and differences in the amount and timing of sediment delivery to streams.

Glacial History

The current interglacial period, during which civilization arose, is only the latest in a series of warm intervals separating glacial cycles over the past 2 to 3 million years. The most recent glaciation covered portions of the coastal temperate rain forest region with up to a kilometer of ice (Figure 3.4). In some areas glacial processes were minor (as in the northern California coast ranges and the Oregon Coast Range). In other areas, meltwater that flowed beneath the ice sheets incised large valleys into thick accumulations of glacial outwash (the Puget/Fraser lowlands); continental glaciation pock-marked landscapes with lakes and wetlands and carved deep fjords (as in southeastern Alaska); and alpine glaciers carved deep U-shaped valleys (Figure 3.5) that filled with outwash terraces when the glaciers withdrew (as in the northern Cascades). Different geomorphic provinces experienced different combinations of glacial processes, each leaving a distinct signature on the postglacial landscape.

Basic landform characteristics—which differ among areas affected by alpine glaciers, continental glaciers, or those never glaciated at all—influence the geomorphological processes that govern ecological disturbances and habitat

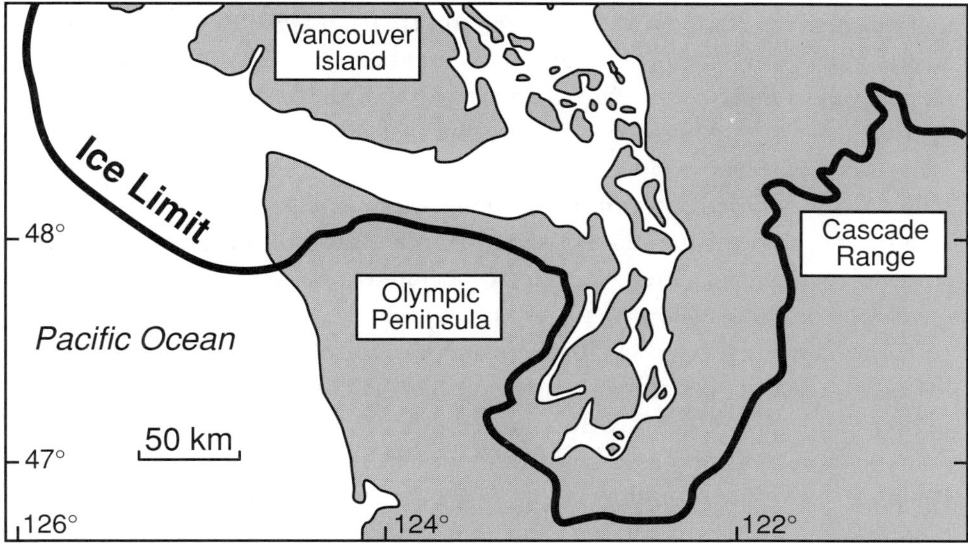

Figure 3.4. Southern extent of the Cordilleran ice sheet during the last glacial maximum. *Source:* After Booth (1994).

Figure 3.5. U-shaped valley carved by alpine glaciers. (Photo courtesy of D. Montgomery.)

characteristics. Unglaciated areas in the coastal temperate rain forest region typically possess steep, well-dissected hillslopes, and generally the processes on these slopes strongly influence stream channel characteristics (Figure 3.6). Shallow landsliding generally initates channels and maintains valleys in steep, soil-mantled landscapes (Montgomery and Dietrich 1988; Benda 1990). Continental glaciation left extensive wetlands that occupy wide valleys, leaving hillslopes only weakly coupled to streams. Most of these wetlands gradually fill with sediment from upslope. During the last glacial maximum, only a few islands of bedrock extended above the ice in many areas of alpine glaciation. Valley walls scraped smooth by alpine glaciers tend to form steep, relatively undissected slopes. Fine-scale dissection is now developing on the steep walls of some of these valleys. Less intensively glaciated areas sometimes exhibit well-developed hanging valleys on upper slopes where ice did not reach.

Tectonic setting, bedrock lithology and structure, and glacial history define at least 12 distinct geomorphic provinces in the coastal temperate rain forest region (Table 3.1). Each province hosts a distinctive suite or arrangement of geomorphological processes that influence hillslope and channel dynamics. Province-scale factors influence, among other things, the type and extent of landsliding, the nature of sediment delivered to stream channels, and the extent of hillslope–channel coupling. This variety of influences on habitat structure and dynamics makes it difficult to evaluate past management practices and to anticipate impacts of future land management decisions.

Figure 3.6. Well-dissected portion of the unglaciated Oregon coast. (Photo courtesy of D. Montgomery.)

Province-level influences on ecosystems involve large-scale physiography and disturbance regimes. Variations in temperature and precipitation related to elevation gradients, for example, strongly influence the structure of plant communities in Oregon and Washington (Franklin and Dyrness 1988). Provinces affected by continental glaciation generally host more abundant wetlands than unglaciated or alpine areas of the coastal temperate rain forest region. Province-level differences in geological and geomorphological processes also determine how different types of disturbance influence plant and animal population dynamics. Province-level processes influence the formation of species and the assembly of plant communities. Slug populations isolated by Pleistocene glacial advances on the Olympic Peninsula, for example, diverged genetically to form a variety of endemic species (Branson 1977). Similarly, salamander populations isolated by volcanic activity and extensive glacier-fed rivers in Oregon and Washington evolved into four distinct species of torrent salamanders (Good and Wake 1992). Province-level differences in environmental conditions also appear to influence the distribution of fish communities in Oregon streams (Hughes et al. 1987).

Watersheds and Lithotopo Units

The geomorphic provinces can be subdivided according to watersheds or according to landform assemblages of similar topography and geology

Table 3.1. Geomorphic Provinces of the West Coast of North America

Geomorphic Province	Tectonic Setting	Lithology	Glacial History
California Coast Ranges	accreted terrains	marine sediments, metamorphosed oceanic crust	unglaciated
Klamath Mts.	accreted terrains and batholiths	granitics, intrusives, marine sediments	minor alpine
Oregon Coast Range	sedimentary wedge	marine sediments	unglaciated
Willamette Valley	fore-arc basin	alluvium	unglaciated
Southern Cascades	active volcanic arc	extrusive and intrusive volcanics	alpine
Northern Cascades	active volcanic arc	extrusive and intrusive volcanics	alpine and continental
Olympic Mts.	sedimentary wedge	marine sediments, basalt, glacial outwash	alpine and continental
Puget/Fraser Lowland	fore-arc basin	glacial outwash, alluvium	continental
Vancouver Island	accreted terrain	marine sediments	continental
Queen Charlotte Is.	sedimentary wedge	marine sediments	continental
British Columbia Coast Range	batholiths and accreted terrains	granite, marine sediments	continental
Alexander Archipelago	accreted terrain and batholiths	marine sediments, granite, volcanics	continental

distinguished by their geomorphological processes and habitat characteristics. Although watersheds provide a logical framework for some forms of resource management, they often cross significant geological boundaries. Distinguishing lithotopo units—areas with similar topography and lithology—stratifies landscapes into areas with similar geomorphological processes—thereby identifying areas in which human activities might influence landscape-scale ecological processes in similar ways. Resource managers can use both approaches: planning, analyzing, and managing on a watershed basis but using lithotopo units to manage across watershed boundaries. Stream channels, for example, may have more in common within a lithotopo unit that crosses several watersheds than within the various lithotopo units making up a single watershed (Figure 3.7). The coastal temperate rain forest region is far too large and complex to specify all of its lithotopo units, but a simple example from Washington's Olympic Peninsula illustrates the process (Figure 3.8). The Olympic Peninsula consists of at least five lithotopo units: high-relief areas underlain by basalt; low-relief areas underlain by Tertiary marine sedimentary rocks; high-relief alpine areas underlain by Tertiary sandstone; lower-elevation areas of moderate to high relief underlain by Tertiary sandstone; and low-relief glacial deposits and outwash terraces.

Watersheds provide a logical framework for assessing geomorphological impacts on ecological processes. Soil and water move down slopes into tributaries and eventually to rivers and estuaries. Hence a watershed context is crucial to link upstream causes with downstream effects, as well as to examine the "cumulative effects" of land management distributed across a landscape. Smaller watersheds can frame detailed analyses of areas of interest to resource managers within a larger basin. Watersheds provide an ideal management and

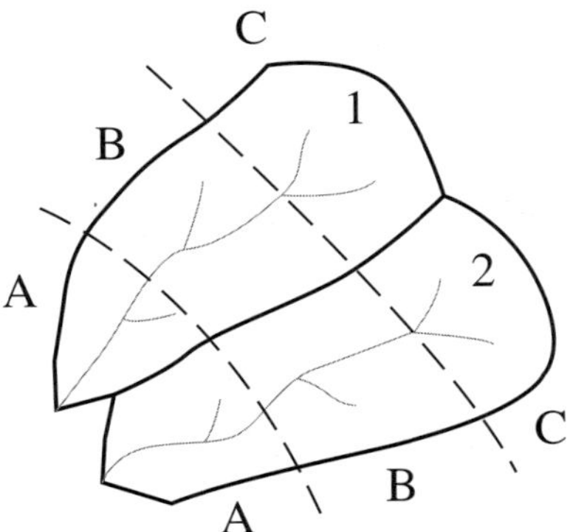

Figure 3.7. Relation of lithotopo units (A, B, C) to watersheds (1, 2).

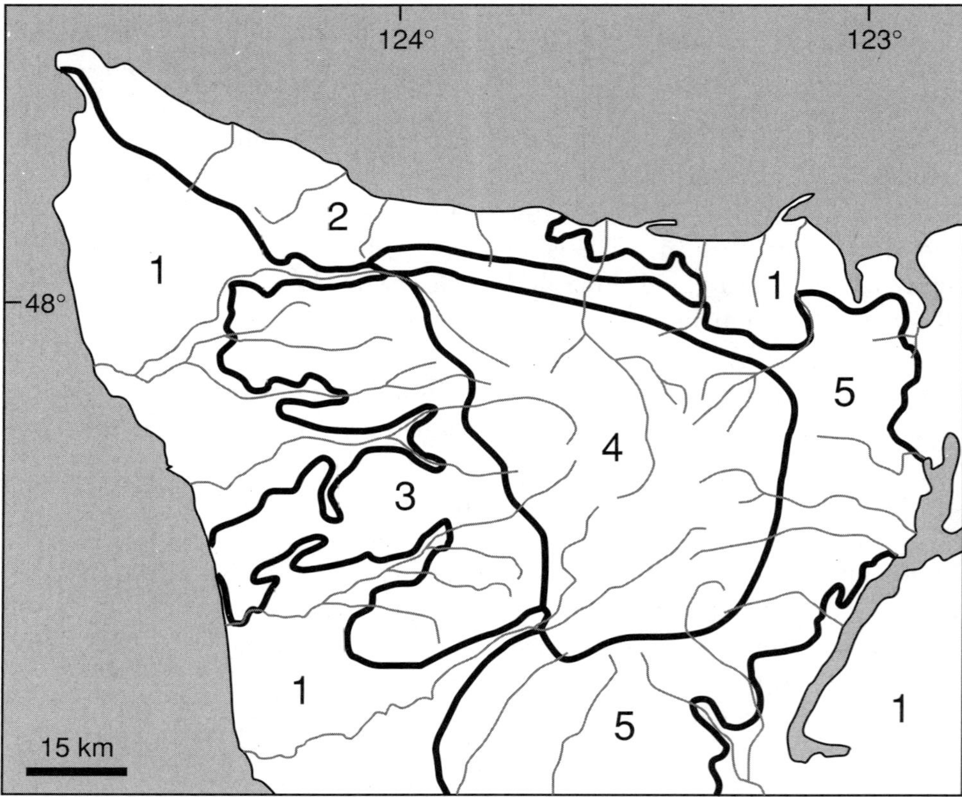

Figure 3.8. Generalized lithotopo units of the Olympic Peninsula: (1) low-relief glacial deposits and outwash; (2) low-relief Tertiary marine sediments; (3) low- to moderate-relief Tertiary sandstone; (4) high-relief Tertiary sandstone; (5) high-relief Tertiary basalt.

analysis framework for many aspects of natural resource management (Schramm 1980; Newson 1992; Montgomery et al. 1995a).

The landform assemblages that define lithotopo units control the abundance, type, and distribution of habitat at the landscape scale. Lithotopo units also influence properties of the landscape at the watershed and larger scales. Patterns of bedrock hydraulic conductivity at the scale of lithotopo units can control groundwater recharge. This may lead, for example, to greater summer flow in areas with more porous rocks. Differences in soil geochemistry at the scale of lithotopo units also influence stream chemistry and vegetation composition, as in the serpentine grasslands of northern California.

Landscape Elements

A natural set of landscape elements frames examination of the link between geomorphological and ecosystem processes at even finer scales. Hillslopes, hollows, channels, and floodplains define the building blocks of watersheds in the simplest such division for the coastal temperate rain forest region. Various combinations and arrangements of these landscape elements influence ecological processes in different landscapes (Figure 3.9). Each fundamental landscape element can exhibit distinct characteristics in different environments, landscapes, or lithotopo units. These differences reflect bedrock distribution, disturbance history, links between landscape elements, and the influences of routing processes, past climates (such as glaciation), and changes in tectonic boundary conditions (uplift and fault offset).

Hillslopes

Hillslopes produce sediment and transport it downslope to stream channels. Sediment transport on soil-mantled hillslopes is limited by transport capacity, while that from bedrock hillslopes is limited by sediment production. Bedrock becomes exposed at the surface if transport exceeds production, whereas a soil profile develops on a slope where production equals or exceeds transport. The relative importance of soil-forming processes—including physical (cyclical freeze-thaw or wetting-drying), chemical, and biological processes (tree throw and burrowing activity)—varies among geomorphic provinces, rock types, and climate.

On soil-mantled slopes, the production and transport of sediment acts like a conveyor belt continuously fed from upslope and the underlying bedrock. Gradual downslope transport of sediment characterizes many hillslopes. On steep hillslopes, however, both shallow and deep-seated landsliding influence sediment transport. Shallow mass wasting involving soil and fractured bedrock typically occurs during intense or sustained rainfall (Caine 1980) and may reflect the frequency of storms or fires or the time required to develop a soil profile thick enough to raise plant roots above bedrock (Okunishi and Iida 1981).

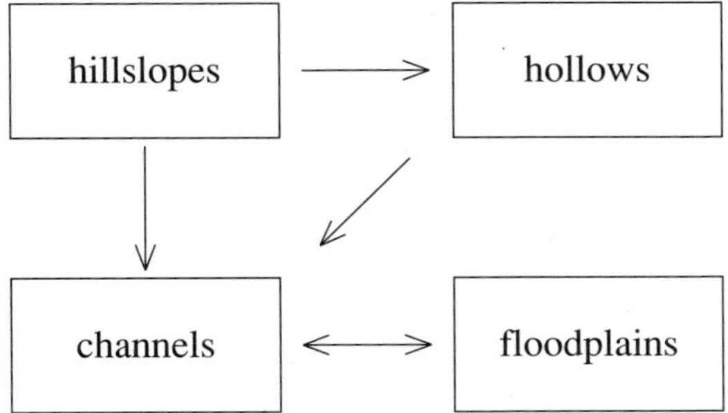

Figure 3.9. Sediment transport links among hillslopes, hollows, channels, and floodplains in mountain drainage basins of the coastal temperate rain forest region.

Earthquakes also can trigger bedrock landslides. Deep-seated landsliding occurs on slopes too steep to be supported by the mechanical strength of the slope-forming materials (Schmidt and Montgomery 1995). Whether shallow or deep-seated, the episodic nature of landslides means that sediment transport in steep landscapes varies significantly through time.

Human activity dramatically affects hillslope processes in many regions of the world. Clearing vegetation and altering drainage can dramatically increase rates of landsliding from steep hillslopes (see the review in Sidle et al. 1985). Soil compaction caused by equipment used in logging or farming can result in overland flow in locations where rainfall previously infiltrated into the soil, altering runoff and streamflows. Human actions can also influence soil-forming processes. Windthrow, for example, is a primary soil-forming process in the Pacific Northwest: the uprooting of trees during serious windstorms mixes bedrock into the soil profile, generating new sediment. Clearcut logging effectively eliminates windthrow and thus may slow soil production. Such changes in hillslope processes can significantly alter the amount and rate of sediment delivered from hillslopes to streams.

Hollows

Hollows—unchanneled valleys—generally occur at the head of the smallest channels in a drainage basin, so-called "first–order" streams. The finest-scale valleys in a landscape, hollows represent an important link between sediment production on hillslopes and transport to stream channels. Scoured intermittently by shallow landslides or incised by upslope extension of stream channels, hollows resist the tendency to accumulate sediment delivered from surrounding hillslopes. Steep hollows experience a cycle of gradual infilling followed by eventual catastrophic excavation by landsliding, whereas shallow hollows may record past responses of the channel network to climate change (Dietrich et al. 1986). Colluvium accumulates for thousands of years between landslides within individual hollows in the coastal temperate rain forest region (Reneau and Dietrich 1990). Gradual accumulation and periodic discharge of sediment from hollows indicate that they function as "sediment capacitors." The frequency of large storms or fires, the root strength of vegetation, the rate of infilling and time since previous failure—all influence the frequency of sediment excavation from hollows.

Human modifications of the landscape can dramatically accelerate sediment delivery from hollows. Weakened root strength following logging, for example, can increase landslide rates from both steep hillslopes and hollows. The common practice of concentrating road drainage into hollows can trigger landslides, even along roads built on ridgetops to minimize road-related landsliding (Figure 3.10).

Figure 3.10. Road-related landslides in the northern Cascade Range, Washington. (Photo courtesy of D. Montgomery.)

Channels

Stream channels receive and transport the sediment eroded from a landscape. Their character reflects the distribution, intensity, and history of erosional processes both in the immediate area and throughout their catchment area, or watershed. Channel morphology and processes reflect sediment supply, transport capacity, and external influences such as bank-forming materials, the presence and abundance of large woody debris, bridge piers, and abandoned cars. Stream channels in mountain watersheds of the Pacific Northwest typically exhibit an orderly pattern of downstream variations in channel morphology and processes that result in distinct habitat patterns (Montgomery and Buffington 1993). While many channel properties vary through time, the variability of channel morphology in space provides a framework in which the influence of geomorphological disturbances and processes on ecological systems can be investigated. The gradient of a channel reach and its position in the channel network, for example, strongly control both channel morphology and types of disturbance (Grant et al. 1990; Montgomery and Buffington 1993). Hence larger-scale (plate margin, province, or lithotopo unit) factors that shape the profile of a stream network from headwaters to outlet exert a major influence on the type, distribution, and variability of aquatic habitat in mountain drainage basins.

In-stream habitat is influenced by the size and type of sediment delivered to a channel, the discharge regime, and the size and abundance of potential flow obstructions. The morphology, dynamics, and habitat attributes of channels in forested mountain drainage basins—as in the coastal temperate rain forest region, for example—are strongly influenced by large woody debris, as well as channel width and type (Swanson and Lienkaemper 1978; Lisle 1986; Montgomery et al. 1995b). Channel morphology, processes, and dynamics reflect disturbance events that occur in the context of the watershed-scale template of geological processes.

Human activity that changes the storage or movement of sediment, water, or woody debris can dramatically alter stream channels. Channel responses to changes in discharge or sediment supply range from textural alteration of the bed surface to aggradation or degradation, channel widening, and changes in reach-level channel morphology (Leopold et al. 1964). Debris flows originating from hillslopes, hollows, or steep channel reaches scour (Figure 3.11) and aggrade (Figure 3.12) downstream channels. Delivery of fine sediment eroded from road surfaces may alter the composition of the bedload—and thus the subsurface sediment carried by mountain streams. Reduction or removal of woody debris can dramatically reduce the spacing between pools and lead to changes in reach-level channel morphology. The wide range of channel types, the spatial and temporal variations in channel morphology, and the influence

Figure 3.11. Stream channel scoured by debris flow on the Olympic Peninsula, Washington. (Photo courtesy of D. Montgomery.)

Figure 3.12. Stream channel aggraded by debris flow on the Olympic Peninsula, Washington. (Photo courtesy of D. Montgomery.)

of disturbance history indicate that a diagnostic approach, rather than simple standardized criteria, is essential for assessing channel conditions and understanding how channel processes influence aquatic ecosystems in the coastal temperate rain forest region.

Floodplains

Floodplains develop where sediment is deposited above streambanks during high flows. Interaction among fluvial processes, geological structure, and hillslope processes control the distribution of floodplains. The broad floodplains of tectonically formed troughs, such as the Willamette Valley and Puget Sound, reflect long-term infilling with sediment delivered from neighboring mountains. Although floodplains tend to widen downstream through channel networks, abrupt decreases in channel slope can widen floodplains locally in mountain drainage basins. In contrast to low-gradient channels in which fluvial processes dominate floodplain development, bedrock outcrops together with hillslope and tributary processes control the morphology of valley bottoms in steep mountain channels (Grant and Swanson 1995).

Floodplain habitats vary greatly between confined floodplains in steep mountain channels and the wider floodplains characteristic of lower-gradient, unconfined channels. Wide floodplains tend to include significant off-channel water bodies and side channels that provide important aquatic habitat and refugia for mobile species during high-flow events. The extent of hyporheic habitat (the subsurface zone associated with stream channels) in mountain watersheds also corresponds to areas with localized floodplain deposits. Uncoupling a channel from its associated floodplain (Figure 3.13) can profoundly affect both channel morphology and aquatic ecosystems.

Landscape Patterns

Viewing landscapes as a modular array of hillslopes, hollows, channels, and floodplains provides a framework for assessing geomorphological influences on ecosystem organization and dynamics. Landscapes with varying degrees of glacial influence illustrate how different arrangements of these landscape elements influence the variability of habitat in space and through time. The routing of debris flows, in particular, illustrates the influence of interactions among landscape elements on watershed processes. V-shaped valleys with narrow valley bottoms—and thus small floodplains—are typical of unglaciated landscapes. Hollows form at the upslope ends of a well-defined dendritic channel network, and debris flows can travel long distances down steep to moderate-gradient streams; there is a strong degree of hillslope–channel coupling.

Figure 3.13. Channel isolated from its floodplain in the Puget lowlands, Washington. (Photo courtesy of D. Montgomery.)

Areas that experienced the most recent continental glaciations, by contrast, tend to have a more gentle topography including numerous depressions, small lakes, and wetlands through which debris flows do not propagate. Areas influenced by alpine glaciation typically possess wide U-shaped main valleys and poorly dissected valley walls. These wide valley bottoms favor deposition of extensive alluvial fans at the base of tributaries, restricting debris-flow-related disturbance to headwater stream channels and largely decoupling major downstream channels from hillslope processes. Such landscape-scale patterns influence the ways in which ecosystems respond to human impacts.

Process Domains: A Link to Ecosystem Dynamics

Geomorphological processes at a variety of scales influence the availability and suitability of habitat for individual species. The definition of habitat, however, depends on both the life form in question and the nature of the inquiry. The distribution of spawning habitat for certain species of salmon and trout, for example, corresponds to reach-level channel types (Montgomery 1994), while habitat for benthic macroinvertebrates is associated with finer-scale channel units (Huryn and Wallace 1987). Although the spatial hierarchy proposed here may prove useful for understanding disturbance and habitat-forming processes, ecological systems involve a plethora of biological processes and interactions (such as interspecies competition and predation). Nonetheless, the spatial distribution of areas in which geomorphological processes shape habitat function and structure provide a fundamental link to ecosystem dynamics.

Subdividing mountain drainage basins, for example, into general process domains reveals correlations with disturbance processes for riparian and upland forests. A typical forested watershed in the Olympic Peninsula can be divided into areas in which the disturbances that initiate development of new forest stands are dominated by avalanches, debris flows, flooding, channel migration, and fire or windthrow (Figure 3.14). Avalanche disturbance occurs frequently on steep slopes in the high-elevation portions of a basin, and downslope-oriented bands of early-successional vegetation characterize these areas. Debris flow disturbances occur only infrequently in steep, confined headwater channels in pristine watersheds of the coastal temperate rain forest region; in areas with intensive logging, however, they occur more frequently. Deciduous forest cover is rarely found in headwater tributaries in pristine forests, whereas a narrow network of alder often traces steep tributaries in intensively managed areas. Flooding dominates disturbance along channel margins in confined stream channels not subject to debris flows. Narrow bands of early-

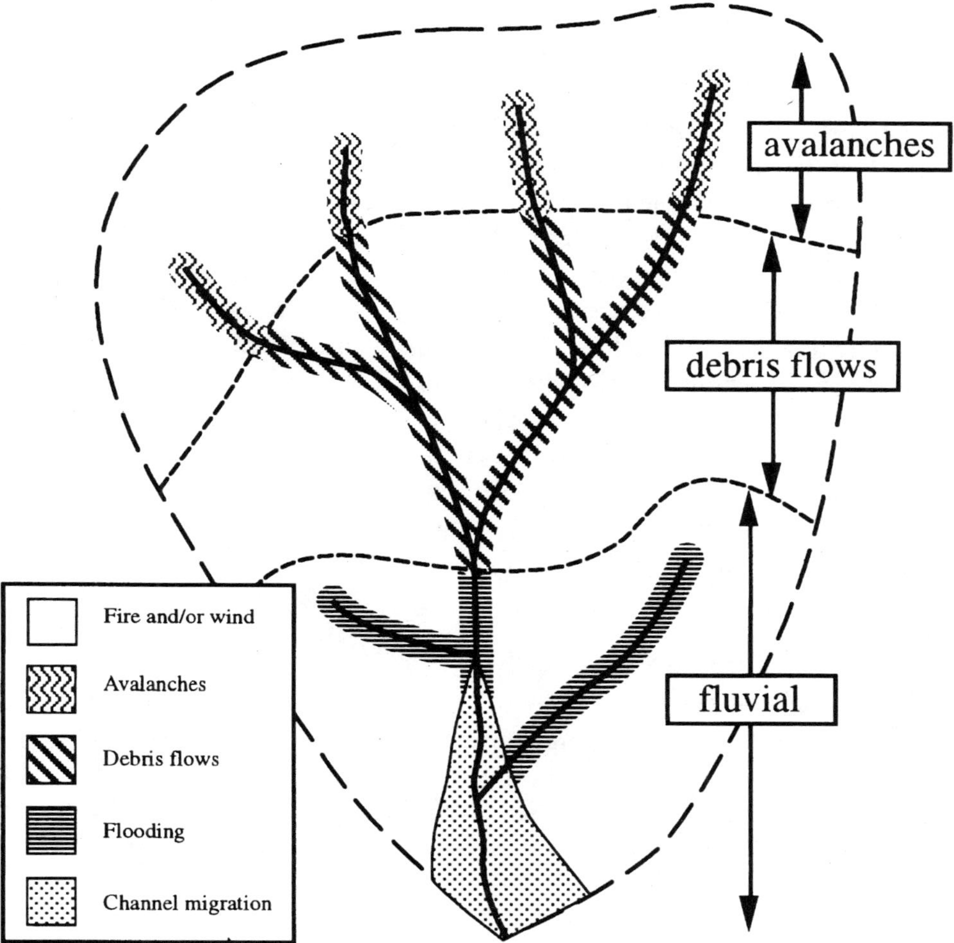

Figure 3.14. General process domains of Olympic Peninsula watersheds.

successional forest also grow along such reaches. Channel migration is a primary disturbance process in gently sloping, unconfined channels with wide floodplains. Distinctive patches and bands of riparian vegetation of different ages typically grow along unconfined channels. On the Olympic Peninsula, intermittant fire and windthrow in the remainder of the landscape result in a matrix of old-growth forest interspersed with large openings created by windstorms or fires, in which young trees regenerate. Simple models for the geomorphological processes that drive the disturbances responsible for forest death and regrowth can predict the size and distribution of geomorphological process domains throughout landscapes of the the coastal temperate rain forest.

Confronting the Issues

Land use planning often neglects the patterns that physical environmental processes impose on landscapes. Because much of this template lies beyond human control, this neglect often results in decisions made by default—decisions that usually encourage resource degradation. Civilization is unlikely to develop more sustainable land and resource use practices without social institutions capable of addressing complex environmental issues head on. Directly examining the links between human actions and ecological processes over many scales of time and space should help us to understand our choices. Evading such issues as habitat loss and unsustainable rates of resource use amounts to a default decision in favor of resource depletion and environmental degradation. In contrast, understanding how human activities fit into the array of geological processes that create the landscape can illuminate two of the most fundamental ethical questions of our time: What kind of environment do we owe future generations? To what extent are we willing to modify present practices in order to preserve their inheritance?

Acknowledgments

Preparation of this chapter was supported by USDA Forest Service Cooperative Agreement PNW 93-0441 and grant TFW-FY94-SH10-004 from the Washington State Timber/Fish/Wildlife agreement. The author wishes to thank Gordon Grant, Harvey Kelsey, Jonathan Stock, Tim Abbe, Kevin Schmidt, and John Buffington for critiques of an earlier draft.

References

Atwater, T. 1970. "Implications of plate tectonics for the Cenozoic tectonic evolution of western North America." *Geol. Soc. Am. Bull.* 81:3513–3536.

Benda, L. 1990. "The influence of debris flows on channels and valley floors in the Oregon Coast Range, U.S.A." *Earth Surface Processes and Landforms* 15:457–466.

Booth, D. B. 1994. "Glaciofluvial infilling and scour of the Puget lowland, Washington, during ice-sheet glaciation." *Geology* 22:695–698.

Branson, B. A. 1977. "Freshwater and terrestrial mollusca of the Olympic Peninsula, Washington." *Veliger* 19:310–330.

Caine, N. 1980. "The rainfall intensity-duration control of shallow landslides and debris flows." *Geografiska Annaler* 62A:23–27.

Davis, W. M. 1899. "The geographical cycle." *Geog. J.* 14:481–504.

Dietrich, W. E., C. J. Wilson, and S. L. Reneau. 1986. "Hollows, colluvium, and landslides in soil-mantled landscapes." In A. D. Abrahams, ed., *Hillslope Processes*. Boston: Allen & Unwin.

Franklin, J. F., and C. T. Dyrness. 1988. *Natural Vegetation of Oregon and Washington.* Corvallis: Oregon State University Press.

Good, D. A., and D. B. Wake. 1992. *Geographic Variation and Speciation in the Torrent Salamanders of the Genus* Rhyacotriton (*Caudata: Rhycotritonidae*). Publication in Zoology 126. Berkeley: University of California Press.

Grant, G. E., and F. J. Swanson. 1995. "Morphology and processes of valley floors in mountain streams, Western Cascades, Oregon." In J. E. Costa, A. J. Miller, K. W. Potter, and P. R. Wilcock, eds., *Natural and Anthropogenic Influences in Fluvial Geomorphology.* Geophysical Monograph 89. Washington, D.C.: American Geophysical Union.

Grant, G. E., F. J. Swanson, and M. G. Wolman. 1990. "Pattern and origin of stepped-bed morphology in high-gradient streams, Western Cascades, Oregon." *Geol. Soc. Am. Bull.* 102:340–352.

Hack, J. T. 1960. "Interpretation of erosional topography in humid temperate regions." *Am. J. Sci.* 258A:80–97.

Holt, R. D. 1993. "Ecology at the mesoscale: The influence of regional processes on local communities." In R. E. Ricklefs and D. Schluter, eds., *Species Diversity in Ecological Communities: Historical and Geographical Perspectives.* Chicago: University of Chicago Press.

Hughes, R. M., E. Rexstad, and C. E. Bond. 1987. "The relationship of aquatic ecoregions, river basins, and physiographic provinces to the ichthyogeographic regions of Oregon." *Copeia* 1987:423–432.

Huryn, A. D., and J. B. Wallace. 1987. "Local geomorphology as a determinant of macrofaunal production in a mountain stream." *Ecology* 68:1932–1942.

Jones, D. L., A. L. Cox, P. Coney, and M. Beck. 1982. "The growth of western North America." *Sci. Am.* 247(5):70–84.

Leopold, L. B., M. G. Wolman, and J. P. Miller. 1964. *Fluvial Processes in Geomorphology.* San Francisco: W. H. Freeman.

Likens, G. E., and F. H. Bormann. 1974. '"Linkages between terrestrial and aquatic ecosystems." *BioScience* 24:447–456.

Lisle, T. E. 1986. "Stabilization of a gravel channel by large streamside obstructions and bedrock bends, Jacoby Creek, northwestern California." *Geol. Soc. Am. Bull.* 97: 999–1011.

Montgomery, D. R. 1994. "Geomorphological influences on salmonid spawning distributions." *Geol. Soc. Am. Abstracts with Programs.* 26(7):A439.

Montgomery, D. R., and J. M. Buffington. 1993. "Channel classification, prediction of channel response, and assessment of channel condition." Report TFW-SH10-93-002. Olympia: Washington State Department of Natural Resources.

Montgomery, D. R., and W. E. Dietrich. 1988. "Where do channels begin?" *Nature* 336:232–234.

Montgomery, D. R., G. E. Grant, and K. Sullivan. 1995a. "Watershed analysis as a framework for implementing ecosystem management." *Water Res. Bull.* 31:369–386.

Montgomery, D. R., J. M. Buffington, R. Smith, K. Schmidt, and G. Pess. 1995b. "Pool frequency in forest channels." *Water Res. Res.* 31:1097–1105.

Newson, M. 1992. *Land, Water and Development: River Basin Systems and Their Sustainable Management.* London: Routledge.

Okunishi, K., and T. Iida. 1981. "Evolution of hillslopes including landslides." *Trans. Jap. Geomorph. Union* 2:291–300.

Omernik, J. M. 1987. "Ecoregions of the coterminous United States." *Ann. of the Assoc. of Am. Geog.* 77:118–125.

Reneau, S. L., and W. E. Dietrich. 1990. "Depositional history of hollows on steep hillslopes, coastal Oregon and Washington." *Nat. Geog. Res.* 6:220–230.

Schmidt, K. M., and D. R. Montgomery. 1995. "Limits to relief." *Science* 270:617–620.

Schramm, G. 1980. "Integrated river basin planning in a holistic universe." *Nat. Res. J.* 20: 787–806.

Sidle, R. C., A. J. Pearce, and C. L. O'Loughlin. 1985. *Hillslope Stability and Land Use.* Water Resources Monograph 11. Washington, D.C.: American Geophysical Union.

Southwood, T.R.E. 1977. "Habitat, the templet for ecological strategies?" *J. Animal Ecol.* 46:337–365.

Stearly, R. F. 1992. "Historical ecology of salmonidae, with special reference to *Oncorhynchus*." In R. L. Mayden, ed., *Systematics, Historical Ecology, and North American Freshwater Fishes.* Palo Alto: Stanford University Press.

Swanson, F. J., and G. W. Lienkaemper. 1978. *Physical Consequences of Large Organic Debris in Pacific Northwest Streams.* General Technical Report PNW-69. Portland, Ore.: Pacific Northwest Forest and Range Experiment Station, USDA Forest Service.

Swanson, F. J., T. K. Kratz, N. Caine, and R. G. Woodmansee. 1988. "Landform effects on ecosystem patterns and processes." *BioScience* 38:92–98.

Turner, B. L., W. C. Clark, R. W. Kates, J. F. Richards, J. T. Mathews, and W. B. Meyer, eds. 1990. *The Earth as Transformed by Human Action.* Cambridge: Cambridge University Press.

Turner, M. G. 1989. "Landscape ecology: The effect of pattern on process." *Ann. Rev. Ecol. and Syst.* 20:171–197.

Umhoefer, P. J. 1987. "Northward translation of 'Baja British Columbia' along the Late Cretaceous to Paleocene margin of western North America." *Tectonics* 6:377–394.

Yanev, K. P. 1980. "Biogeography and distribution of three parapatric salamander species in Coastal and Borderland California." In D. M. Power, ed., *The California Islands.* Santa Barbara, Calif.: Museum of Natural History.

4. Vegetation from Ridgetop to Seashore

Paul Alaback and Jim Pojar

Sailing up the Pacific coast of North America, from the towering redwoods of northern California through the granitic fjordland of British Columbia's Inside Passage to the lush moss-draped spruce forests of southern Alaska, you see an immense dark green cloak over the land. Lighter shades of green color mountaintops, indicating subalpine meadows, recent avalanche paths, and forest clearings. But the prevailing plant cover throughout the region is a dark green coniferous forest. Upon closer examination the forest becomes a complex mosaic of shades of green, each representing a species or life form—lichen, moss, spruce, hemlock, fir, or cedar. The canopies of these forests are rugged: broken or dead treetops and wind-whipped branches reflect a long history of battering wind and rain. As you walk into these forests from the beach, the environment quickly changes from open, bright, and noisy to cool and serene, humid and dark. The only direct evidence that the seashore is nearby may be seashell fragments dropped by otters or by birds from tree or shrub branches.

Elements of Diversity

There is great richness and diversity in these forests. You can see some of this diversity by comparing the luxuriant carpets of mosses and liverworts on the forest floor to thickets of herbs and shrubs, the massive conifer canopy above to mats of epiphytes on branches and bark of trees throughout the forest. This chapter offers a broad overview of the unique and lush vegetation that characterizes the North American coastal temperate rain forest and the closely allied vegetation types that surround it (Figure 4.1; Plates 2 and 3). This broad

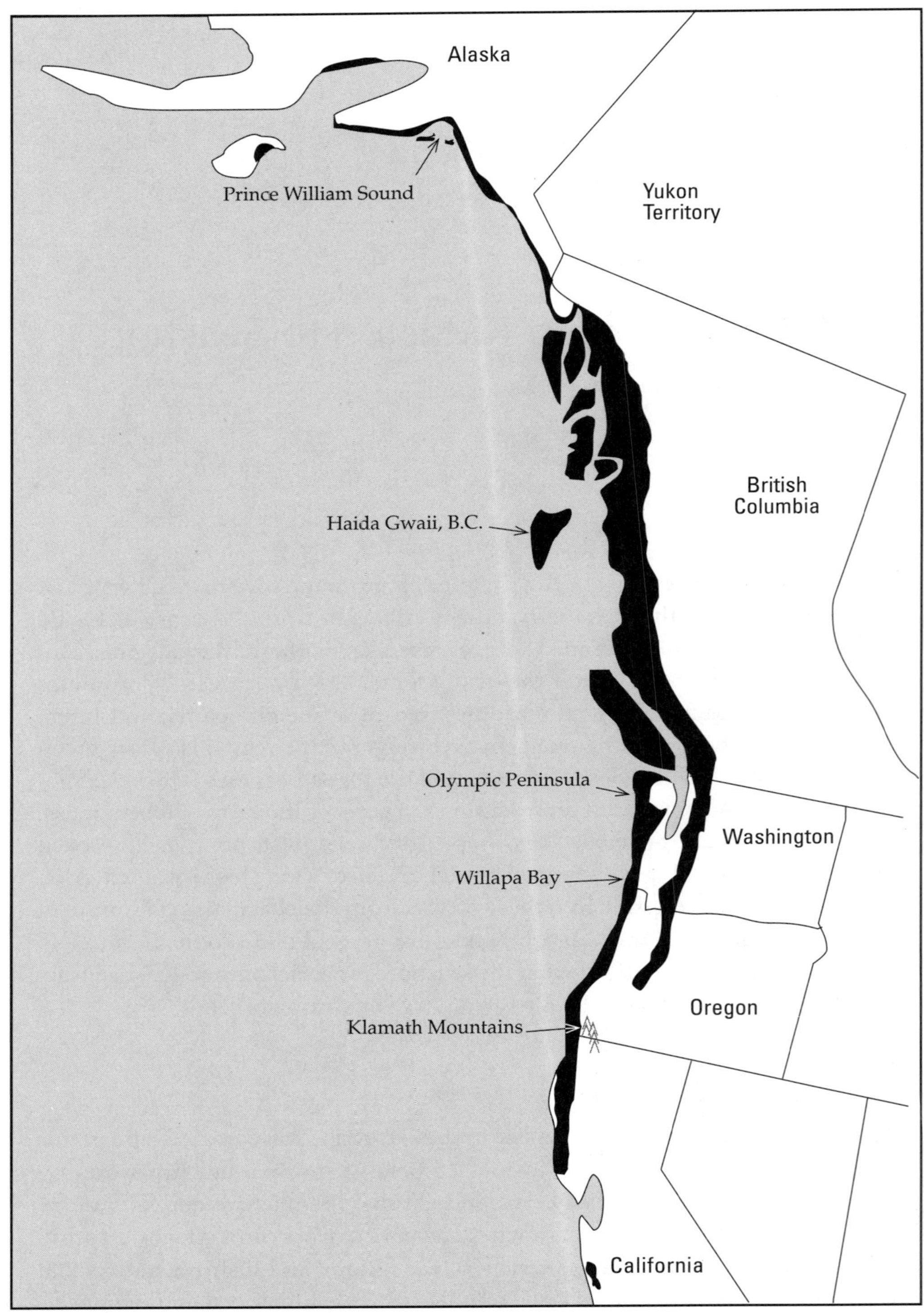

Figure 4.1. Selected locations mentioned in Chapter 4.

geographic scale is needed if we are to understand how these unique forests fit into the larger picture of coastal landscapes, for this is the scale upon which their migration, evolution, and exploitation has occurred.

Understory Plants

The region's wet climate and the even more sheltered microclimate within the forest provide ideal conditions for mosses and liverworts (bryophytes)—little plants that have poorly developed water and food conducting systems and generally fare poorly if they dry out for prolonged periods. Over 600 species of mosses occur in coastal Alaska and British Columbia (Schofield 1968). Epiphytes abound throughout the forest, cloaking or festooning tree trunks and branches. Many are bryophytes (such as *Antitrichia curtipendula* and *Isothecium spiculiferum* (*I. myosuroides*)) and lichens (*Usnea longissima* and *Alectoria sarmentosa*); a few are vascular plants, such as licorice fern (*Polypodium glycyrrhiza*) and *Selaginella oregona*.

The understory plants in these forests often form a nearly impenetrable thicket under the towering conifers. As you crash through this jungle of shrubs and tree seedlings you will see blueberries, hemlocks, vine maples, salmonberry bushes, elderberries, and the most characteristic and notorious of all—the devil's club (*Oplopanax horridum*), complete with spines both on stems and under the leaves, perfectly angled to catch your skin if you try to push through. These understory shrubs and herbs are as lush and dense as in any conifer forest in the temperate zone, aside from the bamboo thickets in the temperate rain forests of the Southern Hemisphere, providing further testament to the salubrious climate for plant growth here.

Heaths and Heathers

Besides the general density of the vegetation, one striking feature of the temperate rain forest shrub layer is the ascendancy of the Ericaceae, the heather family (Pojar and MacKinnon 1994). Many of the most typical and abundant shrubs of the region belong to this family. Ericaceous shrubs not only dominate the understory of many of our forests; they are also abundant in nonforested habitats such as peatlands and high-elevation heaths. These shrubs range from sea-level salal (*Gaultheria shallon*) to timberline mountain-heathers (*Phyllodoce* and *Cassiope* spp.), from bog laurel and Labrador tea (*Kalmia microphylla* ssp. *occidentalis* and *Ledum groenlandicum*) to rock-outcrop bearberry and manzanita (*Arctostaphylos* spp.), from trees (Pacific madrone, *Arbutus*) to tall shrubs (*Vaccinium* spp.) to dwarf trailing shrubs (bog cranberry, *Oxycoccus*). Heather family shrubs include all of our blueberries, huckleberries, bog and mountain cranberries, rhododendrons and false azalea (*Menziesia*), copperbush (*Cladothamnus*), and the mountain heathers.

The closely related wintergreen family (Pyrolaceae, considered by some taxonomists as part of the Ericaceae) extends the realm of this group of plants to evergreen dwarf semishrubs (pipsissewas; *Chimaphila* spp.) and herbs, such as wintergreens (*Pyrola* spp.) and single delight (*Moneses uniflora*). The group also includes several peculiar saprophytes—such as candystick (*Allotropa virgata*), gnome-plant (*Hemitopes congestum*), and pinedrops (*Pterospora andromedea*)—that lack chlorophyll but sport other, often lurid colors, and the waxy white, partial parasite Indian-pipe (*Monotropa uniflora*). Note that some other nongreen, fleshy, parasitic herbs such as ground-cone (*Boschniakia* spp.) and broomrape (*Orobanche* spp.) belong to other families.

Why is the heather family so abundant and diverse in this region? The thick, moist, acidic, organic layers that develop on the ground surface of many of the ecosystems is probably one of the primary reasons. The heather family in general seems to prefer acid soils rich in organic matter, as any rhododendron or azalea gardener knows. One reason for this preference is that the Ericaceae have developed a close relationship to a unique group of mycorrhizal fungi (called ericoid mycorrhizae) that facilitates nutrient uptake on these acid soils. As for the saprophytes, by definition they require decayed organic matter to live. Although it would be difficult to single out any vascular plant as being strictly an old-growth species (that is, completely dependent on old-growth forest for its survival), it could very well be that old forests with well-aged humus are critical habitat for some saprophytes (Schoonmaker and McKee 1988).

Ferns

Another distinctive feature is the profusion of ferns. Although the fifty or so species of ferns is not an unusually large number, the abundance and luxuriance of forest ferns is striking. Many of our moist forests have herb layers dominated by ferns, most notably sword fern (*Polystichum munitum*), spiny wood fern (*Dryopteris austriaca*), giant chain fern (*Woodwardia fimbriata*, especially in the redwood forest), deer fern (*Blechnum spicant*), and oak fern (*Gymnocarpium dryopteris*).

Biogeographic Patterns

While some biogeographers consider the whole region to be but one forest type—Sitka spruce–western hemlock forest—growing evidence suggests that the coastal rain forest should be divided into four distinctive zones, each with its characteristic climate, vegetation, soils, and natural disturbance regimes (Alaback in press). Each of these zones (Plate 2) is defined primarily on the basis of summer climate and prevailing type and intensity of natural disturbance regime. Although many species of plants occur in more than one zone, their ecological role often changes from zone to zone as the physical environment changes.

Warm Temperate Rain Forest

The first major type of forest occurs from San Francisco Bay north to the southern Oregon coast. In this zone, mild wet winters are combined with cool dry summers on the coast or warm dry summers in the interior. Generally less than 5 percent of the annual rainfall occurs during the summer. Snow is rare along the coast, but it may be very deep and persistent on ocean-facing mountaintops. Ground fires are common in summer and fall, but generally they have little impact on dominant trees. Catastrophic fire occurs along the coast only every 100 years or more (Agee 1993). Coastal redwood dominates forest near the coast in association with heavy annual rainfall and persistent summer fog (Waring and Major 1964).

Many of the species that are abundant further north such as Sitka spruce, western redcedar, western hemlock, salmonberry, and blueberry are also found here, but generally in more restricted habitats: along streams, in cool foggy valleys, at higher elevations, or on cool north-facing slopes. These forests have the most diverse canopies of any coastal forest type. There are widely distributed northern trees in addition to many local warm-temperate trees such as California laurel (*Umbellularia californica*), chinquapin (*Castanopsis chrysophylla*), and tan oak (*Lithocarpus densiflorus*). In addition many drought-tolerant species that are more widespread in the interior West—such as white fir (*Abies concolor*), sugar pine (*Pinus lambertiana*), Kellogg's oak (*Quercus kelloggii*), incense cedar (*Calocedrus decurrens*), and ponderosa pine (*Pinus ponderosa*)—play a key role in drier microclimates or on coarse-textured soils in the warm temperate region.

Seasonal Rain Forest

From southern Oregon to central Vancouver Island occurs a lush and productive forest that is usually dominated by western hemlock and Douglas-fir (Krajina 1965; Franklin and Dyrness 1973). Winters are cool to mild; snow is uncommon and transient. Most of the rainfall occurs in the fall and winter, and less than 10 percent occurs during summer months. Summers are dry and cool, but soils are generally moist most of the year. Because of the dry summers and the importance of fire, these forests are often excluded from the strict definition of coastal temperate rain forest. Given their close climatic and geographic tie to the area we include them here as a special type.

Immediately adjacent to the shoreline and in cool fog-drenched valleys, Sitka spruce and western hemlock form a lush towering forest with thickets of big-leaved maple, salmonberry, currants, elderberries, and devil's club underneath. Above this cool fog-influenced zone or further inland, Douglas-fir dominates the forest, generally as a result of large-scale catastrophic fires that burn every 100 years or more. Western redcedar also forms a major forest type in this region, particularly on wet sites. In large valley bottoms these trees can be quite statuesque, commonly reaching 60 to 75 meters (200–250 feet) in height and with trunks 3 meters (10 feet) or more in diameter (Franklin and Dyrness 1973).

Montane and subalpine forests are also well represented in the seasonal rain forest zone (Brooke et al. 1970; Fonda and Bliss 1969; Henderson et al. 1989). Noble fir and Pacific silver fir are characteristic where winter snowpack is persistent, generally from 600 to 1500 meters (2000–5000 feet) in elevation. These cool wet forest zones have understories dominated by northerly species, especially members of the heather family. Mosses and lichens are more characteristic of these forests than in Douglas-fir dominated forests, but not so much so as in coastal perhumid forests. Fires play a key role in these forests, too, although they are generally less frequent than in the Douglas-fir–western hemlock forests. On mountaintops that extend above 1200 meters (4000 feet), subalpine forests composed of mountain hemlock, yellow-cedar, and subalpine fir may also occur. These forests are generally quite open and stunted and represent the extreme limit of forest growth—limited by the length of the growing season and by shifting snow and high winds. These forests often occur as a mosaic of forest patches in a matrix of alpine heaths and meadows.

Perhumid Temperate Rain Forest

From northern Vancouver Island to southeastern Alaska, perhumid rain forests encompass a diverse landscape of wetlands (especially peat bogs) forest and lush subalpine meadows reflecting a distinctly wet and cool climate that persists throughout the year (Alaback in press; Banner et al. 1988). Only one-third of the landscape has dense forests. Summer rainfall comprises 10 to 20 percent of annual rainfall. Snow is common but relatively transient near the shoreline. Summers are very cool and wet. Alpine areas and subalpine forests have heavy snow accumulations, and this snowpack is the key to defining the alpine environment in this climatic zone. Low-elevation forests are dominated by western hemlock, Pacific silver fir, and Sitka spruce with a lush understory dominated by various blueberry or huckleberry species (Alaback and Juday 1989). Western redcedar and yellow-cedar also form a distinctive forest type in this region, usually on sites with low nutrient availability, poor drainage, and open, shrubby forest. Fires are quite rare. Indeed, they are generally considered to be of minor importance to the ecology of this zone, thus distinguishing it ecologically from most other forest types in North America. Pacific silver fir characterizes the montane zone to the southern tip of Alaska, but at its northern extreme it mostly grows in mixture with western hemlock and Sitka spruce. The subalpine zone, generally 450 to 900 meters (1500–3000 feet) in elevation, is dominated, like the seasonal rain forest region, by mountain hemlock and Pacific silver fir. Sitka spruce and yellow-cedar can also be well represented in upper-elevation forests, perhaps due to less competition with western hemlock and the more open nature of these forests, which allow for better regeneration of Sitka spruce.

Subpolar Rain Forest

The highest-latitude coastal forests have a distinctly subalpine character (Alaback and Juday 1989). They commonly have mountain hemlock and Sitka spruce (as in the subalpine type in the perhumid forest zone) and occur in a landscape dominated by alpine meadows, muskeg wetlands, and glaciers. Only about 12 percent of the landscape is forested. Forests are most common on protected sites, especially on mountain slopes where soil drainage is best. Forests are much shorter—often less than 30 meters (100 feet) in height—and more open than at lower latitudes, reflecting short growing seasons, poor soil conditions, and an excessively cool wet climate, often including an annual precipitation greater than 3800 millimeters (150 inches) a year.

Special Forest Types

Forests beside streams and those in the dry "rain shadow" of coastal mountains maintain important similarities across two or more of the coastal rain forest zones. Both streamside ("riparian") forests and rain-shadow forests possess features that distinguish them from prevailing zonal patterns: variations in species composition, disturbance regime, and susceptibility to human impact.

Riparian Forests

Riparian forests include trees and associated understory plants that occur along streams and extend to the limit of flooding or terrace development (or to the bases of trees that can fall into the flooding zone). Riparian forests are often considered the "keystone" ecosystem or the central hot spot of diversity in coastal ecosystems because of the large number of specialized species that occur there and the key linkages between physical and biotic processes that occur in these areas (Chapter 6). These forests are often biologically productive as well as rich in species because of large inputs of nutrient-rich sediments and organic debris and complex patterns of disturbance. Deciduous trees are an important element in diversity because of the high density and diversity of insect communities associated with them.

Many herb and moss species are unique to the riparian zone—especially in floodplains or immediately adjacent to the active stream channel (*Pleuroziopsis ruthenica, Climacium dendroides, Conocephalum conicum, Viola glabella, Prenanthes alata, Trauttveteria caroliniensis*). Because of the unique microclimates and natural migration corridors along streams, riparian areas also have many distinctive tree and shrub species. For example, subalpine or boreal species such as subalpine fir (*Abies lasiocarpa*) and Prince's-pine (*Chimaphila umbellata*) are principally found along mainland rivers in northern southeast Alaska. Riparian systems often are maintained by complex disturbance regimes that vary over many scales of time and space. This is why they are altered in more significant ways by clearcut logging than are upland forest ecosystems—for example, logging can change a spruce–cedar forest into a red alder–salmonberry-dominated forest for many decades. Studies in Alaska suggest that many centuries may be required for some riparian systems to recover their biodiversity after catastrophic natural disturbance or widespread clearcut logging.

Rain-Shadow Forest

This ecosystem occurs in seasonal and warm temperate rain forest latitudes. Rain-shadow forest has the greatest regional diversity of plant species. The Rogue, Umpqua, and Willamette valleys are good examples of rain-shadow forests and grasslands. The northernmost areas with this character are in the Strait of Georgia region of southwestern British Columbia. Dominant species usually include Douglas-fir, grand fir, Pacific madrone, and Garry oak. These ecotonal areas include both rain forest species and dry interior species and even some desert species like prickly-pear cactus (*Opuntia*) and some nasty ones like poison oak (*Rhus diversiloba*). Ponderosa pine can be found on dry sites with coarse-textured soils, a seeming refugee from the extensive fire-dependent forests east of the Cascades. Because of their ecotonal nature, small differences in soils and microclimates provide habitats for a wide range of specialized species in these rain-shadow forests.

Frequent fire disturbance historically created habitats for many species of plants in rain-shadow forests. Aboriginal burning played a key role in maintaining the open nature of many of these valleys and adjacent hillslopes. Exclusion of wildfire (and control of fire set by humans) has dramatically changed the character of this forest, both in terms of its composition and function. Grand fir and western redcedar have colonized extensive areas of this forest in the past century, principally because of this alteration of natural disturbance regimes. The greatest threat to these forests, however, has been urbanization, clearing, and the introduction of exotic species. These forests generally grow in what are considered some of the most pleasant climates for people in the region—warm dry summers and low annual rainfall.

Nonforested Vegetation

Although forests cover most of the region, nonforested areas such as bogs and alpine tundra are significant components of the regional vegetation. Nonforested vegetation occurs where it is too wet, too salty, or too cold and snowy for trees to survive or in areas prone to disturbances, such as avalanches and fires, that destroy stands of trees. In the coastal rain forest it is almost never too dry or too steep and rocky for trees to live. Fire or some other disturbance is usually the limiting or precluding factor.

Maritime Habitats

The northern Pacific coastal region has a very long coastline with a variety of both protected and exposed shores. The British Columbia portion of the coast alone is 2700 kilometers in length. The coastline south of Puget Sound stretches out in long beaches and dune systems, punctuated by rugged exposed headlands, with only a few major rivers. From Puget Sound north through British Columbia and southeastern Alaska, the coastline is an intricate network of islands, long steep-walled fjords, inlets, rivers and their estuaries, and tidal flats. Between the forest and the sea is a strip of maritime vegetation adapted to exposure to salt water, wind, and surf. Few species of vascular plants can tolerate salt spray or immersion in seawater, but those that can are often confined to such habitats. The different types of shoreline differ primarily in their substrate and in the degree to which they are exposed. Each supports a distinctive vegetation type and, taken together, these maritime plant communities can be quite diverse. All beach habitats in exposed localities front a maritime forest community—typically Sitka spruce with a dense salal understory.

Rocky shores are the most common type of shoreline on this coast. Terrestrial plant cover is sparse, especially on exposed rocky headlands, sea stacks, and cliffs. Adaptations to the incessant wind and salt spray and to moisture stress include cushion or matted growth forms and thick, waxy, succulent, or densely hairy leaves. Sea plantain (*Plantago maritima*), hairy cinquefoil (*Potentilla villosa*), ocean strawberry (*Fragaria chiloensis*), northern riceroot (*Fritillaria camschatcensis*), and salal are typical of the hardy vegetation of exposed rocky shores.

Shingle beaches, composed of large gravels or cobbles, are also widespread. They usually support clumps of sea rocket (*Cakile edentula*), dunegrass (*Elymus mollis*), beach pea (*Lathyrus japonicus*), giant vetch (*Vicia gigantea*), ocean strawberry, springbank clover (*Trifolium wormskjoldii*), sea-watch (*Angelica lucida*), and cleavers (*Galium aparine*). Such plants are especially common on the upper beach and among the jumble of driftwood at the furthest reach of winter storm tides and waves.

Sand beaches are widespread in Oregon and Washington, but they are uncommon and local in most of the rest of the region. Notable exceptions are on the northeast Haida Gwaii (Queen Charlotte Islands) and the west coast of Vancouver Island. Elsewhere beaches are mainly in small pockets between the headlands of exposed outer coastal areas. Vegetation is sporadic, but showy species of the driftwood belt include sea rocket, beach carrot (*Glehnia littoralis*), and beach pea. Further up the beach (in the driftwood zone) or on dunes, large-headed sedge (*Carex macrocephala*), dune grasses, paintbrushes (*Castilleja* spp.), lupines (*Lupinus* spp.), dune tansy (*Tanacetum bipinnatum*), and silver burweed (*Ambrosia chamissonis*) become common.

The most productive of the maritime plant communities are tidal marshes—especially those with brackish or low-salinity water, as in estuaries. Soils are usually fine-textured (silts) and nutrient-rich, and they support lush meadow vegetation. Grasses and sedges—especially tufted hairgrass (*Deschampsia caespitosa*) and Lyngby's sedge (*Carex lyngbyei*)—dominate, but these marshes also harbor showy species such as Pacific silverweed (*Potentilla anserina*), springbank clover, and Nootka lupine (*Lupinus nootkatensis*), as well as salt-tolerant oddities like glasswort (*Salicornia virginica*) and sea arrow-grass (*Triglochin maritimum*).

Freshwater Wetlands

Wetlands are common in the coastal rain forest. They are extensive on the outer coast, especially on the coastal lowlands from northern Vancouver Island north to Prince William Sound. Nonforested wetland types include bogs, fens, and marshes. Classic bogs are acid peatlands with stagnant waters that originate as rain or snow falling directly onto the bog. We consider wetlands dominated by *Sphagnum* (peat) moss as bogs. Fens and marshes are less acid and have more nutrients available for plants. Fens and marshes dominated by sedges and grasses, as well as shrubby fens with hardhack (*Spiraea douglasii*), sweet gale (*Myrica gale*), Pacific crabapple (*Malus fusca*), and willows, are infrequent and localized along flowing water, around lake margins, and at river mouths. Throughout the region there are numerous, usually small, topographically controlled bogs in basins, on the margins of small lakes, and on level, poorly drained terrain. The flat bogs of Haida Gwaii (Queen Charlotte Islands) and the domed bogs of the Fraser lowland are extensive examples of the last type.

The colloquial term "muskeg" refers to the complex mosaic of fens, bogs, pools, streams, exposed rock, and scrubby forest that becomes increasingly common as you proceed north (Banner et al. 1988). Muskeg is widespread over the north coastal lowlands and foothills, which feature an unusual landscape of low rocky hills covered with scrub forest and open peatlands. You can see these areas from ferries and boats traveling north to Prince Rupert and southeastern Alaska as you thread your way through the archipelagoes and fjords of the Inside Passage. Stunted and gnarled shore pine, western redcedar, yellow-cedar, and both species of hemlock are scattered throughout muskeg. The pine, in particular, can assume bonsai-like forms. Pit ponds and larger pools contain skunk cabbage (*Lysichiton americanum*), yellow pond-lily (*Nuphar polysepalum*), and various pondweeds (*Potamogeton* spp.). Rills and streams drain the ponds, snaking between thick lawns of *Sphagnum* moss islands of stunted forest and low evergreen shrubs such as Labrador tea, bog laurel, and common juniper (*Juniperus communis*). In the cool, humid, oceanic climate, peatlands can cover both subdued terrain and slopes of considerable steepness. Such "blanket" or "slope" muskeg is extensive on the west coast of Haida Gwaii (Queen Charlotte Islands) and some of the exposed nearshore islands in the Hecate and Alexander depressions. Sometimes the blanket muskeg is continuous from sea level to alpine elevations and has a very diverse flora with low-elevation wetland species intermingled with such typically subalpine species as subalpine daisy (*Erigeron peregrinus*), Indian hellebore (*Veratrum viride*), alpine-azalea (*Loiseleuria procumbens*), and partridgefoot (*Luetkea pectinata*).

High-Elevation Habitats

Nonforested plant communities occur at high elevations above much of the coastal rain forest. The highest elevations are the domain of rock, ice, and permanent snow. The timberline occurs where subalpine forest thins out and becomes stunted and clumped, due to very deep snowpacks that melt slowly. This boundary can be as low as 150 meters (492 feet) in Prince William Sound, around 500 to 700 meters (1500–2000 feet) in southeastern Alaska, 800 to 1000 meters (2500–3500 feet) on the northern British Columbia coast, and 1500 meters (5000 feet) or more in southern British Columbia. The timberline reaches 2200 meters (7200 feet) in coastal southwest Oregon and northern California. Picturesque clumps of trees—usually mountain hemlock (*T. mertensiana*), yellow-cedar (*Chamaecyparis nootkatensis*), silver fir (*A. amabilis*), whitebark pine (*Pinus albicaulis*), and sometimes subalpine fir (*A. lasiocarpa*)—and open heath and meadows form an attractive mosaic termed subalpine parkland.

Heaths characteristically are dominated by dwarf, shrubby, evergreen members of the heather family (especially the mountain-heathers), with other common species such as black crowberry (*Empetrum nigrum*), dwarf blueberries, partridgefoot, and bird's-beak lousewort (*Pedicularis ornithorhyncha*). The ground cover forms a springy carpet often so dense it obscures everything underfoot, including rocks and holes.

Coastal mountain meadows are less extensive than heaths but are lush, intensely green, and dominated by herbs. Typical species include arrow-leafed groundsel (*Senecio triangularis*), subalpine daisy, Sitka valerian (*Valeriana sitchensis*), lupines (*Lupinus arcticus, L. lepidus*), Indian hellebore, gentians, sedges, and grasses such as purple mountain hairgrass (*Vahlodea atropurpurea*). These meadows put on a spectacular wildflower display in summer.

Wet meadows, fens, and marshes characterize poorly drained sites at high elevations. Sedges (especially *Carex aquatilis* and, in snowbeds, *C. nigricans*), cottongrasses, and wetland mosses (including *Sphagnum* spp.) dominate such habitats, which also have numerous other distinctive herbs, graminoids, and shrubs.

Avalanche tracks are very common. Most often they are occupied by slide alder (*Alnus sinuata*) communities—jungles of bowed alder stems with a lush understory of ferns, herbs (including stinging nettle, *Urtica dioica*), sedges, grasses, and other shrubs like red elderberry (*Sambucus racemosa*), salmonberry (*Rubus spectabilis*), currents (*Ribes* spp.), and devil's club. These berry-rich areas often provide ideal habitat for both black and grizzly bears.

Alpine rocklands and steeplands include rock outcrops, cliffs, boulder fields, fellfields, talus and scree slopes, wet runnels and gullies, and avalanche tracks. The plant cover is usually sparse and discontinuous, but it includes various saxifrages as well as ferns, cinquefoils, buttercups, grasses, sedges, and many lichens and bryophytes. Though relatively barren, these rocky areas have a rich flora and are good places to look for some of the most distinctive—and often rare—species in the region.

Patterns of Biological Diversity

The coastal forests of the Pacific coast form a rich center of biological diversity for animals but have considerably fewer plant species than cold-temperate and semiarid forests in the interior (Chapter 5; Pojar et al. 1992). The moderating influence of the ocean provides a usually mild winter and long growing season for the latitude. Several species of plants can persist in these forests only because of the mild climatic conditions. Moreover, many cool-temperate species such as Sitka spruce and shore pine are able to persist because of the acid soils and cool summer temperatures, which dampen the growth of their competitors. The prevailing climate conditions influnce the severity and scale of natural disturbance from wind and fire, which has important implications for biodiversity.

Latitudinal Patterns

As in most other temperate forest regions, plant species diversity declines rapidly with increasing latitude. The warm temperate zone has a particularly rich flora because of its mild climate, the converging of both Mediterranean and cool-temperate species, and a long geological history compared with other parts of the West (Whittaker 1960). In the Klamath Mountains this pattern is most pronounced because dry microsites and local climates often come close to the coast, allowing for the merging of interior and coastal flora. The highest proportion of endemic species (unique to this region) occurs in the warm temperate zone.

Few endemic species occur in the seasonal rain forest zone, but a relatively high species richness is maintained. Douglas-fir, hemlock, and redcedar dominate most sites, but maple, oak, and grand fir may dominate local dry areas. Coastal headlands have high species richness and several endemic species. From the Olympic Peninsula north to Alaska, Douglas-fir declines in importance and dense hemlock–spruce–fir forests become a significant component of the landscape.

Western hemlock trees cast such deep shade that few understory plants can flourish beneath them. In northern forests, therefore, the most diverse upland habitats generally are in areas with low nutrient availability or otherwise suboptimal soils—conditions that lead to more open canopies. Cedar and mountain hemlock forests are good examples of this phenomenon. Peat bogs and riparian areas are particularly species-rich systems in this zone. Riparian areas tend to have a much higher diversity of vascular plant species than peatlands, but peatlands are often rich in nonvascular plants (Figure 4.2).

North of the boundary of Pleistocene continental glaciation, a few areas escaped being covered by ice and may have served as refugia for elements of the regional flora and possibly as theaters for microevolution. These glacial refugia often have a richer array of plant species than surrounding areas, and some have developed local endemic species or subspecies. Well-known examples of such refugia are Haida Gwaii (Queen Charlotte Islands), the outer coast of Glacier Bay, Dall Island, and Kodiak Island. Limestone sites in alpine refugia often have a conspicuous component of regionally rare species. On Haida Gwaii and Prince of Wales Island, limestone areas host several endemic alpine meadow species in addition to several isolated populations of species common elsewhere, such as subalpine fir and reticulated willow (*S. reticulata;* Roemer

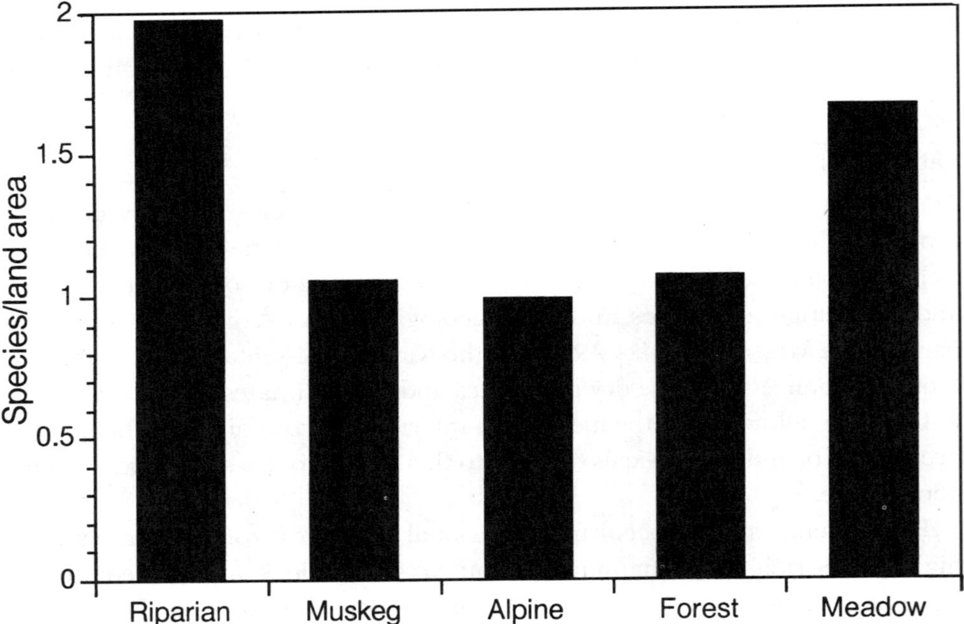

Figure 4.2. Vascular plant species richness by ecosystem type in coastal Alaska. Data represent total flora known to occur on each ecosystem type. "Muskeg" includes peatlands, fens, and bogs. *Source:* Adapted from Alaback (in press).

and Ogilivie 1983). The odd species distributions could also be explained by long-distance dispersal following deglaciation. More detailed paleobotanical work is needed to clarify this history (Chapter 9).

Biodiversity and Ecosystem Productivity

Biological diversity is generally assumed to increase with ecosystem productivity (Huston 1994). Conservation efforts throughout the region have tried to emphasize the most productive sites, because they are thought to be very diverse and are poorly represented in protected areas (Noss and Cooperrider 1994). Coastal forests are very productive. Indeed, some of the greatest productivity estimates for temperate climates have come from the warm temperate and seasonal temperate rain forest zones. The phenomenal productivity of Pacific Northwest conifers compared to all closely related species that evolved elsewhere is generally considered to be a consequence of a high degree of genetic adaptation (less gene depletion than other places), mild winters, long growing seasons, and a lack of summer drought (Waring and Franklin 1979). Due to a combination of shorter growing seasons, less energy available for growth, and declines in nutrient availability, productivity wanes to the north throughout the region.

Diversity and productivity do not always go together in coastal temperate rain forests. As has been demonstrated in some grasslands, increased productivity may lead to greater competitive exclusion and less species richness (Tilman 1982). Dense western hemlock forests are among the most productive temperate forests studied—they can produce over 30 metric tons per hectare per year—but often have low species richness. It appears that because of the limited solar radiation in coastal forests, species richness generally is less in dense, productive forests than in open areas. Heavy annual rainfall, low clouds, and low solar angles all contribute to reducing the light available for plant growth. Understory plants are shaded out in perhumid rain forests at tree canopy densities that would allow a rich diversity of plants to survive at lower latitudes. In the redwood region, by contrast, productive dense forests appear to retain a high level of species richness. The heterogeneity of these forests may encourage diversity by providing a richer complex of habitats and greater penetration of solar radiation (Figure 4.3).

Disturbance and Biodiversity

Natural disturbance plays a key role in maintaining vegetation diversity in temperate rain forests. Dense coniferous canopies often capture 90 percent or

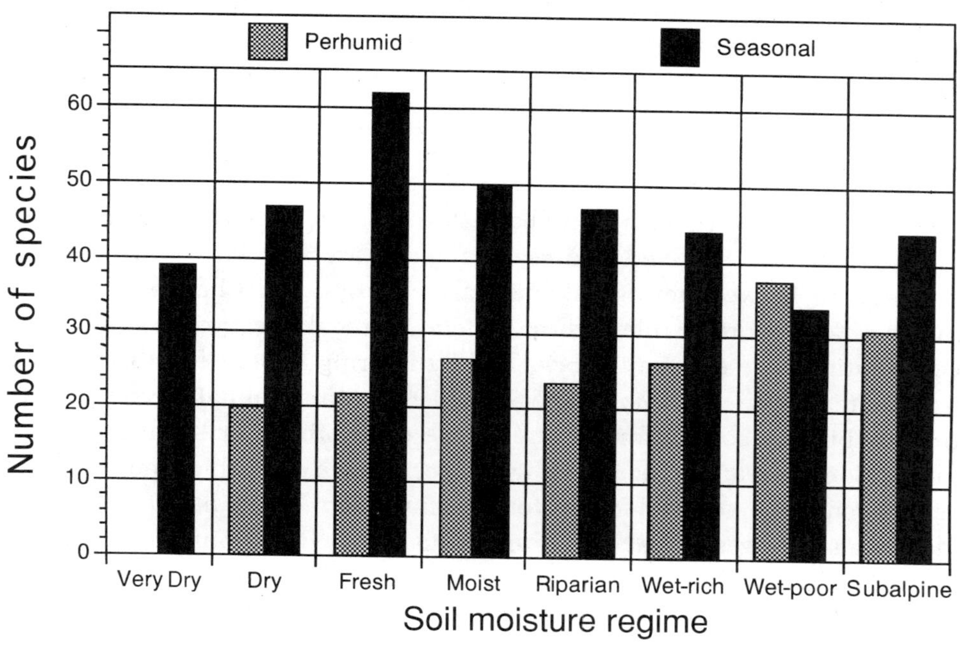

Figure 4.3. Understory vascular plant and bryophyte diversity patterns in two types of northern temperate rain forests, compared according to soil moisture regimes. Species tallies are from 400-square-meter plots and do not include epiphytes or species restricted to decaying wood. Seasonal forests—with generally greater structural diversity and more light penetration—sustain more species in the groups across nearly all soil conditions.

more of the light coming through the canopy, leaving little for understory vegetation. Disturbances that open up the forest canopy enough so that shade intolerant plants can colonize, while still allowing shade-tolerant species to persist, result in a significant increase in species diversity. Windthrow can play this role, although it usually stimulates the growth and reproduction of established plant species rather than encouraging recruitment of new species. The greatest effect of windthrow on species diversity is the creation of new microhabitats such as root-throw mounds, pits, and organic debris. Mosses in the family Polytrichaceae, for example, often colonize root-throw mounds. Even fifty years after tree windthrow, these moss species can persist on root-throw mounds. Some ferns are associated with these microhabitats as well.

Fires play a key role in maintaining species diversity in the seasonal and warm temperate rain forest regions (Agee 1993). Many ephemeral species, including exotic species and annuals, flourish in the aftermath of wildfire. This may be why there is such a limited distribution of these species in perhumid forests, where fire is rare. Moreover, the complex mosaic that fire creates adds to regional biodiversity by increasing heterogeneity at multiple scales. Fire also creates unique germination sites, as it can alter nutrient availability, microclimate, and physical characteristics of the litter and soil layer. In the warm temperate zone, species such as *Ceanothus* require fire for germination. Many drought-tolerant species are also the most aggressive colonists of the forest following wildfire.

Marine/Terrestrial Links

The shoreline provides a variety of habitats for plants, but it also creates an important link between land and sea. Estuaries are well known for being a mixing zone between terrestrial and marine ecosystems (Chapter 7). Forests may play an important role in these systems by regulating cycles of nutrients, contributing coarse woody debris, and structuring habitat for anadromous fish and marine species (Maser and Sedell 1994). Many animal species find optimal habitat in coastal forests and forage on shoreline or marine environments. Some threatened ocean-foraging species such as marbled murrelets are very dependent on coastal forests for breeding and roosting habitat. Some evidence also suggests that anadromous fish may contribute a significant amount of nutrients to riparian forests (Bilby et al. in press). Riparian forests—the most species-rich and most specialized habitats in the coastal zone—are a key link between land and sea. Much more attention needs to be focused on understanding these multifaceted relationships in order to predict the effects of pollution, land clearing, and other human activities on these diverse and sensitive landscapes.

Conservation Implications

Clearcut logging and associated human disturbances have had a profound influence on vegetation patterns and biological diversity throughout the coastal temperate rain forest (Chapter 14; Veblen and Alaback in press). In coastal Alaska it may take three or four centuries for structurally diverse old-growth ecosystems to develop following clearcut logging or other catastrophic disturbances. Since forest vegetation has developed mostly under a disturbance regime of frequent but small-scale windthrow in the perhumid and subpolar rain forest zones, clearcutting represents a drastic contrast to natural disturbance patterns. In seasonal and warm temperate rain forest types, this contrast is less extreme since large-scale catastrophic fire does strike every 100 to 300 years, depending on local site conditions (Agee 1993). Nevertheless, catastrophic fire and clearcutting are different disturbance types. Clearcutting usually leaves much less legacy from the previous stand—such as snags, logs, green trees, and microbes—and is a more spatially uniform disturbance. In the seasonal rain forest zone, old-growth forest may require two or more centuries to develop after a catastrophic disturbance (fire or clearcutting). Even greater recovery times are generally required on poorly drained sites and at higher elevations.

Few exotic species have been introduced in northern rain forest zones, and their role is relatively minor. In the seasonal and warm temperate zone, however, many exotic species are well established in open areas and disturbed habitats. Grasslands, meadows, and wetlands have been particularly heavily affected by the introduction of exotic species. (See the spartina case study on page 96.) In many cases these exotic species appear to be displacing native grass species. Without explicit efforts to reintroduce, reestablish, and restore these grass species it is unlikely that these ecosystems will recover, even when natural disturbance regimes are restored.

Fragmentation of forest habitats has played an important role in changing the biodiversity of coastal forests. While large-scale wildfires and windstorms have ravaged the coast in many areas, greater landscape integrity was generally associated with these events than exists today after clearcut logging. A key factor appears to have been the high degree of ecosystem legacy (surviving propagules, sprouts, logs, dead trees, and microbes), which influenced patterns of ecosystem recovery. While the effects of fragmentation are well established for many animal species along the coast, little research has been published on its effects on plant species. Conifers are generally well dispersed in coastal forests, except during poor cone crop years. Many understory plants produce fleshy fruits, which also tend to be fairly well distributed. The greatest effects of logging on regional plant biodiversity may be on canopy epiphytes, gap-strategist understory herbs, hemisaprophytes, and other specialized species that have poor seed dispersal and require older, more intact habitats than typically exist in human-dominated coastal landscapes (Gustafsson and Hallingbäck 1988).

The rich flora of the North American coastal rain forest offers an unparalleled opportunity to improve our understanding of the effects of climate, marine influences, glacial history, and human activity on biological diversity. While there is still a significant component of pristine forest in the northern portion of the coastline, effects of human encroachment are increasingly evident throughout the region. The most species-rich and unique elements of the landscape—wetlands, shorelines, and rain-shadow forests—have had the greatest human impacts: changing of natural disturbance regimes, clearing, and changing hydrological and soil conditions. Introduction of exotic species has had a disproportionate effect on these ecosystems as well. A well-coordinated conservation strategy is needed to maintain the full richness of habitats and ecosystem structure in the region as well as to preserve functional connectivity for dispersal and genetic interchange of plant and animal populations.

References

Agee, J. K. 1993. *Fire Ecology of Pacific Northwest Forests.* Washington, D.C.: Island Press.

Alaback, P. B. In press. "Biodiversity patterns in relation to climate in the temperate rain forests of North America." In R. Lawford, P. Alaback, and E. R. Fuentes, eds., *High Latitude Rain Forests of the West Coast of the Americas: Climate, Hydrology, Ecology and Conservation.* Berlin: Springer-Verlag.

Alaback, P. B., and G. Juday. 1989. "Structure and composition of low elevation old-growth forests in research natural areas of Southeast Alaska." *Nat. Areas J.* 9:27–39.

Banner, A., R. J. Hebda, E. T. Oswald, J. Pojar, and R. Trowbridge. 1988. "Wetlands of Pacific Canada." In National Wetlands Working Group, ed., *Wetlands of Canada.* Ottawa: Polyscience Publications.

Bilby, R. E., B. R. Fransen, and P. A. Bisson. 1996. "Incorporation of nitrogen and carbon from spawning coho salmon into the trophic system of small streams: Evidence from stable isotopes." *Can. J. of Fish. and Aqua. Sci.* 53(1):164–173.

Brooke, R. C., E. B. Peterson, and V. J. Krajina. 1970. "The subalpine mountain hemlock zone." In V. J. Krajina, ed., *Ecology of Western North America.* Vancouver: University of British Columbia Press.

Fonda, R. W., and L. C. Bliss. 1969. "Forest vegetation of the montane and subalpine zones, Olympic Mountains, Washington." *Ecol. Mono.* 39:271–301.

Franklin, J. F., and C. T. Dyrness. 1973. *The Natural Vegetation of Washington and Oregon.* Corvallis: Oregon State University Press.

Gustafsson, L., and T. Hallingbäck. 1988. "Bryophyte flora and vegetation of managed and virgin coniferous forests in south-west Sweden." *Biol. Cons.* 44:283–300.

Henderson, J. A., D. H. Peter, R. D. Lesher, and D. C. Shaw. 1989. *Plant Associations of the Olympic National Forest.* R6 Ecol. Tech. Paper 001-88. Portland: USDA Forest Service.

Huston, M. A. 1994. *Biological Diversity: The Coexistence of Species on Changing Landscapes.* Cambridge: Cambridge University Press.

Krajina, V. J. 1965. "Biogeoclimatic zones and biogeocoenoses of British Columbia." In V. J. Krajina, ed., *Ecology of Western North America.* Vancouver: University of British Columbia Press.

Maser, C., and J. Sedell. 1994. *From the Forest to the Sea.* Delray Beach, Fla.: St. Lucie Press.

Noss, R. F., and A. Y. Cooperrider. 1994. *Saving Nature's Legacy.* Washington, D.C.: Island Press.

Pojar, J., and A. MacKinnon. 1994. *Plants of Coastal British Columbia.* Vancouver: Lone Pine Press.

Pojar, J., E. Hamilton, D. Meidinger, and A. Nicholson. 1992. "Old growth forests and biological diversity in British Columbia." In G. B. Ingram and M. R. Moss, eds., *Landscape Approaches to Wildlife and Ecosystem Management.* Ottawa: Polyscience Publications.

Roemer, H. L., and R. T. Ogilvie. 1983. "Additions to the flora of the Queen Charlotte Islands on limestone." *Can. J. Bot.* 61:2577–2580.

Roemer, H. L., J. Pojar, and K. R. Joy. 1988. "Protected old-growth forests in coastal British Columbia." *Nat. Areas J.* 8:146–159.

Schofield, W. B. 1968. "A selectively annotated checklist of British Columbia mosses." *Syesis* 1:163–175.

Schoonmaker, P. K., and A. McKee. 1988. "Species composition and diversity during secondary succession of coniferous forests in the western Cascade mountains of Oregon." *Forest Sci.* 34:960–979.

Tilman, D. 1982. *Resource Competition and Community Structure.* Princeton: Princeton University Press.

Veblen, T. T., and P. B. Alaback. In press. "A comparative review of forest dynamics and disturbance in the temperate rain forests in North and South America." In R. Lawford, P. Alaback, and E. R. Fuentes, eds., *High Latitude Rain Forests of the West Coast of the Americas: Climate, Hydrology, Ecology and Conservation.* Berlin: Springer-Verlag.

Waring, R. H., and J. F. Franklin. 1979. "Evergreen coniferous forests of the Pacific Northwest." *Science* 204:1380–1386.

Waring, R. H., and J. Major. 1964. "Some vegetation of the California coastal redwood region in relation to gradients of moisture, nutrients, light, and temperature." *Ecol. Mono.* 34:167–215.

Whittaker, R. H. 1960. "Vegetation of the Siskiyou Mountains, Oregon and California." *Ecol. Mono.* 30:279–338.

Bears in the Greater Kitlope Ecosystem

KEN MARGOLIS

The million-acre greater Kitlope ecosystem on the north coast of British Columbia is the largest intact coastal temperate rain forest watershed on earth. It is also the heart of the traditional territory of the Haisla First Nation and, as well, the last part of their territory that remains intact. The Kitlope includes a continuous series of habitat types—from salt water to high grasslands, ridges, and interior valleys—that in addition to its size make the Kitlope one of the last places along the coast capable of supporting viable populations of grizzly bears and other top predators. Protected by a historic 1994 agreement between the Haisla Nation and the provincial government of British Columbia, a pact supported by the voluntary relinquishment of logging rights by a private timber company, the Kitlope ecosystem is an irreplaceable part of the territory that grizzlies will need if they are to survive in North America over the long term.

Haisla elders have information about the number of grizzlies in the Kitlope both through personal experience and through the cultural transmission of generations of Haisla knowledge. For the past decade, elders have expressed concern about encountering few grizzlies in the Kitlope in places where once they saw many and where Haisla oral traditions record a long history of grizzly presence. The Haisla consider the grizzly an animal of great spiritual power. They have traditionally hunted black bears for food, but Haisla tradition forbids the killing of a grizzly bear in any circumstance except self-defense.

The Haisla were certain that both grizzly and black bear populations had declined alarmingly in the Kitlope, and they strongly suspected that trophy hunting was responsible. In fact, commercial hunting guides licensed by the provincial government maintained a permanent camp in Haisla territory at A-go-yoo-wa on Kitlope Lake, in the heart of the ecosystem. Since only the province has legal authority to enforce a hunting ban, Haisla leaders knew they would have to persuade the provincial government to eliminate bear hunting in the Kitlope.

Wayne McCrory, the most prominent grizzly biologist in western Canada, reviewed recorded kills for the Haisla. He found that the black bears taken by hunters were smaller and younger than in the past and that in recent years hunters had been unable to kill their quota of grizzly bears, even though grizzlies are the preferred species. These indications of population decline were reinforced by McCrory's fieldwork, which confirmed far fewer grizzlies in the Kitlope than the forty-two that available habitat could theoretically support. McCrory concluded that overhunting was the most likely explanation.

Former Haisla Chief Councilor Gerald Amos explains how this could happen: "Jet boats can go right up that river. If you want to kill something, you just have to go down to 'Joe's Diner' at 5:30. Every afternoon they always come there. It's easy to get one. That's what these guys do. They sit beside a river where the sockeye salmon spawn. That's 'Joe's Diner' to them bears. . . . There's nothing challenging about what they call hunting in the Kitlope. It was slaughter, not hunting. Fly in, shoot, and let's have a party.

The tendency of both grizzly and black bears to funnel onto river feeding sites during salmon spawning season has a particular implication for conservation of the species. The funneling effect means that the hunting can be good until the last bear has been shot. Since all bears concentrate at a few spots, there is no early warning of population decline.

McCrory delivered his report to the Haisla in early February 1994. On February 22, the council of the Haisla village at Kitamaat and the Nanakila Institute (a community organization devoted to conservation, stewardship, and appropriate development of the ancestral lands of the Haisla) organized a feast. Over 300 village residents showed up to eat a meal prepared by village elders and served by the young people and to participate in a public forum about the Kitlope. At the climax of the evening, Chief Councilor Rob Robinson and Gerald Amos rose together to speak.

"Our people have noticed in the recent past that we don't see grizzly bears in the Kitlope any more, although once we saw many," Robinson began. "The study done by Wayne McCrory shows that the grizzly population in the Kitlope may be nearly wiped out, and black bears may be in trouble too. Therefore, we are telling our own people, there will be no more hunting of any bears in the Kitlope, at least until we get a better handle on what is going on." Amos added pointedly: "We are sending this out to our own people first to set an example. We are asking others to join in this moratorium. If they do not do so voluntarily, we will determine what steps to take next."

Over the next few weeks, the Haisla pressed the provincial government. It was simple to show that the government actually had no idea how many grizzly or black bears there were in the Kitlope—and that hunting quotas were being set without that crucial information. The McCrory study, while preliminary, presented convincing evidence that the grizzly population was in serious trouble. Stephen Hume, a prominent columnist in the *Vancouver Sun*, wrote a powerful column about the situation, and the wildlife branch of the government was getting several calls a day from Kitamaat village residents. There was talk of Haisla volunteers, carrying whistles and fireworks, following hunters up the river.

On March 19, the government announced a one-year ban on grizzly bear hunting in the Kitlope, effectively canceling the 1994 grizzly bear hunting season. In the same announcement, the provincial minister of environment, lands, and parks acknowledged concern about black bears and made a commitment to conducting a province-wide assessment of black bear populations. In this instance, indigenous and European knowledge systems came together in support of conservation.

In late 1995, the ban on grizzly bear hunting remained in effect. The number of black bears that could be killed in the Kitlope, previously unrestricted, is now limited to six per year. The Haisla consider this arrangement satisfactory—short-term progress toward a long-term solution for black bears. They have every expectation that the ban on grizzly bear hunting will be extended indefinitely, giving grizzlies a chance to survive and Haisla culture a chance to continue its millennia-long relationship with the complete fabric of life in the Kitlope.

Ken Margolis is currently executive director of the Nanakila Institute in Kitamaat, B.C. He served as Ecotrust's director for British Columbia and was a board member of the Nanakila Institute at the time the hunting moratorium was enacted.

□ □ □ □ □ □ □ □ □ □ □ □ □ □

Restoration as Practice in the Mattole

SETH ZUCKERMAN

In the Mattole watershed, a small river basin in northwestern California, a committed group of citizens has been working since 1981 to aid the recovery of the salmon runs and enhance the area's biological productivity and diversity. Like many Pacific Northwest streams, the Mattole River faced a dramatic decline in its salmon populations in the 1970s. From tens of thousands each year, when salmon were a significant source of food for human residents and wild creatures, the number of returning spawners dwindled to hundreds in the aftermath of careless logging, road building, and conversion of forest to pasture.

Our efforts were inspired by the realization that increased erosion was clogging spawning beds with silt, burying salmon eggs, and suffocating them before they could hatch. Residents then launched a project to trap the fish as they return, take their eggs, incubate them in streamside hatchboxes, and return healthy fry to the river at a rate that far exceeds what would be realized in the degraded wild. Small groups of people who live on particular creeks band together to tend these hatchboxes, raise the fish, and release them, often with the help of schoolchildren.

As this effort proceeded, we learned about the habitat salmon need. We came to realize that much damage would have to heal in the hills and valleys before the river would welcome the salmon as it once had and perhaps could again. Road washouts and ruts became more than simply local inconveniences; they came to be seen as one of the causes of the salmon's decline. Clearcuts meant more than ugly scars marring the view; they meant increased erosion and a lack of cover to shield the water from the sun's warming rays. The restoration effort expanded to address these issues. Residents now work to keep the river from carving chunks of the bank and depositing silt in the river where it interferes with salmon procreation. We intercede with regulators to moderate the rate and severity of timber harvest in the watershed. And we are working to preserve remaining stands of old-growth forest, whose streams provide refuges for the remnant fish populations.

The Mattole Restoration Council was founded in 1986 to address these ecosystemwide issues. One of its first projects was a map comparing the watershed's old-growth forest cover in 1947 (before the logging boom began) and in 1988. The map revealed that more than 90 percent of the ancient Douglas-fir forest had been cut during that time and, moreover, showed just how isolated and fragmented the remaining stands were. The map has helped us make the case for preserving the few remaining blocks of ancient forest in the watershed and has been circulated widely as an example of how forests have been treated in our region.

The council has gone on to other projects, including habitat enhancement in streams, landslide stabilization, reforestation, a survey of erosion sources in the basin, and a plan to improve the habitat value of the Mattole estuary. One common theme in the council's efforts is that people who live in the area initiate and direct the action themselves. Restoration is not the work of experts from the university or a government agency—though we draw on experienced advisers whenever we need to—but the practice of those who live here. As a result, our work has a home-grown flavor and is more accessible to our neighbors, on whose participation the success of broad-based restoration depends.

The Mattole is home to several groups of people: ranchers, retired people, new settlers who were part of the 1970s back-to-the-land movement, and other shades too numerous to list. Though we come from varied backgrounds, we all care about the fish and view them as part of everyone's heritage. In the best of times, our shared love for the salmon has brought disparate segments of the human community together. At other times, fears for people's livelihood or the ecosystem have created distrust and kept us from making common cause on behalf of the fish.

One such polarization occurred after Redwood Summer, a series of pro-forest protests in 1990. Tensions soared in the wake of a campaign against old-growth logging in our region and a recommendation by California's Department of Fish and Game that all logging plans in our watershed include provisions to prevent any net increase in sedimentation. The agency suggested that when logging washes extra silt into the river, timber operators should compensate for it by actions to keep sediment out of the river, such as stabilizing existing problem areas. Since just about every activity in the watershed, from homebuilding and road grading to logging and farming, causes sedimentation, this recommendation made a lot of people very nervous.

Just when tensions reached a peak, a small group of residents—ranchers, timber company representatives, restoration workers, and land preservationists—began to meet in order to sort out their differences. Only one government employee was invited, representing the federal Bureau of Land Management, entrusted with the 55,000-acre King Range National Conservation Area, much of which lies in the Mattole basin. By discussing their divergent perspectives, the members of this group gained respect for one another and opened channels of communication. When they invited watershed residents and landowners to meet for the same purpose, a tenth of the valley's population showed up to form the Mattole Watershed Alliance.

We used the alliance as a forum to educate ourselves about the issues that matter to the Mattole; when we were able to reach consensus, we took positions as a group. Our first action was to urge significant restrictions on the sportfishing season to protect dwindling salmon stocks. More than forty people drafted and signed the letter, including homesteaders, resort operators, local merchants, ranchers, and timber workers—a list that demonstrated the broad support the idea enjoyed. The state Fish and Game Commission went on to adopt almost all of our recommendations.

The channels of communication that the alliance opened served the restoration community well in other fields, too. When a local rancher planned a major timber harvest on his and his neighbors' land, he invited members of the restoration council to ride horseback through his forest so he could explain his plan and hear their views about it. As a result, he adopted measures to lessen the impact on streams and hired the council to monitor the harvest's ecological effects. One mitigation was streamside willow planting to shade the creeks and stabilize their banks, work that was farmed out to students at the local high school.

On the way to the worksite, one of the teenage planters looked at the forest and asked the forester, "Is that where you're going to log?"

"Actually," the forester replied with a grin, "that's where we already have logged."

Change happens slowly. Not all of the logging that occurs in the Mattole watershed is quite that benign, nor are all the timber owners quite so open about their plans. But we are laying a foundation. At the local elementary school, before the salmon restoration project began, students painted a mural that showed fish swimming every which way in the river; the only human stood on the bank, fishing. When the hatchbox program was a few years old, another group of students painted a new mural. This time the big fish were all swimming upstream to spawn, and the people in the picture were releasing fingerlings to swim down to the ocean. From such experiences and realizations we are attempting to forge a new way of relating to the watershed we live in.

Seth Zuckerman is a restoration consultant and chairs the board of the Mattole Restoration Council, based in Petrolia, California.

□ □ □ □ □ □ □ □ □ □ □ □ □ □

Crossing the Land of Chalayuck

KURT RUSSO

The Lummi Indians of the Pacific Northwest are people of the salmon and the cedar. The cultural identity, beliefs, and traditions of this tribe are inseparable from the landscape created and transformed by Xales (the Transformer) countless generations before the Great Flood. The Lummi Indians are direct descendants of the *xwalxw'eleqw* ("the people who survived the flood") and have lived for thousands of years in what is now Whatcom County in northwestern Washington.

In the brief span of a few generations, the once vast cedar forest in which the Lummi and other tribes along Puget Sound and the Georgia Strait lived has been reduced to a few remnant groves. The Nooksack River is contaminated with industrial and agricultural pollutants, and over half the creeks and streams in Whatcom County have been degraded by logging and hydropower development. Wild game has retreated to isolated areas in the distant North Cascades, and the salmon, once so thick you could "walk on their backs," are threatened with extinction. Local and state governments continue to erode the promises made to the Lummi Indians and other Washington tribes in the 1855 Treaty of Point Elliott.

Yet the Lummi Indians remain a resilient and resourceful—even optimistic—people who believe in the past and learn from it. They believe that with hard work, proper planning, and guidance from the past, what should be done can be done, regardless of the odds. In 1984 the Lummi fishers, faced with a direct threat to their fishing rights, imposed a tax on themselves to create the Lummi tribe's Treaty Protection Task Force.

The accomplishments of the task force—in areas as varied as treaty rights, international indigenous rights, cross-cultural communication, protection of the sacred, and public awareness—have been inspired by the vision of task force coordinator Jewell Praying Wolf James (tse-Sealth). Described by Mr. James as "crossing the land of Chalayuck," the vision requires travel in the four directions to cross into the lands of the four races, in common cause with the Creation and with future generations.

His vision is reflected in the work of the task force at the local level and the effort to preserve the remaining old-growth cedar forests in the traditional territory of the Lummi people. In its first year, the task force succeeded in placing in the Washington Wilderness Bill some 12,000 acres of old-growth forest used by the *seyown* (spirit dancers) of western Washington. Closer to home, the task force has focused on Arlecho Creek, a tributary of the Nooksack River in the foothills of Mount Baker in the North Cascades.

The Arlecho Creek basin, though it has been extensively logged, contains one of the largest stands of old-growth forest remaining on private land in western Washington and has both spiritual and cultural value to the Lummi Nation. The 2200-acre basin contains numerous cultural and archaeological sites and provides habitat for rare and endangered species (including marbled murrelets that fly up the Nooksack to nest in the ancient trees beside Arlecho Creek). The watershed is also critical to the operation of the Lummi salmon hatchery, located downstream, which must deal with silt problems originating upstream.

The task force led a two-year campaign to protect Arlecho's remnant old growth from logging. The effort included public protests, legal injunctions, two nationally broadcast television specials, and the involvement of local as well as national environmental and church organizations. The pressure and exposure led the landowner, Mutual Life Insurance Company of New York, to sell its timberland holdings in the basin to Crown Pacific Timber Company in 1995.

Even before the sale, the tribe had approached The Nature Conservancy of Washington to intervene on its behalf with the corporate landowner. In recent meetings facilitated by the conservancy, Crown Pacific and the Lummi Nation discussed plans to preserve the Arlecho basin and ensure the sustainable management of the surrounding 8500 acres. The plan under discussion would incorporate tribal values and beliefs in the management of the watershed. One step would include discussions facilitated by the Florence R. Kluckhohn Center for the Study of Values to improve cross-cultural understanding among the tribe, Crown Pacific, and the conservancy. Improved understanding would lay the foundation for cooperative management of this vital and vulnerable watershed.

Like all the activities of the task force, the Arlecho Creek effort can be compared to the facets of a diamond. There is the facet of public policy around treaty-related issues. There is the facet of research—trying to understand barriers between cultures and how to overcome them. There is the facet of networking and empowering one group through association with others. There is the facet of public opinion and the need to continue to educate the public about indigenous values and perspectives on the environment. But perhaps most important are the facets of creativity, imagination, passion, and vision. Without them there is no light, no reflection, and, in the end, no way to endure the losses.

Kurt Russo, a member of the Treaty Protection Task Force of the Lummi tribe, is executive director of the Kluckhohn Center in Bellingham, Washington.

□ □ □ □ □ □ □ □ □ □ □ □ □ □

Stopping Spartina's Hostile Takeover of Willapa Bay

WENDY SUE LEBOVITZ AND DAN'L MARKHAM

Surely, but not so slowly, an aggressive nonnative marsh plant called *Spartina alterniflora* is taking over the tidelands of Willapa Bay, Washington. Patches of spartina are already ruining migratory bird habitats and threatening the bay's shellfish industry. Spartina displaces habitat critical for shellfish, native saltmarsh plants, migratory birds, young salmon, and other wildlife, posing a serious and visible threat to this still-healthy bay. In the early 1980s there were about 800 acres of spartina in 80,000-acre Willapa Bay. Today, spartina patches cover more than 4000 acres of the bay, and the invasion is spreading.

The West Coast commercial oyster business got its start in Willapa Bay in the 1850s, and today roughly one-sixth of all the oysters produced in the United States are harvested in this still-fertile bay. But native oysters were nearly exhausted by the 1880s, and enterprising oystermen introduced oysters from the Chesapeake and other eastern bays to Willapa tidelands in 1894. These transplants came packed in a cordgrass native to the eastern seaboard—spartina. Though the eastern oysters planted in Willapa perished in a shellfish epidemic in 1919 (and were later successfully replaced by Pacific oysters from Japan), the cordgrass took root and hung on for nearly six decades before beginning to spread significantly across the bay's intertidal zone.

In the east, spartina is a natural component of estuary ecology. In Willapa, however, spartina grows in circular patches that can reach 6 feet in height by late summer. As the patches expand, the dense "vegetative blanket" displaces native grasses like eelgrass, as well as young Dungeness crabs, clams, and oysters. Spartina also traps sediment at a rate of 6 inches per year—literally raising the level of the intertidal zone to that of the salt marshes above the high-tide line. In time this accumulation could significantly reduce the volume of water the bay can hold. Loss of the bay's intertidal area could lead to severe flooding of the lower valleys of the Willapa, Naselle, and other rivers that drain into the bay. The shrinking of intertidal areas also eliminates critical feeding grounds for hundreds of thousands of shorebirds and waterfowl that travel the Pacific Flyway in a twice-yearly migration.

Concerned by spartina's spread, the local government of Pacific County convened a group including representatives from several state and federal agencies, local landowners, and businesspeople in October 1993 to develop a practical spartina management plan for the 450 private owners of intertidal properties (primarily oysterbeds) in Willapa Bay. This Pacific County Spartina Task Force completed a draft plan in February 1994.

The task force recommended an integrated weed management approach to control and eradicate spartina. Integrated weed management combines several methods of controlling the plant, allowing tideland users and managers to use the one most likely to succeed on each site, depending on its biological and environmental conditions. Such site-specific treatments are preferable to a generic wide-area treatment in a bay with the diversity of conditions found in Willapa.

Because spartina's ecology in western bays is poorly understood, control methods are still largely experimental and vary from mechanical approaches—pulling seedlings by hand, burning patches at low tide, mowing, cutting with hand-held weed-whackers, covering with plastic—to applications of herbicide including wiping individual plants and controlled spraying. Several of these methods require a permit.

Pacific County commissioners and the county's Noxious Weed Control Board applied for permits to initiate control efforts in early 1994. These applications were immediately challenged by the Seattle office of Friends of the Earth, the Shoalwater Bay Indian Tribe, and a local citizens' group opposed to spraying under any circumstances. These groups filed lawsuits against the government of Pacific County, the Pacific County Noxious Weed Board, the U.S. Fish and Wildlife Service (which administers the Willapa National Wildlife Refuge in the bay), the state Department of Natural Resources, and the state Department of Fish and Wildlife. After lengthy negotiation, the parties reached agreement in 1995 giving the green light to an integrated weed management approach on spartina-infested acres in Willapa Bay.

The process, however, exposed deep divisions among members of the local community. Many residents feared that spartina was becoming another "us versus them" environmental issue and that potential use of herbicides had separated the community into opposing camps with irreconcilable differences. The Willapa Alliance, an organization representing local stakeholders including businesspeople, oystermen, commercial fishermen, timberland owners, and others, sought to get beyond the impasse. The alliance's mission is to enhance the diversity, productivity, and health of Willapa's unique environment, to promote sustainable economic development, and to expand the choices available to Willapa residents. Late in 1994, the alliance independently organized a citizens' group called the Spartina Coordinating Action Group (SCOAG) comprising private interests, local businesses, and environmentalists that split into work groups that could get at the roots of the political stalemate.

When the permits were not forthcoming the following spring, state Senator Sid Snyder from Pacific County introduced legislation in the state senate to streamline the permitting process. The Willapa Alliance and SCOAG provided scientific information to Senator Snyder and testified before the state senate. The legislation was adopted and signed into law by Governor Mike Lowry in May 1995. The legislature also allocated funds to the state Department of Agriculture to aid spartina control and eradication.

In 1995, the weed board, the Department of Natural Resources, and the Department of Fish and Wildlife made a start on spartina control. Crews tested all the approved methods, the Washington Conservation Corps contributed work crews, and the county's spartina coordinator organized monthly seedling pulls. The Willapa Alliance continues to provide leadership in public education, consensus building, and coordinating the control and monitoring efforts.

Spartina has not been eradicated from Willapa Bay—it may never be—but the community has taken promising steps toward reining it in.

Throughout the west, exotic plant species have invaded native ecosystems and disrupted the ecology and productivity of hundreds of thousands of acres. In Willapa, there is still a chance that local resolve, supported by the coordinated efforts of local, state, and federal governments in partnership with private tideland owners, can keep Willapa Bay from sacrificing its native diversity and productivity to spartina.

Wendy Sue Lebovitz is spartina coordinator for the Pacific County Noxious Weed Control Board in South Bend, Washington. Dan'l Markham is executive director of the Willapa Alliance, a citizens' group also based in South Bend.

□ □ □ □ □ □ □ □ □ □ □ □ □ □

Restoring Knowles Creek

THOMAS C. DEWBERRY

In the late 1870s, when the first European settlers ferried their possessions up Oregon's Siuslaw River from the coastal town of Florence, many tributaries of the lower Siuslaw including Knowles Creek supported mostly young stands of Douglas-fir and patches of salal and salmonberry. Much of the landscape the settlers encountered had burned in a fire or series of fires during the 1860s. Except for a few grassy "prairies" maintained by Native American burning, the valley floors withstood the fires that had initiated the young forests on the uplands. Along the creek stood centuries-old western redcedars described by a traveler in 1853 as the most magnificent stand of cedars on the Oregon coast.

Coho and chinook salmon, steelhead, cutthroat trout—all are native to Knowles Creek. Historical sources suggest that 75,000 to 100,000 coho salmon smolts, once the most abundant salmonids in the watershed, migrated to the ocean from Knowles Creek spawning gravels each year. A hundred years later, fewer than 1700 smolts were making that journey. What had been done to make the creek so hostile to salmon survival? And could that change be reversed?

Resource managers with the Siuslaw National Forest, Champion International timber company, and the nonprofit Pacific Rivers Council embarked on a cooperative effort to answer these questions. Years of study of how the Siuslaw River basin looked and functioned before European settlement formed the basis of a strategy that sought to restore part of the basin's natural functions. Salmon provided the yardstick of restoration success.

We began by seeking to understand how water, organic materials, nutrients, and sediment move through the Knowles Creek basin on a time scale of decades to centuries, creating a "digestive" process that develops and maintains salmon habitat. The key turned out to be how frequently landslides and debris torrents deliver sediment to the valley floors and stream channels.

The Knowles Creek basin is underlain by a bedrock of Tyee Sandstone, which weathers to sand and gravel and produces exceptional "skipping rocks." The creek's tributaries flow from steep, unstable hillslopes. Each of the 5000 hollows in the 20-square-mile Knowles Creek basin fills slowly with sediment and woody debris, only to release that material in sudden torrents under certain winter storm conditions. Each hollow delivers some or all of its sediment toward the valley about every 6000 years, suggesting that the accumulation and delivery of material are continuous. Certain reaches of the stream were frequent sites of debris torrent deposits, trapped by streamside trees or a gentle gradient. Terraces would form upstream of these deposits, building the valley floor.

These highly productive flats, built on a lattice of logs and mature trees on the valley floor, hold the key to salmon abundance. When the flats trap and accumulate material from debris flows, they are sites of high biological production. Coho salmon abundance, in particular, seems to be associated with periods when the stream channel is highly connected to the surrounding valley. Small increases in streamflow flood large portions of the valley floor, creating an intricate weave of channels, pools, and ponds.

The picture of Knowles Creek that emerged from our early research was a stream system in which a number of flats maintained the creek's biological productivity, but the specific stream reaches where salmon spawned changed over the years. When white settlers arrived in the Siuslaw, many of the flats were highly connected to their valley floors and the system was enormously productive of fish.

Settlement caused an accelerating simplification of the stream system. Early settlers cleared logs from the lower 6 miles of Knowles Creek, allowing the stream to cut to bedrock; banks as high as 18 feet stood in some places along the lower 5 miles. Human activity in the basin declined during the first half of this century, then accelerated after 1950 in a burst of road building and logging. From the early 1950s to the mid-1980s, landslides from slopes destabilized by logging moved as much material into the channel as would have taken 200 years under natural conditions. In effect, this flow of material, combined with logging of the surviving old growth on the valley floor and "stream cleaning" to allow fish passage, destroyed the "digestive" capabilities of Knowles Creek.

By the 1980s, the lower 12 miles of Knowles Creek had cut to bedrock. The only areas that retained spawning gravels were above remnant boulders from earlier debris torrents. Even the water table had dropped without the flats and debris dams that once spread water across the alluvial formations of the valley floor. A portion of the water in a healthy stream travels subsurface through gravels and stays cool. With virtually no subsurface flow, water in parts of Knowles Creek reaches 75°F during midsummer, fatal to young fish. Salmon populations crashed.

Given our diagnosis of Knowles Creek, we devised a four-part restoration strategy. The first part involved identifying and protecting intact areas. The 5 to 10 percent of the Knowles basin that has never been logged serves as a crucial refuge and resource for restoration of the remaining portion of the basin. Second, we sought to identify source areas for debris torrents in the uplands. Third, we sought to accelerate forest recovery on the valley floor. Most of the valley floors within the Knowles basin are now dominated by second-growth stands of red alder. Believing that little significant restoration could be accomplished in the basin in the absence of mature conifers, we placed high priority on planting western redcedar under stands of alder where cedar was not establishing naturally.

These three steps formed the basis of a 50- to 100-year strategy. But the fish could not wait that long. To increase the digestive capacity of the basin and reestablish the process, we embarked on an effort to simulate debris torrent deposits at a few locations in the basin where they would have naturally occurred. In the summer of 1992, we chose five sites near the headwaters of Knowles Creek and five sites along the lower creek. To capture smaller pieces and create logjams, we placed large pieces of woody debris at each location. To reset flats downstream at times of high storm flow, most key pieces were not anchored in place.

After just three years, it is too early to tell how salmon populations will respond to these efforts. But since the native salmon were adapted to a healthy, functioning watershed, we expect to see an increase in the numbers of salmon smolts coming out of the creek. We are monitoring the fish carefully. What happened to our simulated debris flows after a severe storm in January 1995 gives us hope that we do understand the health of the system. During the six days of that storm, about 12 inches of rain fell and the Siuslaw River crested at 22 feet.

Although the storm pulsed smaller pieces of wood downstream from all of them, it failed to wash out any of our simulated debris torrents. All of the jams reset their respective flats, and each flat accumulated 100 to 150 yards of gravel upstream. The flats captured much of the material transported down the mainstem of Knowles Creek. They were beginning to reset the digestive process.

Promising as these results are, restoration is a long-term process. The emergency measures we undertook by reestablishing small debris deposits in a few places in the basin restored only a small portion of the basin's digestive function. The increases are minuscule compared with the digestive capabilities of the basin as a whole in a healthy condition. Our strategy is a means of buying time while the cedars of the valley floor grow larger and begin to restore the digestive function significantly.

Thomas C. Dewberry is restoration coordinator for the Pacific Rivers Council in Eugene, Oregon, and restoration biologist for the Chinook River restoration initiative in southwest Washington. This case study was adapted from a technical report prepared for the Pacific Rivers Council, John Hancock Timber Management Group, and the Mapleton District of the Siuslaw National Forest and published by the Pacific Rivers Council in 1995.

5. Terrestrial Vertebrates

Fred L. Bunnell and Ann C. Chan-McLeod

□ □

The Pacific Northwest hosts a remarkable richness of terrestrial vertebrates. Within the Pacific Northwest the coastal temperate rain forest stands out as being particularly rich in biological diversity. In this chapter we review the reasons why the Pacific Northwest is so rich in its vertebrate fauna. Although most vertebrate species in the coastal temperate rain forest are forest dwelling, some use the estuaries, lakes, or islets that characterize the region. Indeed water in various forms, from fog and rain to streams and lakes, has a dominant influence in the coastal rain forest, and many vertebrate life histories are intimately tied to water. Other species are closely linked to the natural old forest, which is the product of the long intervals between stand-initiating disturbances. About a hundred species depend on the attributes of older forests.

What are the major implications of these faunal patterns for forest management? For much of our discussion we will rely on experience in British Columbia, which contains the largest intact tracts of coastal temperate rain forest in the world. The conclusions we derive, however, appear broadly applicable to the temperate rain forest region and are corroborated by other studies of the Pacific Northwest.

Species Richness

The vertebrates of the Pacific Northwest generally exhibit greater species richness than the rest of North America with the exception of regions bordering subtropical climates such as southern California and Florida (Bunnell and Kremsater 1990). Canada's westernmost province, British Columbia, for example, makes up less than 10 percent of Canada's landmass but hosts 74 percent of the country's terrestrial mammal species and 70 percent of its breeding bird species (Bunnell and Williams 1980). When species unique to only one province or territory are considered, British Columbia again ranks first with 55 percent of unique breeding birds and 73 percent of unique mammals. Both Quebec and Ontario are larger than British Columbia, but they host far fewer vertebrate species. The situation is similar in the U.S. Pacific Northwest. Washington and Oregon, comprising less than 5 percent of the continental landmass of the United States, contain 37 percent of the native bird fauna and 42 percent of the native mammal fauna (Bunnell and Kremsater 1990).

The Pacific Northwest is extremely rich in vertebrates because of its mountains, its weather, and its forests. The terrain is rugged and relief is generally perpendicular to prevailing weather systems. As a result, annual precipitation ranges widely from windward to leeward areas. Vegetation responds to this range in moisture and shows a remarkable diversity of broad community types (Daubenmire 1952; Meidinger and Pojar 1991). High mountain ranges, greater than 2000 meters in places, also encourage a diversity of vegetation and habitat types that, in turn, encourage a rich fauna. Additionally, the climatic regime has resulted in a landscape dominated by coniferous forests (Waring and Franklin 1979). These forests include some of the largest, most long-lived trees anywhere (Norse 1990). The remarkable biomasses accumulated in these forests—500 to 2000 metric tons per hectare (Fujimori 1971; Franklin and Waring 1980)—exceed those of tropical rain forests and contribute to their great economic value. Long-lived, taller forests also are generally well stratified, providing a wide variety of niches that vertebrates can exploit (MacArthur and MacArthur 1969; Willson 1974).

At the broad regional level, the pronounced vertebrate richness of the Pacific Northwest relative to the rest of the continent is a product of the forest cover (Bunnell and Kremsater 1990). British Columbia represents only 9.5 percent of Canada's land area, yet it contains almost 50 percent of Canada's conifer volume (Figure 5.1*a*). Both Ontario and Quebec, in central Canada, are well forested and larger in area than British Columbia, but because the trees are smaller, these provinces contain less conifer volume. Washington and Oregon comprise only 4.6 percent of the U.S. continental landmass, but they contain 6.8 percent of the coniferous forestland and 25.7 percent of the softwood volume (Bunnell and Kremsater 1990). In comparison, the eastern states of Maine, Michigan, and Minnesota comprise 4.8 percent of the U.S. continental landmass and 7.2 percent of the forested land, but only 5.5 percent of the softwood volume. Not only are both the Canadian and American Pacific Northwest relatively rich in vertebrates, but species richness is generally higher where trees are larger, especially for birds (Figure 5.1*b*). Two more distinctions are apparent at the regional

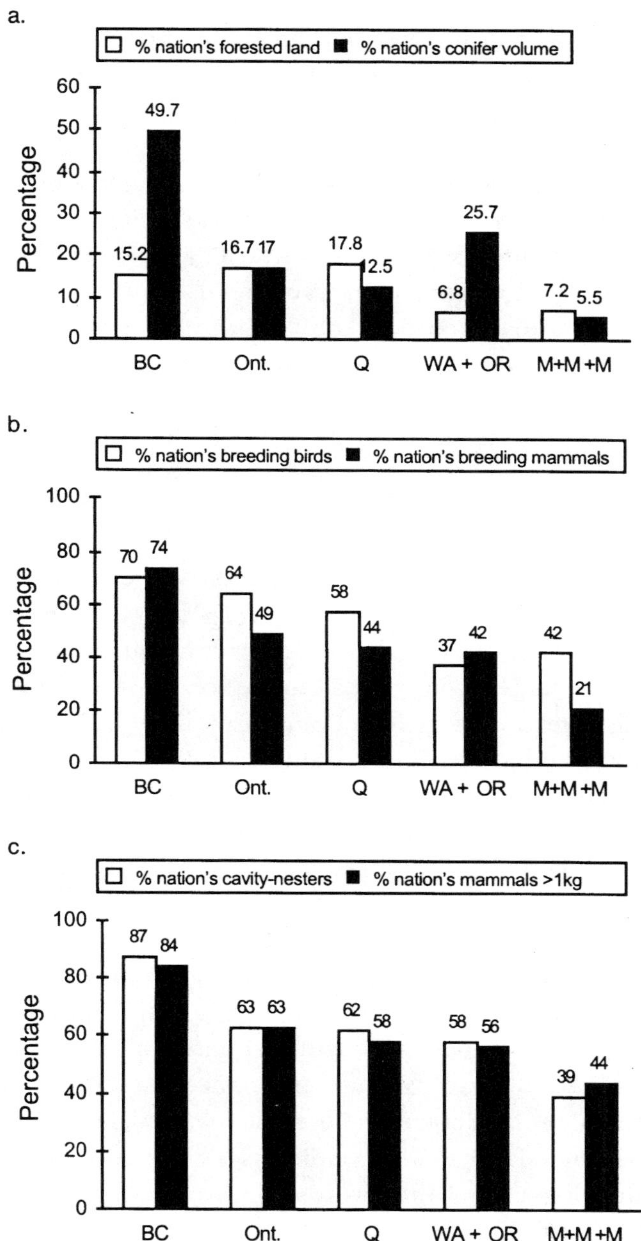

Figure 5.1. Regional comparisons for the Canadian and United States portions of the Pacific Northwest (BC = British Columbia; Ont. = Ontario; PQ. = Quebec; WA + OR = Washington plus Oregon; m+m+m = Maine, Minnesota, and Michigan combined). (*a*) Percentage of the nation's forested land and percentage of the nation's conifer volume. (*b*) Percentage of the nation's native breeding birds and percentage of the nation's native terrestrial mammals. (*c*) Percentage of the nation's cavity-nesting birds and percentage of the nation's large mammals (weighing more than 1 kilogram). *Source:* Bunnell and Kremsater (1990).

level: forests of the Pacific Northwest host an uncommonly large proportion of Canadian and American cavity-nesting bird species and a greater diversity of mammal species larger than 1 kilogram body weight (Figure 5.1*c*).

If one considers biodiversity instead of simple species richness, there are further distinctions at the broad regional level. Definitions of biological diversity recognize variation at the genetic level within species. Although much of this variation is not readily visible, it is suggested by the number of subspecies present. Major factors favoring distinct subpopulations are a wide range of habitats and relative isolation produced by rugged topography. Both features are present in the Pacific Northwest. British Columbia, for example, contains many more subspecies than do other regions of Canada: at least 392 avian subspecies and 281 mammalian subspecies.

Coastal Temperate Rain Forests

Within the remarkable regional richness of the Pacific Northwest, the coastal temperate rain forest hosts one of the most diverse vertebrate faunas found in temperate regions. The pattern is well illustrated in British Columbia, which contains some of the largest contiguous tracts of coastal temperate rain forest in the world. The vegetation of British Columbia has been classified into fourteen biogeoclimatic zones (Meidinger and Pojar 1991). The coastal temperate rain forest is closely represented by the coastal western hemlock zone. Of the twelve forested biogeoclimatic zones in British Columbia, only the interior Douglas-fir zone hosts a greater richness of forest-dwelling vertebrate species (Figure 5.2).

Not all vertebrates found in the coastal temperate rain forest region are forest dwelling, of course. Abundant aquatic and riparian environments host species that are not necessarily dependent on the forest. Some of the estuaries, for example, are globally important staging grounds for migrating shorebirds and waterfowl. These estuaries retain their connection to the forest through downstream transport of sediment and organic material, including whole logs (Chapter 6; Sedell and Maser 1994). Conditions in the estuaries, which may be influenced by the character of the forest well upstream, primarily affect marine species, such as mollusks and spawning salmon. Lakes within the rain forest host ducks and loons and may be influenced by the character of the surrounding forest. Because of its mild climate, the coastal temperate rain forest retains much of its richness all year. Although some species migrate, many others move into the area and its associated waters during winter. Others migrate from breeding habitats in the mountain hemlock zone to winter in the rain forest. In other instances, wintering and breeding habitats differ but both are found within the area.

Excluding very rare occurrences and accidentals, 380 vertebrate species occur in the narrow band of coastal temperate rain forest in the Pacific Northwest. Not all species, as noted, are forest dwelling; the list includes those using lakes, estuaries, and rocky islets. Nearly 69 percent (262) of the total are birds. Distribution of bird species across the coastal temperate rain forest is relatively even, but species richness is distinctly highest in British Columbia (230) and lowest in Alaska (192). Similar numbers of bird species occur in Washington (207), Oregon (212), and California (207). The relatively low species richness in Alaska likely reflects the inhospitable environment of the northern winter. Paradoxically, however, Alaska's highly seasonal climate also provides excellent habitat for breeding birds in spring and summer. A north-to-south gradient in diversity of breeding birds is obvious: the number of breeding species totals 161 in Alaska, 149 in British Columbia, 143 in Washington, 140 in Oregon, and 127 in California. Mammals make up about 21 percent (80 species) of the total terrestrial vertebrate species in the rain forest region. Distribution of mammal species across the area is similar to that for birds: species richness is highest in British Columbia (68), lowest in Alaska (38), and intermediate in Washington (56), Oregon (55), and California (49).

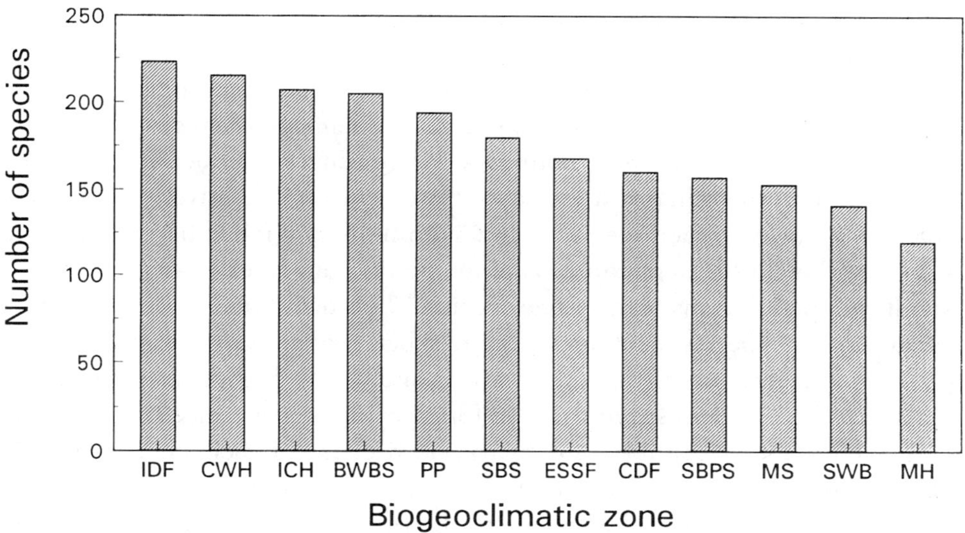

Figure 5.2. Number of native vertebrate species by forested biogeoclimatic zone in British Columbia. (IDF = interior Douglas-fir; CWH = coastal western hemlock or coastal temperate rain forest; ICH = interior cedar–hemlock; BWBS = boreal white and black spruce; PP = ponderosa pine; SBS = subboreal spruce; ESSF = Engelmann spruce–subalpine fir; CDF = coastal Douglas-fir; SBPS = subboreal pine–spruce; MS = montane spruce; SWB = spruce–willow–birch; MH = mountain hemlock.) *Source:* Modified from Bunnell (1995).

Vertebrate species richness in the rain forest is exemplified by Clayoquot Sound. Including lakes and the nearshore environment, Clayoquot Sound encompasses about 350,000 hectares on the west coast of Vancouver Island and contains some of the largest undeveloped tracts of coastal temperate rain forest in the world. At least 304 terrestrial vertebrate species have been recorded in Clayoquot Sound (records of Royal British Columbia Museum; Banfield 1974; Hoyt 1984; van Zyll de Jong 1983, 1985). In an area representing about 0.4 percent of the province, roughly 57 percent of the native terrestrial vertebrate species in the province breed or visit during migration or winter. This richness reflects not only the complex structure of coastal temperate rain forest but also the mild climate and abundant water bodies, including estuaries. Available records also document at least twenty marine mammals in the nearshore environment, including the northern elephant seal, gray whale, and harbor porpoise.

Although the coastal temperate rain forest is extremely rich in vertebrate diversity, relatively few species (11 percent of the total) are restricted to the region (Table 5.1). Without exception, bird species whose ranges correspond most closely to the region also range far north and south of the rain forest boundaries. Amphibian, reptile, and mammal species that occur exclusively within the bioregion (such as the California giant salamander, Coronation Island vole, and Sitka mouse) generally have extremely localized distributions and therefore are not characteristic of the entire rain forest. In all cases, amphibian, reptile, and mammal species largely endemic to the region also occur slightly east of the rain forest boundary. When ranges of endemic species do extend beyond the rain forest, however, the extension is relatively minor, such that the majority of each species occurs within the rain forest borders. Viewed this way, the North American coastal temperate rain forest contains globally significant populations of forty-one species. The true number is, in fact, larger because some species, such as the grizzly bear, range well beyond the temperate rain forest but are most prolific and attain their highest densities there.

As expected, endemism to the rain forest is most pronounced for amphibians, which are susceptible to dehydration because of their permeable skin and low tolerance for dry environments. The heavy reliance of amphibians on the moist environmental conditions of the rain forest is underscored by the fact that twenty-one of twenty-nine amphibian species in the region (72 percent) are largely endemic to the region. The high mobility of birds and mammals, coupled with their relatively low sensitivity to microclimatic conditions, render them even less restricted in geographic distribution than the amphibians. Only 2 percent of bird species and 14 percent of mammal species occurring within the coastal temperate rain forest are largely endemic to the region.

Table 5.1. Vertebrate Species Nearly Exclusive to the Coastal Temperate Rain Forest and Adjoining Pacific Coast

Amphibians	Birds
Taricha granulosa	*Haematophus bachmani*
Taricha rivularis	*Cepphus columba*
Ambystoma gracile	*Brachyramphus brevirostris*
Dicamptodon copei	*Synthliboramphus antiquus*
Dicamptodon ensatus	*Ptychoramphus aleuticus*
Dicamptodon tenebrosus	*Cerorhinca monocerata*
Rhyacotriton cascadae	
Rhyacotriton kezeri	**Mammals**
Rhyacotriton variegatus	*Sorex bendirei*
Aneides ferreus	*Sorex pacificus*
Aneides flavipunctatus	*Neurotrichus gibbsi*
Batrachoseps attentuatis	*Scapanus townsendii*
Ensatina eschscholtzii	*Clethrionomys occidentalis*
Plethodon dunni	*Microtus coronarius*
Plethodon elongatus	*Microtus townsendi*
Plethodon stormi	*Phenacomys albipes*
Plethodon vandykei	*Phenacomys longicaudus*
Plethodon vehiculum	*Peromyscus sitkensis*
Rana aurora	*Zapus trinotatus*
Rana boylii	
Rana cascadae	
Clemmys marmorata	
Reptiles	
Clemmys marmorata	
Contia tenuis	
Thamnophis ordinoides	

Vertebrate/Forest Relations

Habitat use by the vertebrate fauna reflects adaptations to the major characteristics of the coastal temperate rain forest: ubiquitous water and a specific forest cover. Water—as precipitation, fog, and streams and channels—characterizes coastal temperate rain forests. The complexity of habitat use is increased by the diversity of life histories of vertebrates dependent on the coastal rain forest. Habitat use by birds is summarized at the end of this chapter in Appendix A; habitat use by mammals is summarized in Appendix B. A large majority, 72 percent, of the forest-dwelling vertebrates present make significant use of riparian areas (Bunnell et al. 1991). Use by all species of either riparian or shore habitats is even higher (85 percent amphibians and reptiles; 76 percent birds; 85 percent mammals). Even reptiles such as the common garter snake and western terrestrial garter snake are predominantly riparian species. Species not closely associated with forest cover, such as the nine loon and grebe species present, are likewise intimately associated with the water systems that characterize the area. Moreover, loons and other lake users often select water bodies surrounded by undeveloped forests rather than cut-over or open areas. About 150 bird species exploit the estuaries as staging or foraging areas (sandpipers, godwits, dowitchers) or forage and breed in the nearshore environment (cormorants, ducks, mergansers). The moist terrestrial environment favors skin breathers such as the terrestrial salamanders, and the amphibian fauna of coastal temperate rain forest is uncommonly rich for temperate regions (Table 5.1). Of twenty amphibian species found in all of British Columbia, for example, seven are found in the limited area of Clayoquot Sound—again emphasizing the moist, water-dominated character of the environment.

The connections from headwater to estuary and intimate links between forests and water in riparian areas (Sedell et al. 1988; Bunnell and Dupuis 1994; Sedell and Maser 1994) mean, however, that even loons and gulls may be influenced by the character and extent of forest practices. About 75 percent of the vertebrate species (79 percent of amphibians and reptiles, 53 percent of birds, 94 percent of mammals) in the coastal temperate rain forest are forest dwelling and directly influenced by the nature of forest cover. For some species—such as the green-backed heron and the hooded and common mergansers—both tree cover and proximity to water are requisites for successful breeding.

A dominant feature of the coastal temperate rain forest is the natural disturbance regime. The natural fire cycle in coastal temperate rain forest is about 250 years or longer for British Columbia (Bunnell 1995) and about 230 years or longer in Washington and Oregon (Fahnestock and Agee 1983). Agee (1993), for example, summarizes data for the coastal temperate rain forest region indicating fire-initiated stands of 450 to 750 years of age. Forest-dwelling vertebrate fauna appear adapted to the size and frequency of natural disturbances. In British Columbia, for example, the proportion of species using early seral stages to breed is positively correlated with mean fire size whereas the proportion breeding in late seral stages is negatively correlated (Bunnell 1995).

Wind is the major agent of natural disturbance that renews and modifies temperate rain forests. Indeed, in British Columbia the classification of temperate rain forest recognizes two phases based on disturbance by wind. The two dominant tree species of the hemlock–amabilis fir phase appear to benefit from a disturbance regime in which patches of forest ranging from small clumps to stands of several hundred hectares are periodically blown down by wind. Blowdowns release advance regeneration of both tree species and expose mineral soil favoring amabilis fir regeneration. The phase appears to be self-perpetuating, too, because the stand structure resulting from windfall (dense canopies with high "sail" area and shallow-rooted trees) makes the forest susceptible to further windfall and subsequent recruitment of the same species.

The redcedar–hemlock phase is dominated by redcedar with a subcanopy of western hemlock. Stands are more open than those in the hemlock–amabilis fir phase, crowns are less dense, and many cedars are spike-topped. The consequent lower resistance of the canopy to wind, plus redcedar's tenacious roots, make redcedar–hemlock stands more windfirm than hemlock–amabilis fir stands. Although wind is still a major agent of disturbance, the trees blown down tend to be isolated individuals.

The nature of this disturbance regime has several consequences. First, because of the longevity of the trees and the long intervals between stand-initiating disturbances, most natural stands are dominated by old trees. In pristine areas, even-aged young stands are small and rare. Large-scale disturbances, when they occur, usually do not create extensive tracts of even-aged forest but leave pockets of unaltered vegetation and numerous isolated survivors. Second, the natural growth rates, longevity, and low frequency of disturbance allow large biomass to accumulate. Much of this biomass exists as large live trees, large-diameter snags, and large logs on the ground. Downed wood on the lower slopes of the coastal western hemlock zone, for example, can exceed 400 cubic meters or 100 metric tons per hectare. Third, the shade-tolerant nature of the tree species present and the generally small-scale effects of common agents of disturbance ensure a complex structure in the stands. The canopy is usually uneven with gaps where old trees have died or been windthrown. In these gaps the understory is often well developed and includes young conifers. As a result, natural stands, although dominated by old trees, contain trees of a wide range of ages.

Bunnell (1995) has shown that native forest-dwelling vertebrate faunas are adapted to the major natural disturbance regime and the resulting forest structure. In coastal temperate rain forests, we expect the fauna to be adapted to complex structure including large trees, abundant snags, and abundant downed wood. And this is indeed the case. The richness of vertebrate fauna is itself a product of large, long-lived trees creating well-stratified canopies. Because trees are large and structurally complex, the same tree can provide breeding habitat for several species: a brown creeper may be under a bark slab a few meters up; a pileated woodpecker may use a large cavity at about 15 meters; a marbled murrelet may occupy a broad limb 40 meters above the ground. If these birds are unlucky, a red squirrel or marten may also have found a home there. The gaps and abundant understory permit numerous shrub nesters, such as Wilson's warbler or gray-cheeked thrush, as well as tree and cavity nesters such as woodpeckers and nuthatches that breed in the surrounding trees or snags. Winter wrens, marten, and bears nest or den in the root wads or hollows of wind-thrown trees.

At least fifty-six species (nineteen mammals and thirty-seven birds) use cavities of large trees, live or dead, to hibernate or raise their young (Figure 5.3). Several amphibians also exploit snags or large bark slabs falling from snags, especially large snags (Bunnell and Dupuis 1994). Within British Columbia's coastal temperate rain forest (the coastal western hemlock zone) forty-two species, more than in any other zone, use downed wood as breeding sites (Bunnell et al. 1991). Both abundant cavity sites and downed wood are characteristics of old forests.

Some species that are largely restricted to the coastal temperate rain forest, such as the marbled murrelet, appear very closely associated with old-growth forests. Others, such as bald eagles and great blue herons, require large structurally complex trees each year as breeding sites, while black-tailed deer profit during severe winters from closed, older forests with small openings (Bunnell 1985). Some species, such as the western red-backed salamander, require at least limited areas of closed forest tall enough to maintain a humid microclimate (Dupuis 1993).

Comparisons between the coastal temperate rain forest (coastal western hemlock zone) and four other broad forest types in British Columbia are illustrated in Figure 5.3. Only forest-dwelling vertebrates are included; the biogeoclimatic zones illustrated here are described by Meidinger and Pojar (1991).

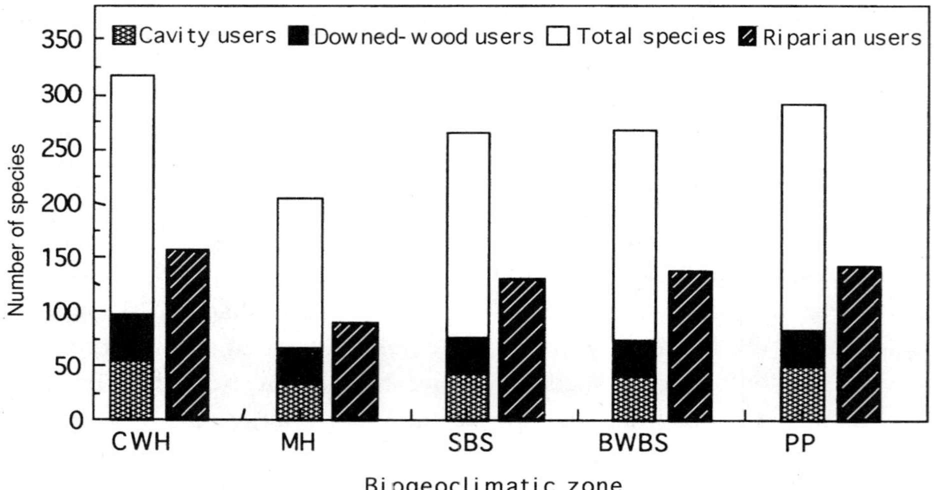

Figure 5.3. Features of the native vertebrate fauna in five forested biogeoclimatic zones of British Columbia. (CWH = coastal western hemlock or coastal temperate rain forest; MH = mountain hemlock; SBS = subboreal spruce; BWBS = boreal white and black spruce; PP = ponderosa pine.) *Source:* Data of Bunnell et al. (1991).

The mountain hemlock zone typically occurs directly above temperate rain forest in British Columbia. Two points merit emphasis. First, although a significant portion of species in all forest types use riparian areas (Raedeke 1988; Bunnell et al. 1991; Bunnell and Dupuis 1994; Chapters 4, 6, 14), the coastal temperate rain forest hosts the greatest number of species using riparian areas. Second, the coastal temperate rain forest also hosts the greatest number of species relying on attributes commonly associated with older forests, such as cavities and downed wood. This association between the vertebrate fauna and elements of older forests is a product of the long interval between stand-initiating disturbances in the temperate rain forest.

Many vertebrates presently occur in managed stands largely because of legacies remaining from unmanaged forests. This point is illustrated by data on black bear dens from coastal British Columbia (unpublished data of B.C. Ministry of Environment, Lands and Parks, Canadian Forest Products, and TimberWest Ltd.). Of sixty-nine reported dens, all were in live trees, snags, or logs larger than 1 meter in diameter. Of the forty-five dens from radio-collared bears, nineteen occurred in second growth but within structures remaining from the previous old-growth forest. The point is simple: although black bears forage actively in areas created by logging (roadsides and recent clearcuts), they are nonetheless dependent on biological legacies remaining from older forests. Similar findings of dependency on old-growth attributes while exploiting conditions created by logging exist for a variety of species in the temperate rain forest, including the marten (Baker 1992), fisher (Jones and Garton 1994), black-tailed deer (Bunnell 1985), and western red-backed salamander (Dupuis 1993).

Although these points about vertebrate/forest relations derive largely from data collected in British Columbia, they apply throughout the North American coastal temperate rain forest. Data of Thomas (1979) and Brown (1985) for Washington and Oregon reveal patterns similar to those of Figure 5.3. Similarly, the apparent dependence on attributes or legacies of older forests is evident for a number of species including black-tailed deer in Alaska (Schoen et al. 1981, 1984), several amphibians (both frogs and salamanders) in Oregon (Corn and Bury 1991a), several birds (Carey et al. 1991), rodents, and shrews (Corn and Bury 1991b), plus larger species such as marten (Buskirk and Powell 1994; Thompson and Harestad 1994) and fisher (Buskirk and Powell 1994; Jones and Garton 1994).

Management Implications

As the large majority of vertebrates in the coastal temperate rain forest are forest dwelling, forest practices can have substantial impacts on native vertebrates. Major implications derive from the two dominant features of the environment: the ubiquitous presence of water and the naturally long interval between stand-initiating disturbances.

Relations between forest practices and the quality, temperature, and amounts or flow rates of water often manifest themselves most clearly in their impact on aquatic vertebrates such as salmon, trout, or the juvenile stages of certain frogs and salamanders. Their impact on subterranean invertebrate species living in the phreatic and hyporheic zones is potentially more profound but largely invisible. These potential impacts themselves merit careful treatment of riparian and adjacent upslope areas to guard against increasing peak rates of discharge, sedimentation, or stream warming. Restriction of activities, such as harvesting or road building, in riparian areas also benefits terrestrial vertebrates.

Terrestrial vertebrates use riparian areas for a variety of reasons (Raedeke 1988; Bunnell et al. 1991; Bunnell and Dupuis 1994). In some instances, it may be the simple juxtaposition of two distinct habitats providing different needs. This benefit is important to a diverse range of species including the mink, river otter, western terrestrial garter snake, several bats, many birds, Pacific giant salamander, and tailed frog. For many of these species the signifiance of riparian areas has nothing to do the with presence of anadromous fish in a stream, and areas well upstream of potential spawning areas are important. This is particularly true among amphibians. Many bird species and some mammals prefer large, deciduous trees as cavity sites, and these are frequently more abundant in riparian areas. The richness and productivity of the riparian area itself confers advantages, as does the structural complexity and provision of snags and downed wood under natural conditions. Some species appear to use riparian corridors as travel routes (Forman 1983; Bunnell and Dupuis 1994). In short, riparian areas meet a diverse collection of needs that are best served when the areas are undisturbed. These needs are both diverse and important in the coastal temperate rain forest, where so many species use riparian areas (Figure 5.3). The fundamental implications are threefold: riparian areas are important from headwall to estuaries in *all* streams, not just those bearing anadromous fish; they should be little disturbed by forest practices; and where logging does occur, particular attention should be given to long-term maintenance of these complex structures.

Implications from the long natural interval between stand-initiating disturbances take two forms: the first relates to the relatively continuous natural forest cover; the second relates to specific stand elements. Species adapted to a forest of nearly continuous structure, such as the coastal temperate rain forest, are likely to be more vulnerable to the negative impacts of edge effects, fragmentation, and isolation. (See the reviews of Hunter 1990 and Primack

1993.) There is growing evidence that species of the coastal temperate rain forest are vulnerable to forest practices that disrupt connectivity among populations. The evidence is most clear for rather sedentary amphibians (Welsh 1990; Bunnell and Dupuis 1994), but it is present for more mobile species such as the fisher (Rosenberg and Raphael 1986) and marten (Buskirk and Powell 1994) as well. The major implication for forestry within the coastal temperate rain forest is that logging should not be allowed to disrupt connectivity among populations. Broad riparian buffers that are left unlogged are an obvious mechanism, but retention of significant portions of the canopy within logged areas is likewise important.

The significant portion of species using elements associated with old growth, such as large logs, trees, or snags (Figure 5.3), emphasizes the importance of forest practices retaining significant amounts of these habitat elements after harvest. The retained portions of a stand should be sufficient to provide both a present and future source of large snags and logs or downed wood. Without this retention, a sizable portion of the vertebrates living in temperate rain forests will find critical elements of their habitat seriously depleted or even eliminated (Bunnell and Allaye-Chan 1984; Spies et al. 1988).

A final implication relates to logging roads. Roads and road construction often contribute to serious sedimentation and degradation of the aquatic environment. They also have implications for terrestrial vertebrates. Roads can contribute to habitat fragmentation and isolation of smaller species (Oxley et al. 1974; Mader 1984) and may affect larger, hunted species. In British Columbia local declines in mountain goat, caribou, and grizzly bear are associated with increased access provided by roads, among other things. Bunnell and Kremsater (1990) note that the loss or dramatic declines of such forest-dwelling creatures as cougars, wolverines, and fishers in eastern North America demonstrate two points: first, the importance of human access to habitats of larger carnivores and, second, the role of roadless forests in providing some refuge from human access or land-use practices. Refuge may be particularly important in coastal temperate rain forests where the vertebrate fauna appears adapted to a nearly continuous, long-lived forest cover. Thus the density of road networks should be kept low and roads should be deactivated as soon as possible after their construction.

Protecting the Diversity

North America's coastal temperate rain forest hosts one of the most diverse vertebrate faunas in temperate regions. Some 380 terrestrial vertebrates occur within the region's boundaries as delineated by Alaback (1990) and Ecotrust et al. (1995). About forty-one of these vertebrate species are largely restricted to the coastal rain forest and adjoining Pacific coast. Amphibians, with their heavy reliance on moist and humid environments, comprise nearly half of these coastally restricted species.

Vertebrates of the coastal temperate rain forest are apparently adapted to the ubiquitous water habitats and forest cover types characterizing this biome. A large majority use riparian or seashore environments for at least part of the year. Similarly, most have some requirement for forest cover. The native fauna also shows adaptations to the natural disturbance regime of the coastal temperate rain forest and the forest structure resulting from that regime. Relative to other regions, vertebrates in the coastal temperate rain forest show heavy use of the products of infrequent stand-initiating disturbances (large trees, abundant large snags, downed wood). The native fauna also thrives in the well-stratified and patchy forest cover produced by frequent but small-scale wind disturbance.

The remarkable species richness of the coastal temperate rain forest, coupled with the heavy dependence of these species on unique riparian and rain forest habitats, underscore the need for land-use practices that minimize disturbance of all types in riparian areas, maintain landscape connectivity, retain attributes of old-growth stands, and minimize the density of road networks.

Acknowledgments

Alton Harestad and Laurie Kremsater reviewed the manuscript. Linda Dupuis and Bruce Bury evaluated our list of amphibians and reptiles in Table 5.1. R. C. Campbell and Alton Harestad assisted in compiling the estimates of species richness for Clayoquot Sound. A. N. Hamilton and Dave Lindsay provided unpublished data on black bears. We are grateful to all of them for their assistance.

References

Agee, J. K. 1993. *Fire Ecology of Pacific Northwest Forests.* Washington, D.C.: Island Press.

Alaback, P. 1990. *Comparative Ecology of Temperate Rain Forests of the Americas Along Analogous Climatic Gradients.* Juneau: USDA Forest Service, Pacific Northwest Research Station.

Armstrong, R. H. 1990. *Guide to the Birds of Alaska.* Anchorage and Seattle: Alaska Northwest Books.

Baker, J. 1992. "Habitat use and spatial organization of pine marten on southern Vancouver Island, British Columbia." M.Sc. thesis, Simon Fraser University, Burnaby, B.C.

Banfield, A.W.F. 1974. *The Mammals of Canada.* Toronto: University of Toronto Press.

Behler, J. L., and F. W. King. 1979. *The Audubon Society Field Guide to North American Reptiles and Amphibians.* New York: Knopf.

Brown, E. R., ed. 1985. *Management of Wildlife and Fish Habitats in Forests of Western Oregon and Washington.* Publication R6-F&WL-192-1985. Portland: USDA Forest Service.

Bunnell, F. L. 1985. "Forestry and black-tailed deer: Conflicts, crisis, or cooperation." *For. Chron.* 61:180–184.

———. 1990. "Biodiversity: What, where, why, and how." In *Wildlife Forestry Symposium: A Workshop on Resource Integration for Wildlife and Forest Managers.* Symposium proceedings, Prince George, B.C., March 7–8, 1990.

———. 1995. "Biodiversity, silviculture, constraints, and opportunities." In *Proceedings of the Silvicultural Conference: Stewardship in the New Forest.* Ottawa: Forestry Canada.

———. 1995. "Forest-dwelling vertebrate faunas and natural fire regimes in British Columbia: Patterns and implications for conservation." *Cons. Biol.* 9:636–644.

Bunnell, F. L., and A. Allaye-Chan. 1984. "Potential of ungulate winter-range reserves as habitat for cavity-nesting birds." In W. R. Meehan, T. R. Merrell, and T. A. Hanley, eds., *Fish and Wildlife Relationships in Old-Growth Forests.* Morehead, N.C.: American Institute of Fishery Research Biologists.

Bunnell, F. L., and L. A. Dupuis. 1994. "Riparian habitats in British Columbia: Their nature and role." In K. H. Morgan and M. A. Lashar, eds., *Riparian Habitat Management and Research.* Ladner, B.C.: Environment Canada's Fraser River Action Plan and Canadian Wildlife Service.

Bunnell, F. L., and L. L. Kremsater. 1990. "Sustaining wildlife in managed forests." *Northwest Env. J.* 6:243–269.

Bunnell, F. L., and R. G. Williams. 1980. "Subspecies and diversity—the spice of life or prophet of doom?" In R. Stace-Smith, L. Joahns, and P. Joslin, eds., *Threatened and Endangered Species and Habitats in British Columbia and the Yukon.* Victoria: B.C. Ministry of Environment.

Bunnell, F. L., D. K. Daust, W. Klenner, L. L. Klenner, and R. K. McCann. 1991. "Managing for biodiversity in forested ecosystems." Report to Forest Sector of the Old Growth Strategy. Vancouver: Council of Forest Industries.

Burt, W. H., and R. P. Grossenheider. 1976. *A Field Guide to the Mammals of America North of Mexico.* Boston: Houghton Mifflin.

Buskirk, S. W., and R. A. Powell. 1994. "Habitat ecology of fishers and American martens." In S. W. Buskirk, A. S. Haresta, M. G. Raphael, and R. A. Powell, eds., *Martens, Sables and Fishers: Biology and Conservation.* Ithaca: Cornell University Press.

Campbell, R. W., N. K. Dawer, I. Cowan, J. M. Cooper, G. W. Kaiser, and M.C. McNall. 1990. *The Birds of British Columbia.* Vols. 1 and 2. Victoria: Royal British Columbia Museum.

Carey, A. B., M. M. Hardt, S. P. Horton, and B. L. Biswell. 1991. "Spring bird communities in the Oregon Coast Range." In L. F. Ruggiero, K. B. Aubry, A. B. Carey, and M. H. Huff, tech. coords., *Wildlife and Vegetation of Unmanaged Douglas-Fir Forests.* Gen. Tech. Rep. PNW-GTR-285. Portland: USDA Forest Service.

Corn, P. S., and R. B. Bury. 1991a. "Terrestrial amphibian communities in the Oregon Coast Range." In L. F. Ruggiero, K. B. Aubry, A. B. Carey, and M. H. Huff, tech. coords., *Wildlife and Vegetation of Unmanaged Douglas-Fir Forests.* Gen. Tech. Rep. PNW-GTR-285. Portland: USDA Forest Service.

———. 1991b. "Small mammal communities in the Oregon Coast Range." In L. F. Ruggiero, K. B. Aubry, A. B. Carey, and M. H. Huff, tech. coords., *Wildlife and Vegetation of Unmanaged Douglas-Fir Forests.* Gen. Tech. Rep. PNW-GTR-285. Portland: USDA Forest Service.

Cowan, I., and C. J. Guiguet. 1978. *The Mammals of British Columbia.* Victoria: B.C. Provincial Museum.

Daubenmire, R. 1952. "Forest vegetation of northern Idaho and adjacent Washington, and its bearing on concepts of vegetation classification." *Ecol. Mono.* 22:301–330.

Dupuis, L. A. 1993. "The status and distribution of terrestrial amphibians in old-growth forests and managed stands." M.Sc. thesis, University of British Columbia, Vancouver.

Ecotrust, Pacific GIS, and Conservation International. 1995. *The Rain Forests of Home: An Atlas of People and Place.* Part 1: *Natural Forests and Native Languages of the Coastal Temperate Rain Forest.* Portland, Ore.

Fahnestock, G. R., and J. K. Agee. 1983. "Biomass consumption and smoke production by prehistoric and modern forest fires in western Washington." *J. For.* 81:653–657.

Forman, R.T.T. 1983. "Corridors in a landscape: Their ecological structure and function." *Ekologiya* (USSR) 2:375–387.

Franklin, J., and R. Waring. 1980. "Distinctive features of the northwestern coniferous forest: Development, structure and function." In R. Waring, ed., *Forests: Fresh Perspectives from Ecosystem Analysis.* Proceedings of the 40th Annual Biology Colloquium. Corvallis: Oregon State University Press.

Fujimori, T. 1971. "Primary production of a young *Tsuga heterophylla* stand and some speculations about biomass of forest communities on the Oregon Coast." Research Paper PNW-123. Portland: USDA Forest Service.

Hoyt, E. 1984. *The Whales of Canada.* Camden East, Ont.: Camden House Press.

Hunter, M. 1990. *Wildlife, Forests, and Forestry—Principles of Managing Forests for Biological Diversity.* Englewood Cliffs, N.J.: Prentice-Hall.

Jones, J. L., and E. O. Garton. 1994. "Selection of successional stages by fishers in north-central Idaho." In S. W. Buskirk, A. S. Haresta, M. G. Raphael, and R. A. Powell, eds., *Martens, Sables and Fishers: Biology and Conservation.* Ithaca: Cornell University Press.

Kessel, B., and D. D. Gibson. 1978. *Status and Distribution of Alaska Birds.* Lawrence, Kan.: Cooper Ornithological Society.

MacArthur, R. H., and J. W. MacArthur. 1969. "On bird species diversity." *Ecology* 42:594–598.

Mader, H. J. 1984. "Animal habitat isolation by roads and agricultural fields." *Biol. Cons.* 29:81–96.

Meidinger, D., and J. Pojar, eds. 1991. *Ecosystems of British Columbia.* Victoria: B.C. Ministry of Forests.

Norse, E. A. 1990. *Ancient Forests of the Pacific Northwest.* Washington, D.C.: Island Press.

Orchard, S. A. 1984. *Amphibians and Reptiles of British Columbia: An Ecological Review.* WHR-15. Victoria: B.C. Ministry of Forests.

Oxley, D. J., M. B. Fenton, and G. R. Carmody. 1974. "The effects of roads on populations of small mammals." *J. Appl. Ecol.* 11:51–59.

Peterson, R. T. 1990. *A Field Guide to Western Birds.* 3rd ed. Boston: Houghton Mifflin.

Primack, R. B. 1993. *Essentials of Conservation Biology.* Sunderland, Mass.: Sinauer Associates.

Raedeke, K. J., ed. 1988. *Streamside Management: Riparian Wildlife and Forestry Interactions.* Contribution 59. Seattle: University of Washington, College of Forest Resources.

Rosenberg, K. V., and M. G. Raphael. 1986. "Effects of forest fragmentation on vertebrates in Douglas-fir forests." In J. Verner, M. L. Morrison, and C. J. Ralph, eds., *Wildlife 2000: Modeling Habitat Relationships of Terrestrial Vertebrates.* Madison: University of Wisconsin Press.

Schoen, J. W., M. D. Kirchoff, and O. C. Wallmo. 1984. "Sitka black-tailed deer/old-growth relationships in southeast Alaska." In W. R. Meehan, T. R. Merrell, and T. A. Hanley, eds., *Fish and Wildlife Relationships in Old-growth Forests.* Morehead, N.C.: American Institute of Fishery Research Biologists.

Schoen, J. W., O. C. Wallmo, and M. D. Kirchoff. 1981. "Wildlife-forest relationships: Is a re-evaluation of old growth necessary?" *Trans. N. Am. Wildl. and Nat. Res. Conf.* 46:531–544.

Sedell, J., and C. Maser. 1994. *From the Forest to the Sea: The Ecology of Wood in Streams, Rivers, Estuaries, and Oceans.* Delray Beach, Fla.: St. Lucie Press.

Sedell, J. R., P. A. Bisson, F. J. Swanson, and S. V. Gregory. 1988. "What we know about large trees that fall into streams and rivers." In C. Maser, R. F. Tarrant, J. M. Trappe, and J. F. Franklin, eds., *From the Forest to the Sea: A Story of Fallen Trees.* Gen. Tech. Rep. PNW-GTR-229. Portland: USDA Forest Service.

Spies, T. A., J. F. Franklin, and T. Thomas. 1988. "Coarse woody debris in Douglas-fir forests of western Oregon and Washington." *Ecology* 69:1689–1702.

Thomas, J. W., ed. 1979. *Wildlife Habitats in Managed Forests.* Agricultural Handbook 533. Washington, D.C.: USDA Forest Service.

Thompson, I. D., and A. S. Harestad. 1994. "Effects of logging on American martens and models for habitat management." In S. W. Buskirk, A. S. Harestad, M. G. Raphael, and R. A. Powell, eds., *Martens, Sables and Fishers: Biology and Conservation.* Ithaca: Cornell University Press.

van Zyll de Jong, C. G. 1983. *Handbook of Canadian Mammals.* Vol. 1: *Marsupials and Insectivores.* Ottawa: National Museum of Natural Sciences.

———. 1985. *Handbook of Canadian Mammals.* Vol. 2: *Bats.* Ottawa: National Museum of Natural Sciences.

Waring, R. H., and J. F. Franklin. 1979. "Evergreen coniferous forests of the Pacific Northwest." *Science* 204:1380–1386.

Welsh, H. H., Jr. 1990. "Relictual amphibians and old-growth forests." *Cons. Biol.* 4:309–319.

Willson, M. F. 1974. "Avian community organization and habitat structure." *Ecology* 55:1017–1029.

Appendix A. Occurrence and Habitat Use of Bird Species in the Coastal Temperate Rain Forest

Seashore	Riparian	Grass/ shrub	Open/ edge	Closed/ mature	Old growth	Species	Common Name	AK	BC	WA	OR	CA
f	fb					*Gavia stellata*	Red-throated loon	a	a	w	w	w
f	fb					*Gavia pacifica*	Pacific loon	a	b	w	w	w
	fb	fb				*Gavia immer*	Common loon	a	a	b	b	w
f	f					*Gavia adamsii*	Yellow-billed loon	w	w			
fb	fb					*Podilybus podiceps*	Pied-billed grebe		a	b	b	b
f	f					*Podiceps auritus*	Horned grebe	w	w	w	w	w
f	f					*Podiceps grisegena*	Red-necked grebe		a	w	w	w
f	f					*Podiceps nigricollis*	Eared grebe		w	w	w	w
f	f					*Aechmophorus occidentalis*	Western grebe		w	w	w	a
			fb			*Oceanodroma furcata*	Fork-tailed storm petrel	a	a		a	
			fb		fb	*Oceanodroma leucorhoa*	Leach's storm petrel	b	a	a	a	
f	f					*Pelecanus erythrorhynchos*	American white pelican		w			w
fb						*Pelecanus occidentalis*	Brown pelican			a	a	a
fb	fb					*Phalacrocorax auritus*	Double-crested cormorant	a	a	a	a	a
fb						*Phalacrocorax penicillatus*	Brandt's cormorant	a	a	a	a	a
fb						*Phalacrocorax pelagicus*	Pelagic cormorant	a	a	a	a	a
fb						*Phalacrocorax urile*	Red-faced cormorant	a				
f	fb	rfrb				*Botaurus lentiginosus*	American bittern	b	w	a	a	a
f	f			fb	rfrb	*Ardea herodias*	Great blue heron	a	a	a	a	a
f	f	rf	rf	fb	rf	*Casmerodius albus*	Great egret				w	a
fb	fb					*Egretta thula*	Snowy egret				w	a
f	f					*Bubulcus ibis*	Cattle egret		w			a
f	f	rfrb	fb	rfrb	rfrb	*Butorides striatus*	Green-backed heron		a	b	b	a
f	f	rf	rf	rf		*Nycticorax nycticorax*	Black-crowned night heron		a	w	w	a
f	f	rf				*Cygnus columbianus*	Tundra swan	w	w	w	w	w
f	f	rf				*Cygnus buccinator*	Trumpeter swan	a	w	w	w	
f	f					*Anser albifrons*	Greater white-fronted goose	a	w	w	w	w
f	f					*Chen caerulescens*	Snow goose	w	w	w	w	w
f						*Branta bernicla*	Brant	w	w	w	w	w
f	fb	rf				*Branta canadensis*	Canada goose	a	a	a	a	w
f	fb	f	fb	fb	fb	*Aix sponsa*	Wood duck		a	a	a	a
f	f	rf				*Anas crecca*	Green-winged teal	b	a	a	a	w

f	fb	rfrb				*Anas platyrhynchos*	Mallard	a	a	a	a	a
f	fb	rfrb				*Anas acuta*	Northern pintail	a	a	a	a	w
f	fb	rfrb				*Anas discors*	Blue-winged teal	b	a	a	a	w
f	fb	rfrb				*Anas cyanoptera*	Cinnamon teal		a	a	a	a
f	fb	rfrb				*Anas clypeata*	Northern shoveler	b	a	a	a	w
f	fb	rfrb				*Anas strepera*	Gadwall	a	a	a	w	w
f	f	rf				*Anas penelope*	Eurasian wigeon	w	w	w	w	
f	fb	rfrb				*Anas americana*	American wigeon	a	w	a	w	w
f	f					*Aythya valisineria*	Canvasback		w	w	w	w
f	f					*Aythya americana*	Redhead	b	w	w	w	w
f	fb	rfrb	rfrb	rfrb	rfrb	*Aythya collaris*	Ring-necked duck	b	a	a	w	w
f	f					*Aythya fuligula*	Tufted duck		w			
f	f					*Aythya marila*	Greater scaup	a	w	w	w	w
f	fb					*Aythya affinis*	Lesser scaup	a	w	w	w	w
f						*Polysticta stelleri*	Steller's eider	w				
f	fb			fb	fb	*Histrionicus histrionicus*	Harlequin duck	a	a	b	b	w
f						*Clangula hyemalis*	Oldsquaw	a	w	w	w	w
f						*Melanitta nigra*	Black scoter	w	w			w
f	f					*Melanitta perspicillata*	Surf scoter	w	w			w
f						*Melanitta fusca*	White-winged scoter	w	w			w
f	f					*Bucephala clangula*	Common goldeneye	a	a	w	w	w
f	fb			fb	fb	*Bucephala islandica*	Barrow's goldeneye	a	a	a	a	w
f	fb			fb	fb	*Bucephala albeola*	Bufflehead	a	a	a	w	w
f	fb			fb	fb	*Lophodytes cucullatus*	Hooded merganser	a	a	a	a	w
f	fb			fb	fb	*Mergus merganser*	Common merganser	a	a	a	a	a
fb	f					*Mergus serrator*	Red-breasted merganser	a	a	w	w	w
f	fb					*Oxyura jamaicensis*	Ruddy duck		a	a	a	a
f	f	fb	f		fb	*Cathartes aura*	Turkey vulture		a	b	b	a
f	f			fb	fb	*Pandion haliaetus*	Osprey	b	a	b	b	b
	f	f	rfrb	rfrb	rfrb	*Elanus caeruleus*	Black-shouldered kite				a	a
f	f		f	f	fb	*Haliaeetus leucocephalus*	Bald eagle	a	a	a	a	a
	fb	f	fb			*Circus cyaneus*	Northern harrier	a	a	a	a	a
	f	f	f	fb	fb	*Accipiter striatus*	Sharp-shinned hawk	a	a	a	a	a
	f	f	fb	fb	fb	*Accipiter cooperii*	Cooper's hawk		w	a	a	a
			f	fb	fb	*Accipiter gentilis*	Northern goshawk	a	a	a	a	a

Continues

Appendix A. *Continued*

Seashore	Riparian	Grass/ shrub	Open/ edge	Closed/ mature	Old growth	Species	Common Name	AK	BC	WA	OR	CA
	fb	f	fb	fb	fb	*Buteo lineatus*	Red-shouldered hawk				a	a
	f					*Buteo swainsoni*	Swainson's hawk		w			a
	fb	f	fb	fb	fb	*Buteo jamaicensis*	Red-tailed hawk	b	a	a	a	a
	f	f				*Buteo lagopus*	Rough-legged hawk		w	w	w	w
		f	fb	f	fb	*Aquila chrysaetos*	Golden eagle	a	a	a	a	a
	fb	f	fb	fb	fb	*Falco sparverius*	American kestrel	b	a	a	a	a
f	f		f			*Falco columbarius*	Merlin	a	a	w	w	
f	f		f	fb	fb	*Falco peregrinus*	Peregrine falcon	a	a	a	a	a
f	f					*Falco rusticolus*	Gyrfalcon	a	w			
			fb			*Dendragapus canadensis*	Spruce grouse	a	a			
			fb	f	f	*Dendragapus obscurus*	Blue grouse	a	a	a	a	a
	fb	fb	fb			*Lagopus lagopus*	Willow ptarmigan	a	a			
	fb		fb	fb	fb	*Bonasa umbellus*	Ruffed grouse	a	a	a	a	a
			fb			*Oreortyx pictus*	Mountain quail		a	a	a	a
	fb	rfrb				*Rallus limicola*	Virginia rail		a	a	a	a
	fb	rfrb				*Porzana carolina*	Sora rail	b	w	b	b	a
f	fb	rfrb				*Fulica americana*	American coot	a	a	a	a	a
fb	fb	fb		fb	fb	*Grus canadensis*	Sandhill crane	a	a			
f	f					*Pluvialis squatarola*	Black-bellied plover	w	w	w	w	w
f	f					*Pluvialis dominica*	Lesser golden plover	w	w			w
fb						*Charadrius alexandrinus*	Snowy plover			a	a	a
fb	f	rf				*Charadrius semipalmatus*	Semipalmated plover	a	a	b	w	w
fb	fb					*Charadrius vociferus*	Killdeer	a	a	a	a	a
fb						*Haematopus bachmani*	Black oystercatcher	a	a	a	a	a
f	f					*Tringa melanoleuca*	Greater yellowlegs	a	w	w	w	w
f	f					*Tringa flavipes*	Lesser yellowlegs	a	w	w	w	w
f	f	rf				*Tringa solitaria*	Solitary sandpiper	b	w	w	w	
f						*Catoptrophorus semipalmatus*	Willet		w	w	w	w
f						*Heteroscelus incanus*	Wandering tattler	a	w	w	w	w
fb	fb					*Actitis macularia*	Spotted sandpiper	b	a	a	a	a
f	f					*Numenius phaeopus*	Whimbrel	w	w	w	w	w
f	f					*Numenius americanus*	Long-billed curlew			w	w	w

f	fb		*Limosa haemoastica*	Hudsonian godwit	b	w			
f			*Limosa lapponica*	Bar-tailed godwit	w				
f			*Limosa fedoa*	Marbled godwit		w	w	w	w
f			*Arenaria interpres*	Ruddy turnstone	w	w	w	w	w
f			*Arenaria melanocephala*	Black turnstone	w	w	w	w	w
f			*Aphriza virgata*	Surfbird	a	w	w	w	w
f			*Calidris canutus*	Red knot	w	w	w	w	w
f	f		*Calidris alba*	Sanderling	w	w	w	w	w
f	f		*Calidris pusilla*	Semipalmated sandpiper	w	w	w	w	w
f	f		*Calidris mauri*	Western sandpiper	w	w	w	w	w
f	f		*Calidris minutilla*	Least sandpiper	a	w	w	w	w
f	f		*Calidris bairdii*	Baird's sandpiper	w	w	w	w	w
f	f		*Calidris melanotos*	Pectoral sandpiper	w	w	w	w	w
f			*Calidris acuminata*	Sharp-tailed sandpiper		w			
f			*Calidris ptilocnemis*	Rock sandpiper	w	w	w	w	w
f	f		*Calidris alpina*	Dunlin	a	w	w	w	w
f	f		*Calidris himantopus*	Stilt sandpiper	w	w			
f	f		*Limnodromus griseus*	Short-billed dowitcher	a	w	w	w	w
f	f		*Limnodromus scolopaceus*	Long-billed dowitcher	w	w	w	w	w
f	fb	rfb	*Gallinago gallinago*	Common snipe	a	a	a	a	w
f	fb		*Phalaropus tricolor*	Wilson's phalarope		w	b	b	
f	f		*Phalaropus lobatus*	Red-necked phalarope	a	w	w	w	w
f	f		*Phalaropus fulicaria*	Red phalarope			w	w	w
f	f		*Larus pipixcan*	Franklin's gull		w	w	w	w
f	f		*Larus philadelphia*	Bonaparte's gull	a	w	w	w	w
f	f		*Larus heermanni*	Heermann's gull		w	w	w	w
f	f		*Larus canus*	Mew gull	a	a	w	w	w
f	f		*Larus delawarensis*	Ring-billed gull	w	w	a	a	w
f	f		*Larus californicus*	California gull	w	w	w	w	w
f	f		*Larus argentatus*	Herring gull	a	w	w	w	w
f	f		*Larus thayeri*	Thayer's gull	w	w	w	w	w
f	f		*Larus occidentalis*	Western gull		w	a	a	a
f	f		*Larus glaucescens*	Glaucous-winged gull	a	a	a	a	w
f	f		*Larus hyperboreus*	Glaucous gull	w	w	w	w	w
fb			*Rissa tridactyla*	Black-legged kittiwake	a	w			
f	f		*Sterna caspia*	Caspian tern	a	w	b	b	b

Continues

Appendix A. *Continued*

Seashore	Riparian	Grass/ shrub	Open/ edge	Closed/ mature	Old growth	Species	Common Name	AK	BC	WA	OR	CA
f	f					*Sterna hirundo*	Common tern		w	w	w	w
f	f					*Sterna paradisaea*	Arctic tern	a		b	w	
fb	fb					*Sterna aleutica*	Aleutian tern	b				
fb						*Uria aalge*	Common murre	a	w	a	a	a
fb						*Uria lomvia*	Thick-billed murre	a				
fb						*Cepphus columba*	Pigeon guillemot	a	a	a	a	a
f	f			fb	fb	*Brachyramphus marmoratus*	Marbled murrelet	a	a	a	a	a
fb						*Brachyramphus brevirostris*	Kittlitz's murrelet	a				
fb						*Synthliboramphus antiquus*	Ancient murrelet	a	a			
fb						*Ptychoramphus aleuticus*	Cassin's auklet	w	a			
fb						*Cerorhinca monocerata*	Rhinocerous auklet	a	a			
f	f		fb	frb	frb	*Columba fasciata*	Band-tailed pigeon		a	b	a	a
	f		fb	fb	fb	*Zemaida macroura*	Mourning dove		a	a	a	a
	f	f	fb	fb	fb	*Tyto alba*	Barn owl		a	a	a	a
	f		fb			*Otus kennicottii*	Western screech owl	a	a	a	a	a
f	fb		fb	fb	fb	*Bubo virginianus*	Great horned owl	a	a	a	a	a
	f	f	fb	fb		*Surnia ulula*	Northern hawk owl	a	w			
	fb			fb	fb	*Glaucidium gnoma*	Northern pygmy owl	a	a	a	a	a
				fb	fb	*Aegolius funereus*	Boreal owl	a				
			f	f	fb	*Strix occidentalis*	Spotted owl		w	a	a	a
				fb	fb	*Strix varia*	Barred owl		a			
	fb	f	f	b	b	*Strix nebulosa*	Great gray owl	a	w			
	fb		f	rb	rb	*Asio otus*	Long-eared owl		w	a	a	a
fb	fb	rf				*Asio flammeus*	Short-eared owl	a	w	a	a	a
			fb	fb	fb	*Aegolius acadicus*	Northern saw-whet owl	a	a	a	a	a
fb	fb		fb	f	f	*Chordeiles minor*	Common nighthawk		a	b	b	b
			fb			*Phalaenoptilus nuttallii*	Common poorwill			w	b	b
f	fb	fb*r*	fb*r*	fb*r*	fb*r*	*Cypseloides niger*	Black swift		w	b		
fb	fb		fb	fb	fb	*Chaetura vauxi*	Vaux's swift		w	b	b	b
		rfrb	fb			*Calypte anna*	Anna's hummingbird		w	b	b	a
		f	fb			*Stellula calliope*	Calliope hummingbird			w	w	b
	f		fb	fb	fb	*Selasphorus rufus*	Rufous hummingbird	a	a	b	b	b

			fb		fb	*Selasphorus sasin*	Allen's hummingbird				b	b
f	fb			f	f	*Ceryle alcyon*	Belted kingfisher	a	a	a	a	a
		f	fb	fb	fb	*Melanerpes lewis*	Lewis' woodpecker		a	w	b	a
	fb			fb	fb	*Sphyrapicus ruber*	Red-breasted sapsucker	a	a	a	a	a
	fb		rfrb	rfrb	rfrb	*Picoides pubescens*	Downy woodpecker	a	a	a	a	a
				fb	fb	*Picoides villosus*	Hairy woodpecker	a	a	a	a	a
	fb		fb	rfrb	rfrb	*Picoides tridactylus*	Three-toed woodpecker	a	w			
	fb		rfrb	rfrb	rfrb	*Picoides arcticus*	Black-backed woodpecker	a				
			fb		fb	*Colaptes auratus*	Northern flicker	a	a	a	a	a
				fb	fb	*Dryocopus pileatus*	Pileated woodpecker		a	a	a	a
			fb	fb	fb	*Contopus borealis*	Olive-sided flycatcher	b	b	b	b	b
	fb		fb	fb	fb	*Contopus sordidulus*	Western wood-pewee	b	b	b	b	b
	fb		f			*Empidonax alnorum*	Alder flycatcher	b	b			
	fb	rf	fb	rfrb	rfrb	*Empidonax traillii*	Willow flycatcher		b	b	b	
			fb	fb	fb	*Empidonax hammondii*	Hammond's flycatcher	b	b	b	b	
	fb		fb			*Empidonax difficilis*	Pacific slope flycatcher	b	b	b	b	b
f	f		fb	fb	fb	*Progne subis*	Purple martin		b	b	b	b
f	f			fb	fb	*Tachycineta bicolor*	Tree swallow	b	b	b	b	
f	f			fb	fb	*Tachycineta thalassina*	Violet-green swallow	b	b	b	b	
	f	fb*	fb*	fb*	fb*	*Stelgidopteryx serripennis*	Northern rough-winged swallow	b	b	b	b	b
fb	f	fb*	fb*	fb*	fb*	*Riparia riparia*	Bank swallow	b				
f	f					*Hirundo pyrrhonota*	Cliff swallow	b	b	b	b	b
f	f					*Hirundo rustica*	Barn swallow	b	b	b	b	b
				fb	fb	*Perisoreus canadensis*	Gray jay	a	a	a	a	a
			fb	fb	fb	*Cyanocitta stelleri*	Steller's jay	a	a	a	a	a
	f	f	fb	rfrb		*Pica pica*	Black-billed magpie	a				
f	fb	f	fb	fb	fb	*Corvus brachyrhynchos*	American crow			a	a	a
fb			fb			*Corvus caurinus*	Northwestern crow	a	a	a		
f		f	fb	fb	fb	*Corvus corax*	Common raven	a	a	a	a	a
	fb		fb			*Parus atricapillus*	Black-capped chickadee	a	a	a	a	
	f			fb	fb	*Parus rufescens*	Chestnut-backed chickadee	a	a	a	a	a
				fb	fb	*Sitta canadensis*	Red-breasted nuthatch	a	a	a	a	a
				fb	fb	*Certhia americana*	Brown creeper	a	a	a	a	a
	fb	rfrb	fb	rfrb	rfrb	*Thryomanes bewickii*	Bewick's wren		a	a	a	a

Continues

Appendix A. *Continued*

Seashore	Riparian	Grass/ shrub	Open/ edge	Closed/ mature	Old growth	Species	Common Name	AK	BC	WA	OR	CA
		rfrb	fb	rfrb	rfrb	*Troglodytes aedon*	House wren		b	b	b	a
	fb		fb	fb	fb	*Troglodytes troglodytes*	Winter wren	a	a	a	a	a
fb		rfrb				*Cistothorus palustris*	Marsh wren		a	a	a	a
fb	fb			*	*	*Cinclus mexicanus*	American dipper	a	a	a	a	a
				fb	fb	*Regulus satrapa*	Golden-crowned kinglet	a	a	a	a	a
			b	fb	fb	*Regulus calendula*	Ruby-crowned kinglet	b	a	w	a	w
		f	fb	rf	rf	*Sialia mexicana*	Western bluebird		b	a	a	a
			fb		fb	*Myadestes townsendi*	Townsend's solitaire	b	a	a	a	a
		fb	fb	f	f	*Catharus minimus*	Gray-cheeked thrush	b				
	fb		fb	fb	fb	*Catharus ustulatus*	Swainson's thrush	b	b	b	b	b
			fb	fb	fb	*Catharus guttatus*	Hermit thrush	b	a	a	a	a
		fb	fb	fb	fb	*Turdus migratorius*	American robin	a	a	a	a	a
	fb		fb	fb	fb	*Ixoreus naevius*	Varied thrush	a	a	a	a	a
		fb	fb			*Dumetella carolinensis*	Gray catbird		b			
f	f	f				*Anthus spinoletta*	American pipit	b				b
			f	f	f	*Bombycilla garrulus*	Bohemian waxwing	a	w	w	w	w
	f	rfrb	rfrb	frb	frb	*Bombycilla cedrorum*	Cedar waxwing	b	a	a	a	a
		f	f			*Lanius excubitor*	Northern shrike	a	w	w	w	w
				fb	fb	*Vireo solitarius*	Solitary vireo		b	b	b	b
		rfrb	fb	rfrb	rfrb	*Vireo huttoni*	Hutton's vireo		a	a	a	a
		rfrb	fb	rfrb	rfrb	*Vireo gilvus*	Warbling vireo	b	b	b	b	b
	fb		rfrb	rfrb	rfrb	*Vireo olivaceus*	Red-eyed vireo	b	b	b		
	fb	f	fb	fb	fb	*Vermivora peregrina*	Tennessee warbler	b				
			fb		fb	*Vermivora celata*	Orange-crowned warbler	b	b	a	a	a
	fb	rfrb	fb	rfrb	rfrb	*Dendroica petechia*	Yellow warbler	b	b	b	b	b
	fb	f	fb	fb	f	*Dendroica coronata*	Yellow-rumped warbler	b	a	a	a	a
				fb	f	*Dendroica townsendi*	Townsend's warbler	b	a	a	w	w
				fb	fb	*Dendroica occidentalis*	Hermit warbler			b	b	b
		fb	fb	fb	fb	*Dendroica striata*	Blackpoll warbler	b	b			
	fb		fb	fb		*Setophaga ruticilla*	American redstart	b				
	fb					*Seiurus noveboracensis*	Northern waterthrush	b				
	fb		fb	fb		*Oporornis tolmiei*	MacGillivray's warbler	b	b	b	b	b

	fb	fb				*Geothlypis trichas*	Common yellowthroat	b	b	b	b	b
		fb	fb	fb	fb	*Wilsonia pusilla*	Wilson's warbler	b	b	b	b	b
			fb	fb	fb	*Piranga ludoviciana*	Western tanager	b	b	b	b	a
	fb		fb			*Pheucticus melanocephalus*	Black-headed grosbeak		b	b	b	b
		fb	rfrb			*Passerina amoena*	Lazuli bunting				b	b
		fb	fb	fb	fb	*Pipilo erythrophthalmus*	Rufous-sided towhee		a	a	a	a
	f		fb			*Spizella arborea*	American tree sparrow	w	w			
		fb	fb	fb	fb	*Spizella passerina*	Chipping sparrow	b	b	b	b	b
fb	fb	fb				*Passerculus sandwichensis*	Savannah sparrow	b	a	a	a	a
		fb	fb	rfrb	rfrb	*Passerella iliaca*	Fox sparrow	a	a	a	w	w
	fb	fb	fb			*Melospiza melodia*	Song sparrow	a	b	b	b	b
	fb	fb	rfrb			*Melospiza lincolnii*	Lincoln's sparrow	b	b		w	w
		f	f		f	*Zonotrichia albicollis*	White-throated sparrow			w	w	w
		f				*Zonotrichia atricapilla*	Golden-crowned sparrow	a	a	w	w	w
		fb				*Zonotrichia leucophrys*	White-crowned sparrow	a	a	a	a	a
		fb	fb	fb	fb	*Junco hyemalis*	Dark-eyed junco	a	a	a	a	a
f		f				*Plectrophenax nivalis*	Snow bunting	a				
f	fb	rfrb	rfrb			*Agelaius phoeniceus*	Red-winged blackbird	b	a	a	a	b
	fb					*Euphagus carolinus*	Rusty blackbird	a	b			
f		fb	fb			*Euphagus cyanocephalus*	Brewer's blackbird		a	a	a	a
		fb	fb	fb	fb	*Molothrus ater*	Brown-headed cowbird	b	a	a	a	a
		rf	fb			*Icterus galbula*	Northern oriole		b			b
			fb	f	frb	*Pinicola enucleator*	Pine grosbeak	a	a			
			fb	fb	fb	*Carpodacus purpureus*	Purple finch		a	a	a	a
			rfrb	fb	fb	*Loxia curvirostra*	Red crossbill	a	a	a	a	a
				fb	fb	*Loxia leucoptera*	White-winged crossbill	a	a	a	a	
		f	fb	fb	fb	*Carduelis pinus*	Pine siskin	a	a	a	a	a
		f	fb			*Carduelis tristis*	American goldfinch		a	a	a	a
		fb	fb	fb	rfrb	*Coccothraustes vespertinus*	Evening grosbeak		a	a	a	a

Seashore = estuary and beach; Riparian = rivers, streams, lakes, ponds, marshes, bogs, and swamps; Open/edge = open forest cover including pole-sapling stands and forest edges; Closed/mature = closed forest cover and mature stands; f = foraging, resting, and other nonbreeding activities; b = breeding habitat; r = use of habitat type restricted to riparian areas; * = use when special habitat available; a = all-year occurrence; b = breeding use; w = nonbreeding use.

Appendix B. Occurrence and Habitat Use of Mammal Species in the Coastal Temperate Rain Forest

Seashore	Riparian	Grass/ shrub	Open/ edge	Closed/ mature	Old growth	Species	Common Name	AK	BC	WA	OR	CA
fb	fb	fb	fb	fb		*Didelphis virginiana*	Virginia opposum		x	x	x	x
fb	fb	fb	fb	fb	fb	*Sorex bendirii*	Pacific water shrew		x	x	x	x
fb	fb	fb	fb	fb	fb	*Sorex cinereus*	Masked shrew	x	x	x		
fb	fb	fb	fb	fb	fb	*Sorex obscurus*	Dusky shrew	x	x	x	x	
	fb	fb	fb	fb	fb	*Sorex pacificus*	Pacific shrew				x	x
	fb	*fb	*fb	*f*b	*f*b	*Sorex palustris*	Northern water shrew	x	x	x	x	
fb	fb			fb	fb	*Sorex trowbridgii*	Trowbridge shrew		x	x	x	x
fb	fb	fb	fb	fb	fb	*Sorex vagrans*	Vagrant shrew		x	x	x	x
	fb	fb	fb	f	fb	*Microsorex hoyi*	Pygmy shrew	x	x			
fb	fb	fb	fb	fb	fb	*Neurotrichus gibbsii*	Shrew-mole		x	x	x	x
	fb	fb	fb	fh	fb	*Scapanus orarius*	Coast mole		x	x	x	x
	fb	fb	fb			*Scapanus townsendii*	Townsend's mole		x	x	x	x
f	fb	f	f	fb	fb	*Eptesicus fuscus*	Big brown bat		x	x	x	x
	f	f	f	fb	fb	*Lasionycteris noctivagans*	Silver-haired bat	x	x	x	x	x
			fb	fb	fb	*Nycteris cinerea*	Hoary bat		x	x	x	x
			fb		fb	*Myotis californicus*	California myotis		x	x	x	x
f	fb	f	f	fb	fb	*Myotis evotis*	Long-eared myotis		x	x	x	x
	f	f	rf	fb	fb	*Myotis keenii*	Keen's myotis		x	x		
f	f	f	f	fb	fb	*Myotis lucifugus*	Little brown myotis	x	x	x	x	x
	f	f	f	fb	fb	*Myotis thysanodes*	Fringed myotis			x	x	x
f	f	f	f	fb	fb	*Myotis volans*	Long-legged myotis	x	x	x	x	x
f	fb	f	f	fb	fb	*Myotis yumanensis*	Yuma myotis		x	x	x	x
f	f	*f*b	*f*b	*f*b	*f*b	*Plecotus townsendii*	Townsend's big-eared bat		x	x	x	x
	fb	fb	fb	fb	fb	*Lepus americanus*	Snowshoe hare	x	x	x	x	
		f	*	*	*	*Ochotona collaris*	Collared pika	x	x			
		f	*	*	*	*Ochotona princeps*	Common pika		x			
	fb	fb	fb	fb	fb	*Aplodontia rufa*	Mountain beaver		x	x	x	x
fb				fb	fb	*Clethrionomys occidentalis*	Western red-backed vole		x	x	x	x
fb	fb			fb	fb	*Clethrionomys gapperi*	Southern red-backed vole	x	x	x		
			fb	fb	fb	*Clethrionomys rutilus*	Northern red-backed vole	x	x			
						Microtus coronarius	Coronation Island vole	x				

fb	fb	fb	fb			*Microtus longicaudus*	Long-tailed vole	x	x	x	x	x
fb	fb	fb	fb	fb	fb	*Microtus oregoni*	Creeping vole		x	x	x	x
	f	f		f		*Microtus pennsylvanicus*	Meadow vole	x				
	fv	b*f* r				*Microtus richardsoni*	Water vole			x	x	
fb	fb	fb	fb			*Microtus townsendii*	Townsend's vole		x	x	x	x
fb	fb	*rf*rb	*rf*rb	*rf*rb	*rf*rb	*Ondatra zibethicus*	Muskrat		x	x	x	x
	f			f		*Phenacomys albipes*	White-footed vole				x	x
	fb	rfb		rfb	rfb	*Phenacomys intermedius*	Heather vole		x			
	f		fb		fb	*Phenacomys longicaudus*	Tree phenacomys		x			x
fb	fb	rfb	rfb	rfb	rfb	*Castor canadensis*	Beaver	x	x	x	x	x
fb	fb	fb	fb	fb	fb	*Neotoma cinerea*	Bushy-tailed woodrat	x	x	x	x	x
	fb	fb	fb	fb	fb	*Neotoma fuscipes*	Dusky-footed woodrat				x	x
fb	fb	fb	fb	fb	fb	*Peromyscus maniculatus*	Deer mouse	x	x	x	x	x
			fb			*Peromyscus oreas*	Columbian mouse		x			
			fb	fb	fb	*Peromyscus sitkensis*	Sitka mouse	x	x			
	fb	fb	fb	fb	fb	*Erethizon dorsatum*	Porcupine	x	x	x	x	x
	f			fb	fb	*Glaucomys sabrinus*	Northern flying squirrel	x	x	x	x	x
	f		f	fb	fb	*Sciurus griseus*	Western gray squirrel			x	x	
	fb	fb	fb	fb		*Eutamias amoenus*	Yellow-pine chipmunk		x	x		
	fb	fb	fb	fb	fb	*Eutamias townsendi*	Townsend's chipmunk		x	x	x	x
	f		fb	fb	fb	*Tamiasciurus douglasii*	Douglas' squirrel		x	x	x	x
	f		fb	fb	fb	*Tamiasciurus hudsonicus*	Red squirrel	x	x			
	fb	f	fb	f	fb	*Zapus hudsonius*	Meadow jumping mouse	x	x			
	fb	rfb				*Zapus princeps*	Western jumping mouse	x				
fb	fb	fb	fb	fb	fb	*Zapus trinotatus*	Pacific jumping mouse			x	x	x
fb	fb	fb	fb	fb	fb	*Canis latrans*	Coyote	x	x	x	x	x
			f	f	f	*Canis lupus*	Gray wolf	x	x			
	f	f	f	f	f	*Felis concolor*	Cougar		x	x	x	x
	f	rf	rf	rf	rf	*Lynx canadensis*	Lynx	x	x			
fb	f	fb	fb	fb	fb	*Lynx rufus*	Bobcat		x	x	x	x
	f	rf	rf	rf	rf	*Gulo gulo*	Wolverine	x	x			
fb	fb	fb	fb	fb	fb	*Lutra canadensis*	River otter	x	x	x	x	x
			f	fb	fb	*Martes americana*	Marten	x	x	x	x	x
	fb			fb	fb	*Martes pennanti*	Fisher		x	x	x	x
fb	fb	fb				*Mephitis mephitis*	Striped skunk		x	x	x	x

CONTINUES

Appendix B. *Continued*

Seashore	Riparian	Grass/ shrub	Open/ edge	Closed/ mature	Old growth	Species	Common Name	AK	BC	WA	OR	CA
	f	f	f			*Mustela erminea*	Ermine	x	x	x	x	x
f	fb	fb	fb	fb	fb	*Mustela frenata*	Long-tailed weasel		x	x	x	x
	f	f	f			*Mustela nivalis*	Least weasel	x				
fb	fb	fb	fb	fb	fb	*Mustela vison*	Mink	x	x	x	x	x
	fb	fb	fb			*Spilogale putorius*	Spotted skunk		x	x	x	
f	fb	fb	fb	fb	fb	*Procyon lotor*	Raccoon		x	x	x	x
	fb	f	fb	fb	fb	*Ursus americanus*	Black bear	x	x	x	x	x
	f	rf	rf	rf	rf	*Ursus arctos*	Grizzly bear	x	x			
		f	f			*Oreamnos americanus*	Mountain goat	x	x			
		f	f			*Ovis dalli*	Thinhorn sheep	x	x			
	f		f	f	f	*Alces alces*	Moose	x	x			
f	f	fb	fb	f	f	*Cervus elaphus*	Elk		x	x	x	x
fb	fb	fb	fb	fb	fb	*Odocoileus hemionus*	Mule deer	x	x	x	x	x
	fb	fb	fb	fb		*Odocoileus virginianus*	White-tailed deer			x	x	

Seashore = estuary and beach; Riparian = rivers, streams, lakes, ponds, marshes, bogs, and swamps; Open/edge = open forest cover including pole-sapling stands and forest edges; Closed/mature = closed forest cover and mature stands; f = foraging, resting, and other nonbreeding activities; b=breeding habitat; r = use of habitat type restricted to riparian areas; * = use when special habitat available; x = occurrence.

6. Streams and Rivers: Their Physical and Biological Variability

Robert J. Naiman and Eric C. Anderson

□ □

Rivers, streams, and associated aquatic ecosystems reflect local geomorphology, climate, the history and scale of disturbances, and the features of the riparian vegetation that grows beside them. As a consequence, the various aquatic ecosystems within a watershed (ponds, lakes, and wetlands, as well as streams and rivers) often acquire unique features that are related to their location in the watershed, patterns of water movement, gradient, and accessibility to colonizing plants and animals. For example, small streams in cold, steep, headwaters regions have biotic characteristics substantially different from comparably small, lowland streams near the ocean. The biophysical diversity of aquatic ecosystems increases when strong latitudinal or longitudinal gradients in geology and climate exist in a region. The ecological consequences of this physical heterogeneity are expressed as variations in life history strategies, evolutionary processes, the composition of plant and animal communities, system-level productivity, ecosystem integrity, and patterns of biodiversity. These consequences, in turn, have strong implications for management and restoration strategies.

This is especially true in the coastal temperate rain forest, a bioregion characterized by strong geological and climatic gradients ranging from northern California to Kodiak Island, Alaska, a distance of over 22 degrees of latitude or 2300 kilometers (Plate 1; Figure 6.1). On a global scale, the coastal temperate rain forest appears to be sufficiently homogeneous to be considered a single "ecoregion" for the interpretation of global biogeochemical processes and responses to climatic change (Neilson and Marks 1994). Similar community

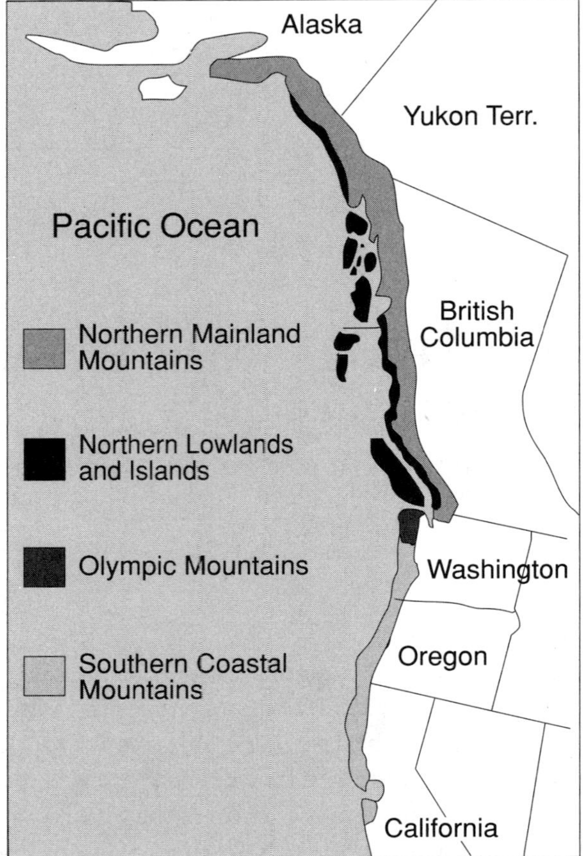

Figure 6.1. Four physiographic subregions in the coastal temperate rain forest region of the Pacific Northwest.

structure in aquatic organisms (especially fishes) and vegetation (Naiman et al. 1992) supports this conclusion. At smaller scales relevant to resource management and conservation, however, ecologically significant variations in genetic diversity, evolutionary processes, life history strategies, and other biotic characteristics are responsive to the wide variations in physical properties and histories of the aquatic ecosystems.

In this chapter we characterize the streams and rivers of the coastal temperate rain forest and illustrate the breadth of that heterogeneity. Streams and rivers are important components of ecosystems and economies throughout the rain forest region. Running waters are evenly distributed throughout the region, and we have a well-developed database on their physical characteristics. The objectives of this chapter are, first, to describe characteristic stream features of coastal temperate rain forest watersheds; second, to describe the variations in watershed size, discharge characteristics, and temperature regimes for major rivers originating along a gradient of latitude; third, to describe the effects of the physical environmental variations on the organisms found in streams and rivers; and finally to discuss the implications of these biophysical variations and adaptations for management, restoration, and conservation of aquatic ecosystems throughout the region.

Earlier chapters have described the latitudinal and altitudinal variation in geomorphological processes (Chapter 3) and climate (Chapter 2). The result—a geological–climatic template of considerable complexity—shapes the heterogeneity of biological communities and the physical character of streams and rivers in the coastal temperate rain forest. The way in which landscapes and watersheds were recolonized by plants and animals after the retreat of the Pleistocene ice sheet adds further complexity to the biotic characteristics of streams and rivers. High mountains to the east limited the movement of freshwater organisms, as did the paucity of postglacial connections between major drainages such as the Fraser and Columbia rivers (McPhail and Lindsey 1970; McPhail 1995). Most aquatic recolonization of the northern rivers occurred via the ocean from southern areas or from northern regions in which ice-free refuges survived the glacial period. As a consequence, total species richness is relatively low, but substantial genetic heterogeneity has arisen through millennia of isolation: populations of numerous aquatic species are highly adapted to local environmental conditions (Ricker 1972; Taylor 1991; Nehlsen et al. 1991; Meffe 1992).

Overview of the Pacific Coastal Ecoregion

Rivers in the coastal temperate rain forest integrate countless physical and biological processes, providing a long-term memory of environmental conditions (Naiman et al. 1992b). As is true for other ecoregions, connections between rivers and the surrounding terrain, and considerable biophysical variability, control the ecological vitality of rivers. But rivers in the North American coastal rain forest appear to possess several characteristics that may be shared only with rivers in other pristine temperate rain forests on geologically young terrain. These characteristics include the presence of large woody debris in the channel, the role played by riparian corridors in maintaining watershed-scale biodiversity, adaptations of organisms to highly fluctuating environments, the ease by which exotic plants can invade the riparian corridor, and the extent and significance of water movements in subsurface channels.

Rivers in the coastal rain forest are linked to surrounding terrain by an intricate network of stream channels with average densities of 2.5 kilometers per square kilometer. Of these, over 80 percent are classified as first- or second-order streams less than about 2 meters wide. In addition, at least forty different types of channels can be distinguished, based on various combinations of biophysical characteristics (Naiman et al. 1992b). In combination with regional variations in climate, this diversity results in a diverse array of habitat conditions exploited by life forms that employ a variety of life history strategies.

Perhaps the most visual evidence of the strong linkage of streams to the surrounding terrain is the presence (and importance) of large woody debris in the channels (Maser and Sedell 1994). Much of this debris remains in channels for as long as seven millennia (Benda et al. 1992; D. Montgomery, pers. comm.)—influencing the channel's physical configuration, the distribution of bed shear stress during high flows, and the ability of riparian vegetation to colonize (Fetherston et al. 1995). In essence, large woody debris acts as a legacy of past disturbances that influence the environmental character of the river far into the future. In small streams, woody debris density can approach 40 kilograms per square meter; in larger rivers the density may only be 5 to 10 kilograms per square meter. The role played by woody debris changes as the channel's size increases (Naiman et al. 1992b).

The presence of large woody debris in river corridors is also central to maintaining watershed-scale biodiversity. The woody debris creates diverse habitats for aquatic life forms (Maser and Sedell 1994) and also for riparian fauna and flora. Lock and Naiman (unpublished manuscript) found that the richness and abundance of spring breeding birds is related to the diversity of deciduous trees and vegetative patch types along river corridors in the coastal rain forest, for example, while Steel et al. (unpublished manuscript) found that piles of large woody debris deposited by fluvial processes along riverbanks significantly increase the abundance of small mammals and birds found on exposed cobble bars. Moreover, Pollock (1995) has demonstrated that microtopographical variations (created by large woody debris), flooding frequency, and site productivity are correlated with plant species richness in southeastern Alaska, where over 90 percent of the vascular plants are associated with riparian wetlands. In each instance, large woody debris interacting with the seasonal flood regime maintains the biological organization of the riparian zone.

Many of the organisms in the coastal rain forest are well adapted to the highly variable (yet seasonally predictable) physical environment. Regional variations in water temperatures, flow regimes, and lithologies have resulted in unique stocks of salmonids and a variety of life history strategies (Groot and Margolis 1991). Similarly, the riparian plants exhibit interesting adaptations. Black cottonwood (*Populus trichocarpa*) seedlings send roots nearly 40 centimeters deep within ten days after germination in response to declining water levels in late spring (John Reed, unpublished data), while both willow (*Salix*) and cottonwood employ different reproductive strategies according to flow regimes.

The intensity and frequency of floods, small-scale variations in topography and soils, variations in climate with altitude, the disturbance regimes imposed on the riparian corridor by upland environments—all create the mosaic of habitats in a nonequilibrium system that allows the wide variety of species and life history strategies to coexist (Naiman et al. 1993). These same characteristics, however, make rivers in the coastal rain forest susceptible to invasion by exotic organisms. DeFerrari and Naiman (1994) have shown that the type of disturbance and time since disturbance are major factors influencing the susceptibility of rivers to invasions on Washington's Olympic Peninsula. Vegetative patch size, position within the watershed (distance to human settlements, major highways, or river mouth), and environmental variables (slope, aspect, or elevation) were not, they discovered, important indicators of susceptibility to invasion by exotic plants. They found fifty-two exotic species, accounting for 23 percent of the total flora at the watershed scale. The total number of exotic species was nearly 33 percent greater in riparian zones than in uplands, and the average number and cover of exotic species were more than 50 percent greater in riparian zones than in uplands. The long-term ecological consequences of these exotic invasions are not known, but it can be expected that coastal rain forest rivers provide a principal corridor for the movement of these organisms.

Finally, subsurface movements of water are not unique to rain forest rivers, but the extent and importance of the hyporheic zone to the overall productivity of the system may be unrivaled elsewhere (Naiman et al. 1992b). The hyporheic zone is the interstitial habitat beneath the streambed—that is, the interface between surface water and the adjoining groundwater. Recent investigations have shown that this habitat can extend throughout the alluvial gravels of floodplains; the most conductive conditions appear to be in buried paleo-channels created soon after the retreat of the glaciers (Poole et al. 1996). On the Flathead River, Montana, the hyporheic zone extends laterally as much as 3 kilometers and spans a vertical dimension of 10 meters (Stanford and Ward 1988).

Hyporheic areas act as regulators of nutrient conditions in streams and provide habitat for a variety of species. As a retention or storage compartment for nutrients, the hyporheic zone provides a space for biotic processing and transformation (Triska et al. 1989; Vervier and Naiman 1992). This biotic activity appears to be strongly influenced by the rate of water exchange between the stream and the hyporheic zone. This exchange, in turn, is regulated by the physical configuration of the channel (Stanford and Ward 1993). The hyporheic zone supports a large variety of invertebrates, many of which were heretofore considered rare in their immature forms (Stanford and Ward 1988, 1993).

Taken together, spatial and temporal variations in geology, lithology, climate, large woody debris, temperature, biodiversity, exotic invasions, hyporheic characteristics, and life history strategies have created a bewildering array of river and community types in the coastal temperate rain forest. Yet, upon careful analysis, it is possible to discern patterns at the regional scale. These patterns, based on a sound understanding of rivers as ecological systems, can be used to improve management activities.

Identifying Subregions

To make sense of the complexity inherent in the coastal temperate rain forest region, we have identified four subregions with contrasting physiographies related to topography, lithology, and glacial history (Table 6.1). The southern coastal mountain subregion encompasses the coastal mountains of California, Oregon, and southern Washington (Figure 6.1). The mountains, although steep, are generally less than 1600 meters in elevation, harbor no active glaciers, and were ice-free during the Pleistocene (Wilson 1958). The Olympic Mountains subregion contains high coastal mountains (up to 2427 meters) possessing active glaciers; the rain shadow east of the Olympic Mountains is excluded. The northern mainland mountains subregion contains the Cascade Mountains north of Snoqualmie Pass, Washington, and the massive ice-mantled coastal mountains of British Columbia and continental Alaska. Finally, the northern lowlands and islands subregion includes all coastal lowlands or islands substantially influenced by oceanic conditions north of the U.S.–Canadian border.

Latitudinal Gradients in the Physical Environment

Watershed size decreases generally from north to south. Ninety-five percent of the 608 watersheds in the northern lowlands and islands subregion are smaller

Table 6.1. Physical Characteristics of the Physiographic Subregions in the Coastal Temperate Rain Forest

Subregion	Topography	Pleistocene Glacial History	Present Glaciology	Climate
Southern coastal mountains	low	unglaciated	no glaciers	rain-dominated
Olympic Mountains	moderately high	partially glaciated	several glaciers	rain and snow
Northern lowlands and islands	very low to moderately high	extensively glaciated	some glaciers	rain and snow
Northern mainland mountains	high	extensively glaciated	many glaciers	snow-dominated

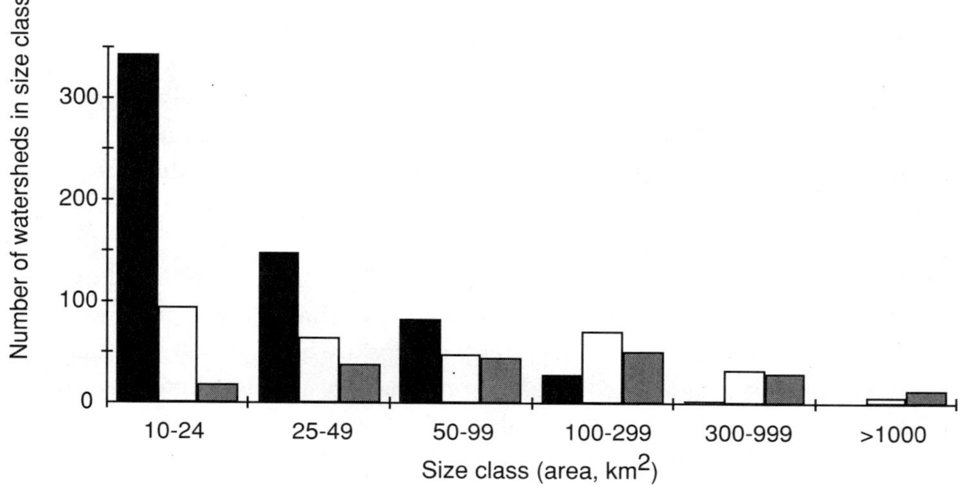

a. Number of primary watersheds

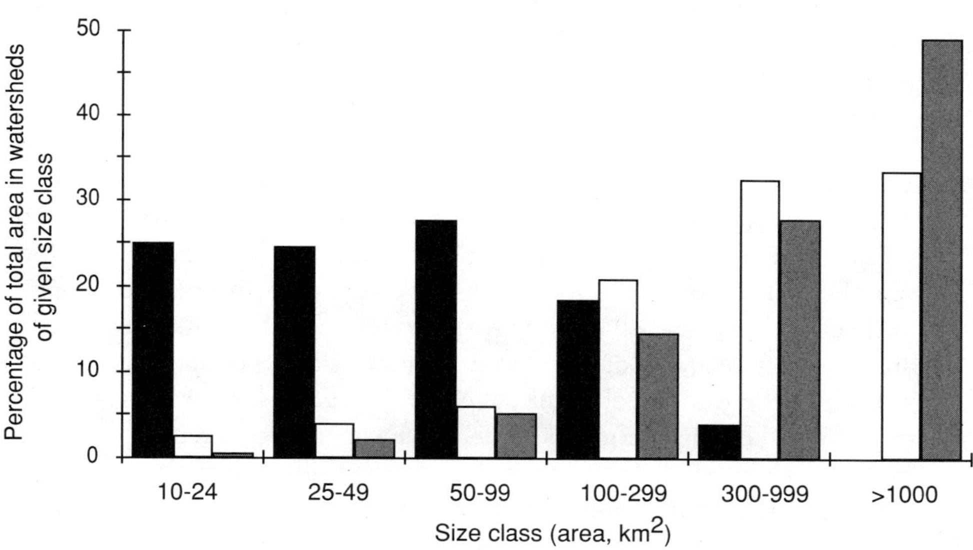

b. Distribution of areal extent

Figure 6.2. (*a*) Number of primary watersheds of different size classes from 21,216 square kilometers of the northern lowlands and islands (black) and 57,884 square kilometers of the northern mainland mountains (white) subregions in the Tongass National Forest of Alaska and from 63,561 square kilometers of the southern coastal mountains and Olympic Mountains subregions (gray). (*b*) Distribution of the areal extent of primary watersheds in the different size classes.

than 100 square kilometers. In the northern mainland mountains (324 watersheds) and the southern coastal mountains (197 watersheds), only 65 percent and 52 percent, respectively, are comparably small. As a result, in the northern lowlands and islands, watersheds smaller than 100 square kilometers account for more than three-quarters of the total area while in the northern mainland and southern coastal mountain subregions the comparable values are 13 percent and 8 percent, respectively (Figure 6.2). In the northern mainland and southern coastal subregions, 66 percent and 77 percent of the total area, respectively, is encompassed by watersheds greater than 300 square kilometers. These subregional differences in watershed size are ecologically significant for two reasons. First, watershed size, in combination with local geology and climate, determines the hydrological disturbance regime, which influences system productivity, patterns of biotic diversity, life history strategies, and other fundamental ecological characteristics. Second, watershed size is highly correlated with stream size (Naiman 1983), and streams of different sizes have different communities and processes (Vannote et al. 1980).

The mean annual runoff from coastal rain forest rivers increases from 0.5 meters a year in central California to 7.4 meters a year in southeastern Alaska. The variability in runoff also increases (Figure 6.3*a*). The full range of annual discharge observed over the entire latitudinal gradient of the coastal temperate rain forest region can be found among the rivers located in northern British Columbia and Alaska. The ecological significance of this order-of-magnitude difference in runoff quantities within the region accounts for many of the differences in community dynamics and ecosystem-level processes observed in the region's rivers. Increasing discharge means faster flow and increasing bed shear stress, all other factors being equal (Dunne and Leopold 1978). This has profound effects on erosion rates, lateral channel migrations, the ability of organisms to colonize and maintain viable populations, and the retention of large woody debris in channels. Indeed, the entire character of rivers is affected by the energy associated with water discharge.

The seasonal patterns of discharge also show significant variation with latitude. Nearly 70 percent of annual discharge occurs in winter in central California; in parts of British Columbia and southeastern Alaska, this proportion declines to 25 percent (Figure 6.3*b*). The northern lowlands and islands rivers generally discharge proportionally more water in winter than do rivers draining the higher-elevation (and colder) northern mainland mountains. The proportion of annual discharge in summer, however, increases with latitude (Figure 6.3*c*). In central California less than 2 percent of the annual discharge occurs in summer, while in British Columbia and Alaska the proportion can be as high as 60 percent. As the snows accumulated at higher levels melt, the

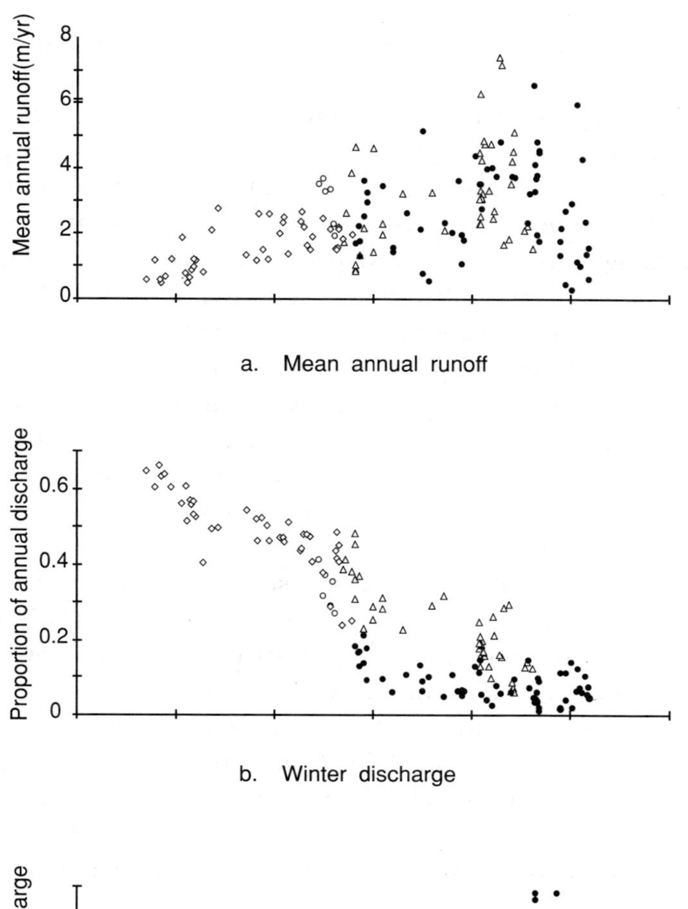

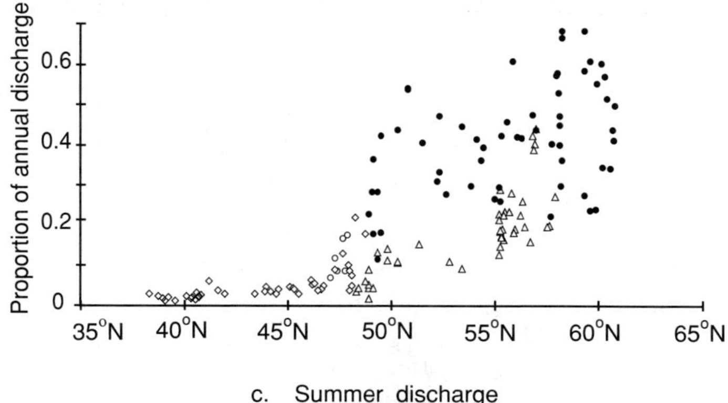

c. Summer discharge

Figure 6.3. Streamflow characteristics for 151 rivers of the coastal temperate rain forest region. Each symbol represents one river within one of four subregions: southern coastal mountains (◇); Olympic Mountains (○); northern lowlands and islands (Δ); and northern mainland mountains (●). (*a*) Mean annual runoff as a function of latitude. (*b*) Proportion of mean annual discharge occurring during the three months of winter. (*c*) Proportion of mean annual discharge occurring during the three months of

northern mainland mountains discharge proportionally more water than the northern lowlands and islands.

The timing of high flows has a strong influence on reproductive strategies, species migrations, and system productivity. A substantial literature (reviewed by Statzner et al. 1988 and Poff and Ward 1989) shows how variations in water hydraulics in space and through time influence behavioral and physiological features including mating, insect drift, orientation, competition, net and case building, schooling, territoriality, and respiration. Organisms seek optimal conditions for reproduction and migration in the short term—and they optimize life history strategies (fecundity, egg size, allocation of energy) in the long term—in response to the discharge patterns and associated factors including water velocity, the mobility of streambed materials, and other hydraulic features. Likewise, discharge patterns influence overall productivity in streams and rivers by modifying such features as light regimes, fluxes of nutrients, and rates of juvenile mortality.

A sharp latitudinal demarcation in the seasonal timing of annual floods within the coastal temperate rain forest occurs at about 48 degrees north (Figure 6.4). Ninety-seven percent of the 1441 documented annual floods on rivers between central California and the Washington–British Columbia border occurred from November to April. In contrast, north of the border only 31 percent of the 1552 documented floods occurred during the same months, and most of those floods occurred in the northern lowlands and islands. This demarcation corresponds roughly to the boundary between the northern subregions and the southern coastal mountains and Olympic subregions and may

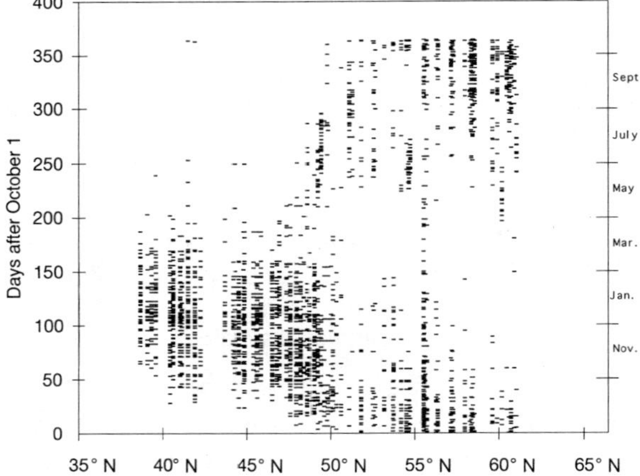

Figure 6.4. Temporal distribution of peak annual discharge. Each point represents the date and latitude of a maximum daily discharge for a given river in a single water year (one point per year; all rivers have at least ten years in record; ninety-three rivers total).

relate to latitudinal patterns in the seasonal timing of precipitation (Chapters 1 and 2) and the greater likelihood of snowfall in the northern mainland mountains. The annual timing of floods has ecological consequences similar to those associated with discharge patterns. In the long term, life history strategies and species migrations are intimately tied to annual floods (Montgomery 1994; Seegrist and Gard 1972; Erman et al. 1988).

Average yearly water temperatures decline from about 15°C in central California to about 3°C in south central Alaska (Figure 6.5). This trend has three ecologically significant aspects. First, the range of temperatures experienced in a river decreases in the northern latitudes, since the freezing point sets a lower limit on temperature for flowing water. Second, mean annual maximum temperatures in the more southerly latitudes approach (and even exceed) 25°C—past the physiological thresholds that many cool-adapted organisms can tolerate. Third, mean annual minimum temperatures reach 0°C north of latitude 48, indicating that ice poses a challenge to aquatic life forms in the winter.

The physiological effects of temperature on species distributions, reproductive success, life history strategies, and community processes are well known (Danks 1978, Ward and Stanford 1992). Accumulated thermal units (degree-days) synchronize the life histories of most aquatic organisms. Depending on the type and duration, ice also strongly determines the character of biotic

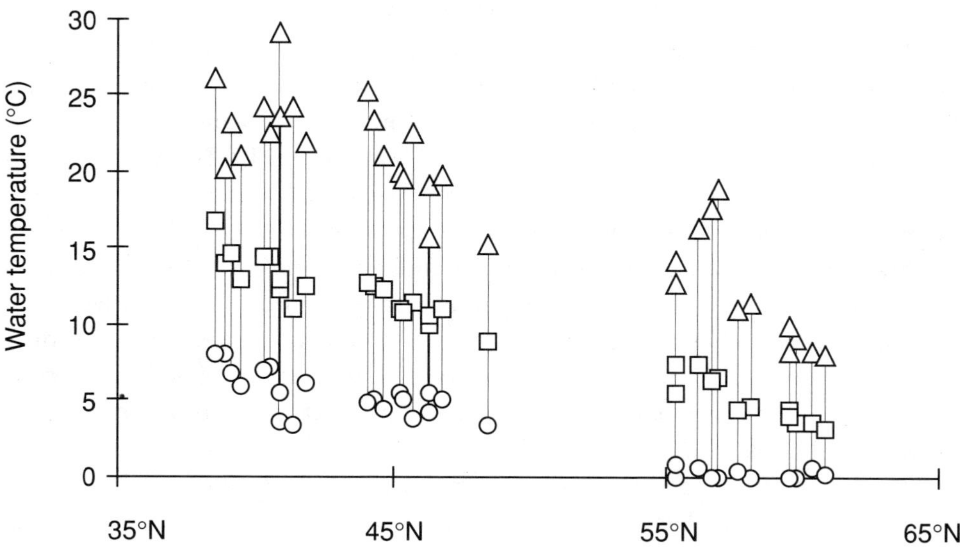

Figure 6.5. Water temperature as a function of latitude for thirty-two rivers. Mean annual temperature and the maxima and minima are date-averaged over 8 to 82 years, depending on the site.

communities (Harper 1981; Olsson 1981) by changing the nature of competition between species or by killing some individuals.

Ecological Implications of Latitudinal Variations

Few biotic studies encompass the entire range of physical variables encountered in the region, but a number of studies in specific subregions of the coastal temperate rain forest, and from other ecoregions, strongly suggest that the biotic heterogeneity of the coastal temperate region may be much greater than previously recognized. This heterogeneity is expressed in community organization and in the behavioral, evolutionary, and physiological adaptations of organisms to water flow and temperature.

River size (as related to watershed size) is strongly associated with certain patterns of biological organization in rivers. This relationship is apparent in the types and patterns of disturbance experienced by aquatic organisms, in the character of riparian vegetation, in the storage and flux of organic materials, and in the attributes of the in-stream community (Gregory et al. 1991, reviewed in Naiman et al. 1992b). In the Cascade Mountains, for example, the significance of aquatic plants as the basis of river food chains increases as streams become larger, while the significance of organic matter washed or blown into the river decreases, the mass of large woody debris decreases, and the invertebrate community composition changes according to the proportions of habitat types available (Naiman and Sedell 1980; Anderson and Sedell 1979; Hawkins et al. 1982; Bilby and Ward 1989).

Many of the rivers in the region are alluvial. Therefore, river size also influences the creation, extent, and long-term maintenance of the hyporheic zone—the unique habitat located beneath the streambed, the zone where surface water and groundwater meet (Stanford and Ward 1988, 1993). The history of channel development, the delivery and routing of sediments from upstream areas, the physical constraints imposed by the surrounding uplands—collectively all determine the extent and depth of the hyporheic zone. All other factors being equal, small watersheds presumably have smaller hyporheic volumes than larger watersheds. The hyporheic zone contributes substantial productivity to otherwise low-productivity surface waters (Stanford and Ward 1993). The evidence suggests that the overall aquatic productivity of a watershed may be intimately linked to the size and accessibility of its hyporheic zone.

The distribution and adaptations of salmonids indigenous to the coastal temperate rain forest best illustrate the nature of biotic responses to the strong latitudinal gradients in the physical features of watersheds (Chapters 7 and 8). Each salmonid species employs a different life history strategy (Everest 1987) and, consequently, uses a different set of habitats (Bisson et al. 1982). These differing life history strategies allow several salmonid species to fully utilize available habitats in a single watershed by using different sections of a river or using the same sections at different times. Salmonids respond to the characteristics and location of spawning areas, postemergent rearing areas, summer (low-water) rearing areas, winter rearing areas, and the estuarine conditions encountered at smoltification (when juvenile salmon undergo the physiological adjustment that allows them to migrate into salt water). For example, juvenile steelhead (*Oncorhynchus mykiss*) and coho salmon (*O. kisutch*) use the rearing habitats in small streams differently. During the summer rearing period, coho tend to occupy pools whereas steelhead prefer riffles and glides (Bisson et al. 1988). During winter, however, when water levels are higher, coho move to wall-base channels (Peterson and Reid 1984) and steelhead occupy terrace tributaries (Scarlett and Cederholm 1984). Spawning areas and timing of spawning may nearly overlap among species such as pink salmon (*O. gorbuscha*), chum salmon (*O. keta*), and coho salmon. But because pink salmon and chum salmon migrate to sea soon after emergence, juveniles of these species do not compete while in streams.

Water temperatures and the presence of ice have equally influential effects on the biotic diversity of rivers in the coastal temperate region. Water temperatures control and synchronize most physiological functions of ectothermic life forms (those whose body temperature matches their environment), including fishes (reviewed by Ward and Stanford 1992). Temperature is a crucial determinant of the incubation rate of salmon eggs (Murray and McPhail 1988), for example, and it also affects the growth of juvenile, stream-rearing salmonids (Holtby 1988). In the colder northern regions of the rain forest, salmon spawn earlier in the autumn, as their eggs require more time to accumulate the number of degree-days required for hatching. Northern chinook salmon (*O. tshawytscha*) populations finish almost half their spawning by August; southern populations may not spawn until several months later (Healey 1991). Some stocks of chum salmon from very cold rivers seem to have physiologically adapted their egg metabolism to require fewer degree-days for hatching (Salo 1991). Moreover, the preponderance of chinook salmon that require two to three years of juvenile growth before smoltification in northern rivers (as opposed to one year or even less in southern rivers) is related to the low trophic productivity associated with the low-temperature rivers (Taylor 1990). The same biotic responses can be expected in fishes other than salmonids as well as the aquatic invertebrates of the coastal rain forest region.

Where seasonal water temperatures are extreme, "patchiness" in temperature throughout the river network becomes very important. In northern California streams, for example, when ambient water temperatures exceed 23°C juvenile steelhead migrate to the cooler depths of nearby pools. Adult, summer-run steelhead congregate in such pools on California's Eel River, remaining in the cool water below the thermocline (Nielsen et al. 1994).

Ice poses special challenges for aquatic organisms (Harper 1981; Olsson 1981). Each of the three basic types of ice—surface ice, anchor ice, and frazzle ice—has different consequences. Anchor ice is probably the most disruptive to bottom-dwelling organisms because of its tendency to exclude organisms from the sediments and to raft sediments if the ice mass is large enough to float. Frazzle ice, in sufficient quantities, can clog fishes' gills and cause additional damage to sensitive tissues. Rivers with substantial ice formation each year tend to be dominated by low-diversity communities and contain numerous insect species that complete their life cycles in less than a year (Diptera, especially Simulidae, and mosquitoes). In contrast, rivers with less regular ice formation tend to have higher diversity and to contain insect species that may take several years to complete their life cycles (Trichoptera, Plecoptera). The sustained periods of freezing in many northern rivers in the rain forest are sufficient to account for fundamental differences in community composition between the northern and southern subregions.

Management, Restoration, and Conservation of Rain Forest Rivers

The ecological consequences of the subregional differences in watershed sizes, as well as the latitudinal gradients in discharge patterns and temperature, have important implications for the management, restoration, and conservation of rivers in the coastal temperate rain forest. The rain forest is a heterogeneous region with substantial biophysical variability both within watersheds and among watersheds. Management philosophies and restoration and conservation strategies that embrace this natural variation will be the most successful in the long term. The most effective strategies will be those that consider the unique behavioral and physiological adaptations of local populations and species. Although aquatic species diversity is relatively low throughout the region, compared to ecoregions further south, the diversity of adaptations to local conditions is impressive and evolutionary divergence continues to expand. The region is highly vulnerable to invasion by exotic species, which cause generally undesirable consequences. This is especially true along river corridors with periodic disturbances caused by fluctuating flow levels. To maintain the integrity of coastal rain forest rivers, continued vigilance against exotic invasions along rivers and streams is required.

Human activities will affect the most sensitive and most important components of the riverine ecosystems first: the hyporheic zone, by clogging interstitial spaces, and the riparian forest, by altering community demographics.

Without comprehensive and sustained monitoring programs, however, widespread changes to these components will be difficult to detect. Monitoring riverine conditions against a background of a highly variable physical environment remains a major challenge for managers and researchers alike—one that must to be resolved if we are to maintain the long-term integrity of the coastal temperate rain forest streams and rivers.

Acknowledgments

The authors thank Stanley V. Gregory for helpful comments and suggestions on the content of this chapter. Randall H. Hagenstein, Andrew P. Mitchell, and Ed Backus generously provided data gathered during research for Pacific GIS and Ecotrust. Our work was supported by the Pacific Northwest Research Station of the USDA Forest Service, by private contributions to the Center for Streamside Studies, and by the H. Mason Keeler Endowment of the University of Washington's School of Fisheries.

References

Anderson, N. H., and J. R. Sedell. 1979. "Detritus processing by macroinvertebrates in stream ecosystems." *Ann. Rev. Entomol.* 24:351–377.

Benda, L. E., T. J. Beechie, R. C. Wissmar, and A. C. Johnson. 1992. "Morphology and evolution of salmonid habitats in a recently deglaciated river basin, Washington State." *Can. J. Fish. Aqua. Sci.* 49(6):1246–1256.

Bilby, R. E., and J. W. Ward. 1989. "Changes in characteristics and function of woody debris with increasing size of streams in western Washington." *Trans. Amr. Fish. Soc.* 118:368–378.

Bisson, P. A., K. Sullivan, and J. L. Nielsen. 1988. "Channel hydraulics, habitat use, and body form of juvenile coho salmon, steelhead, and cutthroat trout in streams." *Trans. Amr. Fish. Soc.* 117:262–273.

Bisson, P. A., J. L. Nielsen, R. A. Palmson, and L. E. Grove. 1982. "A system of naming habitat types in small streams, with examples of habitat utilization by salmonids during low streamflow." In N. B. Armantrout, ed., *Acquisition and Utilization of Aquatic Habitat Inventory Information.* Bethesda, Md.: American Fisheries Society, Western Division.

Danks, H. V. 1978. "Modes of seasonal adaptations in the insects. I: Winter survival." *Can. Ent.* 110:1167–1205.

DeFerrari, C., and R. J. Naiman. 1994. "A multiscale assessment of exotic plants on the Olympic Peninsula, Washington." *J. Veg. Sci.* 5:247–258.

Demarchi, D. 1990. *An Introduction to the Ecoregion Classification Used to Define British Columbia's Regional Ecosystems.* Victoria: B.C. Ministry of Environment, Wildlife Branch.

Dunne, T., and L. B. Leopold. 1978. *Water in Environmental Planning.* San Francisco: W. H. Freeman.

Erman, D. C., E. D. Andrews, and M. Yoder-Williams. 1988. "Effects of winter floods on fishes in the Sierra Nevada." *Can. J. Fish. Aqua. Sci.* 45:2195–2200.

Everest, F. H. 1987. "Salmonids of western forested watersheds." In E. O. Salo and T. W. Cundy, eds., *Streamside Management: Forestry and Fishery Interactions.* Contribution 57. Seattle: Institute of Forest Resources, University of Washington.

Fetherston, K. L., R. J. Naiman, and R. E. Bilby. 1995. "Large woody debris, physical process, riparian forest development in montane river networks." *J. Geomorph.* 13:133–144.

Gregory, S. V., F. J. Swanson, W. A. McKee, and K. W. Cummins. 1991. "An ecosystem perspective of riparian zones." *BioScience* 41:540–551.

Groot, C., and L. Margolis, eds. 1991. *Pacific Salmon Life Histories.* Vancouver: University of British Columbia Press.

Harper, P. P. 1981. "Ecology of streams at high latitudes." In M. A. Lock and D. D. Williams, eds., *Perspectives in Running Water Ecology.* New York: Plenum Press.

Hawkins, C. P., M. L. Murphy, and N. H. Anderson. 1982. "Effects of canopy, substrate composition, and gradient on the structure of macroinvertebrate communities in Cascade Range streams of Oregon." *Ecology* 63:1840–1856.

Healey, M. C. 1991. "Life history of chinook salmon (*Oncorhynchus tsawytscha*)." In C. Groot and L. Margolis, eds., *Pacific Salmon Life Histories.* Vancouver: University of British Columbia Press.

Holtby, L. B. 1988. "Effects of logging on stream temperatures in Carnation Creek, British Columbia, and associated impacts on the coho salmon (*Oncorhynchus kisutch*)." *Can. J. Fish. Aqua. Sci.* 49:682–693.

Lock, P. A., and R. J. Naiman. Unpublished manuscript. "Effects of stream size and riparian vegetation on bird community structure in coastal temperate forests of the Pacific Northwest."

Maser, C., and J. R. Sedell. 1994. *From the Forest to the Sea.* Delray Beach, Fla.: St. Lucie Press.

McPhail, J. D. 1995. "The origin and speciation of Oncorhynchus revisited." In D. J. Stouder, P. A. Bisson, and R. J. Naiman, eds., *Pacific Salmon and Their Ecosystems.* New York: Chapman & Hall.

McPhail, J. D., and C. C. Lindsey. 1970. "Freshwater fishes of northwestern Canada and Alaska." *Fish. Res. Bd. of Canada Bull.* 173.

Meffe, G. K. 1992. "Techno-arrogance and halfway technologies: Salmon hatcheries on the Pacific coast of North America." *Cons. Biol.* 6:350–354.

Montgomery, D. R. 1994. "Geomorphological influences on salmonid spawning distributions." *Geol. Soc. Am. Abstracts with Programs* 26(7):A439.

Murray, C. B., and J. D. McPhail. 1988. "Effect of incubation temperature on the development of five species of Pacific salmon (*Oncorhynchus*) embryos and alevins." *Can. J. Zoo.* 66:266–273.

Naiman, R. J. 1983. "The annual pattern and spatial distribution of aquatic oxygen metabolism in boreal forest watersheds." *Ecol. Mono.* 53:73–94.

Naiman, R. J., and J. R. Sedell. 1980. "Relationships between metabolic parameters and stream order in Oregon." *Can. J. Fish. Aqua. Sci.* 37:83–847.

Naiman, R. J., H. Décamps, and M. M. Pollock. 1993. "The role of riparian corridors in maintaining regional biodiversity." *Ecol. App.* 3:209–212.

Naiman, R. J., D. G. Lonzarich, T. J. Beechie, and S. C. Ralph. 1992a. "General principles of classification and the assessment of conservation potential in rivers." In P. Boon, P. Calow, and G. Petts, eds., *River Conservation and Management.* Chichester, UK: Wiley.

Naiman, R. J., T. J. Beechie, L. B. Benda, D. R. Berg, P. A. Bisson, L. H. MacDonald, M. D. O'Connor, P. L. Olson, and E. A. Steel. 1992b. "Fundamental elements of ecologically healthy watersheds in the Pacific Northwest coastal ecoregion." In R. J. Naiman, ed., *Watershed Management: Balancing Sustainability and Environmental Change.* New York: Springer-Verlag.

Nehlsen, W., J. E. Williams, and J. A. Lichatowich. 1991. "Pacific salmon at the crossroads: Stocks of salmon at risk from California, Oregon, Idaho, and Washington." *Fisheries* 16:4–21.

Neilson, R. P., and D. Marks. 1994. "A global perspective of regional vegetation and hydrologic sensitivities from climatic change." *J. Veg. Sci.* 5:715–730.

Nielsen, J. L., T. E. Lisle, and V. Ozaki. 1994. "Thermally stratified pools and their use by steelhead in northern California streams." *Trans. Am. Fish. Soc.* 123:613–626.

Olsson, T. I. 1981. "Overwintering of benthic macroinvertebrates in ice and frozen sediment in a north Swedish river." *Holarctic Ecol.* 4:161–166.

Peterson, N. P., and L. M. Reid. 1984. "Wall-base channels: Their evolution, distribution and use by juvenile coho salmon in the Clearwater River, Washington." In J. M. Walton and D. B. Houston, eds., *Proceedings of the Olympic Wild Fish Conference.* Port Angeles, Wash.: Fisheries Technology Program, Peninsula College.

Poff, N. L., and J. V. Ward. 1989. "Implications of streamflow variability and predictability for lotic community structure: A regional analysis of streamflow patterns." *Can. J. Fish. Aqua. Sci.* 46:1805–1818.

Pollock, M. M. 1995. "Patterns of plant species richness in emergent and forested wetlands of southeast Alaska." Ph.D. diss., University of Washington, Seattle.

Poole, G. C., R. J. Naiman, J. Pastor, and J. A. Stanford. 1996. "Uses and limitations of ground-penetrating RADAR in two riparian systems." In J. Gibert, J. Mathieu, and F. Fournier, eds., *Groundwater/Surface Water Ecotones: Biological and Hydrological Interactions and Management Options.* Cambridge: Cambridge University Press.

Ricker, W. E. 1972. "Hereditary and environmental factors affecting certain salmonid populations." In R. C. Simon and P. A. Larkin, eds., *The Stock Concept in Pacific Salmon.* Vancouver: Institute of Animal Resource Ecology, University of British Columbia.

Salo, E. O. 1991. "Life history of chum salmon (*Oncorhynchus keta*)." In C. Groot and L. Margolis, eds., *Pacific Salmon Life Histories.* Vancouver: University of British Columbia Press.

Scarlett, W. S., and C. J. Cederholm. 1984. "Juvenile coho salmon fall-winter utilization of two small tributaries of the Clearwater River, Washington." In J. M. Walton and D. B. Houston, eds., *Proceedings of the Olympic Wild Fish Conference.* Port Angeles, Wash.: Fisheries Technology Program, Peninsula College.

Seegrist, D. W., and R. Gard. 1972. "Effects of floods on trout in Sagehen Creek, California." *Trans. Amr. Fish. Soc.* 101:478–482.

Stanford, J. A., and J. V. Ward. 1988. "The hyporheic habitat of river ecosystems." *Nature* 335:64–66.

———. 1993. "An ecosystem perspective on alluvial rivers: Connectivity and the hyporheic corridor." *J. N. Am. Benthol. Soc.* 12:48–60.

Statzner, B., J. A. Gore, and V. H. Resh. 1988. "Hydraulic stream ecology: Observed patterns and potential applications." *J. N. Am. Benthol. Soc.* 7:307–360.

Steel, E. A., R. J. Naiman, and S. West. Unpublished manuscript. "Woody debris piles: Habitat for birds and small mammals in the riparian zone."

Taylor, E. B. 1990. "Environmental correlates of life-history variation in juvenile chinook salmon, *Oncorhynchus tshawytscha* (Walbaum)." *J. Fish. Biol.* 37:1–17.

———. 1991. "A review of local adaptation in salmonidae with particular reference to Pacific and Atlantic salmon." *Aquaculture* 98(1–3):185–208.

Triska, F. J., V. C. Kennedy, R. J. Avanzino, G. W. Zellweger, and K. E. Bencala. 1989. "Retention and transport of nutrients in a third-order stream in northwestern California: Hyporheic processes." *Ecology* 70:1893–1905.

Vannote, R. L., G. W. Minshall, K. W. Cummins, J. R. Sedell, and C. E. Cushing. 1980. "The river continuum concept." *Can. J. Fish. Aqua. Sci.* 37:130–137.

Vervier, P., and R. J. Naiman. 1992. "Spatial and temporal fluctuations of dissolved organic carbon in subsurface flow of the Stillaguamish River (Washington, USA)." *Archiv für Hydrobiol.* 123(4):401–412.

Ward, J. V., and J. A. Stanford. 1992. "Thermal responses in the evolutionary ecology of aquatic insects." *Ann. Rev. Entomol.* 27:97–117.

Wilson, J. T. 1958. *Glacial Map of Canada.* Ottawa: Geological Society of Canada.

Wissmar, R. C., and F. J. Swanson. 1990. "Landscape disturbances and lotic ecotones." In R. J. Naiman and H. Décamps, eds., *The Ecology and Management of Aquatic-Terrestrial Ecotones.* Park Ridge, N.J.: Parthenon.

7. The Terrestrial/Marine Ecotone

Charles A. Simenstad, Megan Dethier, Colin Levings, and Douglas Hay

□ □

Perched at the edge of the North American continent, coastal temperate rain forest ecosystems epitomize the concept that ecologists call "ecotones"—that is, "zones of transition between adjacent ecological systems, having a set of characteristics uniquely defined by space and time scales and by the strength of the interactions between the adjacent ecological systems" (Holland 1988; Risser 1990). A multitude of energies converge at this terrestrial/marine ecotone, as the heavy precipitation that intercepts the coastal mountains for eight to nine months of the year discharges through major and minor rivers, permeating the sea/land interface with the movement of fresh water, sediments, nutrients, organic matter and debris of all sizes, and animals. At the land/sea ecotone of the coastal temperate rain forest, rivers, tides, weather, and organisms are in a constant flux.

This chapter describes some of the unique features of the terrestrial/marine ecotone, as well as the dynamic large-scale and long-term processes that span and shape its strong interactions among adjacent ecosystems. We examine how these processes vary across several spatial scales, from regional to local, and explore how biological communities and particular fauna within them have evolved in response to these interacting scales of variation. We conclude with a perspective on management of human activities at the interface of land and sea that demands understanding, decision making, and further scientific inquiry at multiple scales in time and space. Not only do natural resources respond to large-scale, land-margin ecosystem processes, they in turn influence those processes. Therefore, rational management of the coastal temperate rain forest cannot operate at any scale smaller than the processes that create and sustain the terrestrial/marine ecotone.

The Nature of Terrestrial/Marine Ecotones

The coastal temperate rain forest terrestrial/marine ecotone is not always distinct, although it may seem so when viewing the crashing surf on a coastal headland. The interface can be as narrow as the ribbon of meltwater from a hanging glacier in northern British Columbia or as wide as a tidal-freshwater river delta that intrudes 100 miles into the coastal landmass (Figure 7.1). The terrestrial/marine ecotone of the temperate rain forest includes broad deltas formed by high sedimentation rates, such as those found at the mouth of the Fraser River (Figure 7.2), expansive drowned river valleys like Grays Harbor on the central Washington coast (Figure 7.3), elongated fjords such as Hood Canal and adjoining Puget Sound (Figure 7.4), and rocky beaches and wave-washed headlands such as Point of Arches on the Olympic Peninsula (Figure 7.5).

Geology

The rain forest coast is shaped by the past energies of tectonic activity and glaciation and by present forces of river, wind, wave, and tide that erode rocks and transport unconsolidated sediments such as sand and silts. Glaciation was responsible for most of the physical features of the coastal temperate rain forest—both directly by eroding the land surface and indirectly by causing changes in sea level. River valleys carved out of the coastal plain were filled when sea levels rose at the close of Pleistocene glaciation (1.6 million to 15,000 B.P.), the same event that carved the fjords and inlets of Puget Sound, the Strait of Georgia, and most of the rain forest coastline in British Columbia and southeastern Alaska.

And the region is not yet at rest. It is prone to slow, chronic uplifting of the land relative to the sea by as much as 2 to 3 millimeters per year because the land is still rebounding (termed "isostatic" rebound) from the release of the weight of the glaciers (Chelton and Davis 1982; Lajoie 1986) and prone also to cataclysmic but infrequent earthquakes (every 1000 years or so) of the Cascadia subduction zone (where the Juan de Fuca ocean crustal plate slides under the North American continental plate) that can cause coastal land to subside between 0.5 to 2 meters in a matter of minutes. Excavation in the marshes of coastal Washington estuaries, like that performed by Brian Atwater (1987) in Willapa Bay and along the lower Columbia River, would likely unearth evidence of a sequence of at least six such subsidences and burials of salt marsh over the last 7000 years, each marking an abrupt end of marsh

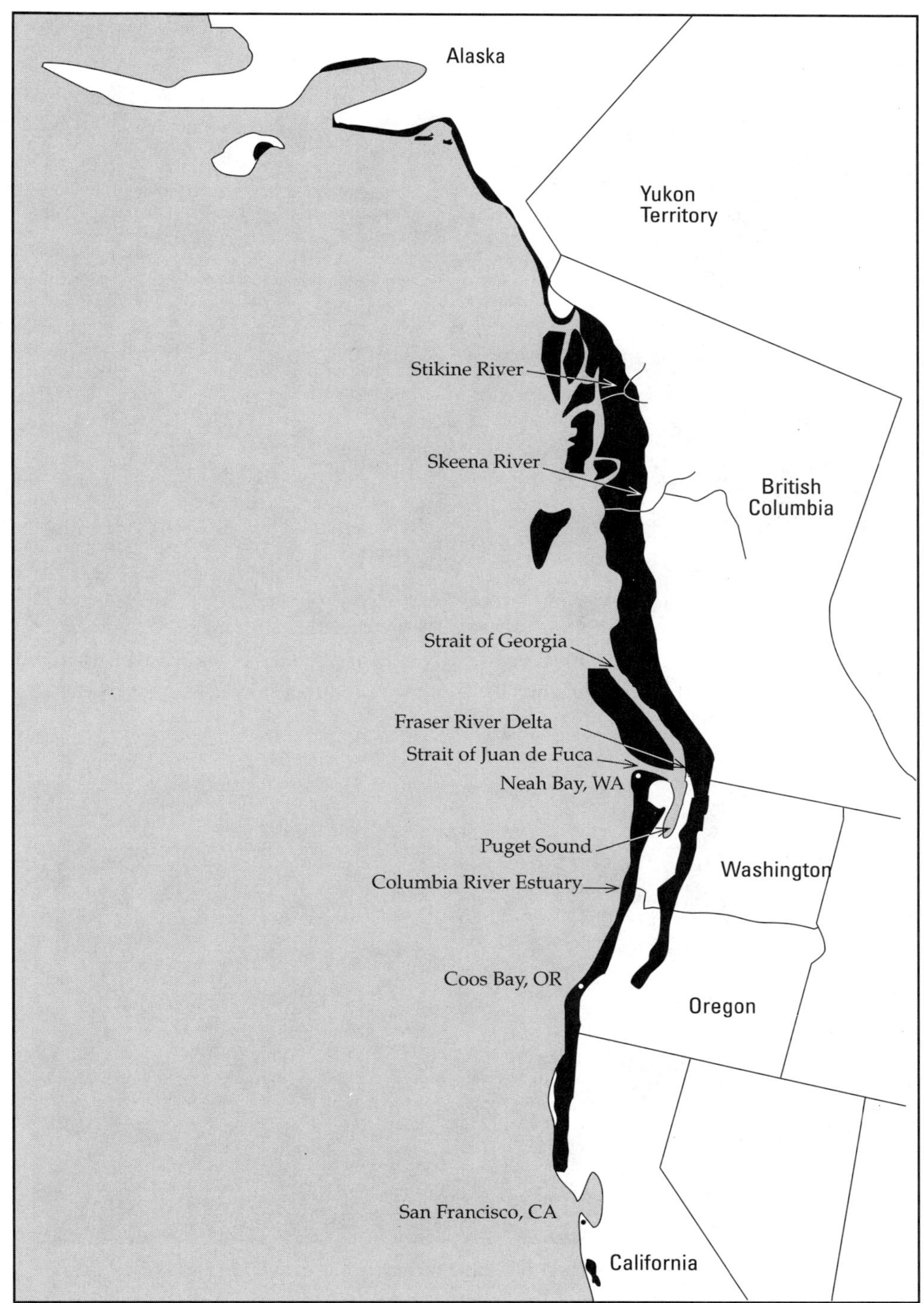

Figure 7.1. Selected locations mentioned in Chapter 7.

Figure 7.2. Aerial view of the lower Fraser River near the city of Vancouver, British Columbia, a broad delta in which the river's present channel is constrained by urban development. (Photo courtesy of D. Levy.)

Figure 7.3. Aerial view of Grays Harbor—an example of a drowned river valley estuary at the terrestrial/marine ecotone of the coastal temperate rain forest ecosystem on the central Washington coast. Developed land surrounds the upper estuary (mid-foreground), but the outer estuary is comparatively unaltered. (Photo courtesy of K. Kunz, U.S. Army Corps of Engineers, Seattle District.)

Figure 7.4. Aerial view of Hood Canal—a fjord adjoining the eastern margin of the coastal temperate rain forest on the Olympic Peninsula, Washington. The view is southwest from above the Kitsap Peninsula; the Olympic Mountains are on the center-right horizon. (Photo courtesy of C. Simenstad, University of Washington.)

Figure 7.5. Exposed coastal beach and rocky headland at Point of Arches on the Olympic Peninsula in Washington. This exquisite example of the highest-energy ecotone in the coastal temperate rain forest is now protected within the Olympic Coast National Marine Sanctuary. (Photo courtesy of M. Dethier, University of Washington.)

salt marsh over the last 7000 years, each marking an abrupt end of marsh development and a return to low intertidal mudflats.

The rock and soil that form the coast are also prone to erosion, and ample mineral sediments, accompanied by considerable organic matter, are flushed from the land by the rivers. Much of the finer suspended material settles in the estuary and continuously rebuilds intertidal platforms until they ultimately attain elevations that can be colonized by marsh vegetation. On time scales that exceed our cultural perception, the land/sea ecotone itself is going through "lifetimes" in which earthquake disruption alternates with natural restoration; estuaries are perpetually rising, getting shallower, becoming progressively more vegetated until the next disturbance resets the clock. To view coastal ecosystems as static or predictable—and to attempt to "manage" processes at human scales of time and space—is naive at best.

The exposed shore of the coastal rain forest may be the most geologically and ecologically dynamic in the Pacific Northwest, vulnerable to extreme forces of waves, currents, wind, and rain. The region is geomorphically diverse (Chapter 3). Where present, resistant headlands and seastacks both shape the shoreline and control the rates and patterns of longshore sediment exchange along the coast. On exposed shores (as in southern Washington and much of Oregon and British Columbia), waves encounter the entire "face" of the land, and eroded sediments move freely along the coast, driven by longshore currents. Coastal cliffs erode rapidly (about 100 meters every 100 years), and landslides are common. Consequently, these shores are very responsive to changes in the rate and character of sediment supply (primarily gravel and sand deposits left by the glaciers and river-borne sediments). Seasonal changes are also apparent: finer sands tend to be transported offshore by winter storm waves, leaving beaches coarser and steeper in the winter.

Climate

As described in detail in Chapters 1 and 2, perhaps the most important processes shaping the land/sea margin of the coastal rain forest are associated with its dynamic climate. Important climatic features include the strong seasonal variation in weather intensity, sharp transitions in temperature and precipitation gradients, and the seasonally high discharge of precipitation or snowmelt. Climatic effects are probably most evident in the fall and winter, when storms with relatively moist, dense air move northeasterly across the Pacific, along the subtropical high and North Pacific low pressure gradient, and interact with cooler, drier fronts along the landmass. Sharp temperature and precipitation gradients form along steeply rising terrain along the coast—Olympic Mountains, Coast Range Mountains, and Vancouver Island—forcing storms to rise. The volume of precipitation is tremendous, falling as rain

(30–50 inches over December–February alone) at lower elevations and as snow (up to 100 inches) at the higher elevations. Mild, moist winters, on the other hand, inhibit freezing. Where river valleys and fjords penetrate coastal mountains, milder marine air masses flow particularly far into the continental landmass. Summers are cool and foggy, but only 10 percent or so of the total annual precipitation occurs between June and August. Cloud cover is persistent along the coast: 75 to 90 percent during both winter and summer. Winds are equally persistent and quite cyclic (on the scale of every three days): the maximum wind speeds are in fall and winter, decreasing through spring, and lowest in summer (Thomson 1981).

Rivers and Their Inputs

River flow completes the hydrological cycle begun in the North Pacific and dilutes the sea, forming a salinity gradient that plays a large part in defining the breadth of the land/sea ecotone. Depending on its volume, freshwater flow can affect the salinity of the ocean for as little as the width of the surf zone to as much as hundreds of miles north and south of the river mouth. The freshwater plume of the Columbia River is among the most prominent features of the region's coastal oceanography. Despite the forty-eight major dams and reservoirs on the Columbia and its tributaries, the river's annual average discharge of 5500 cubic meters per second (ranging from 3000 to 17,000) spreads offshore as far north as Vancouver Island and as far south as as northern California, depending on seasonal wind patterns. The Columbia contributes between 60 percent (winter) and 90 percent (summer) of the fresh water diluting the salinity of the sea surface waters over that region (Barnes et al. 1972; Sherwood et al. 1990).

Average annual discharge for rivers in the coastal rain forest can range from less than 90,000 cubic meters per year, for the tributary rivers of Willapa Bay, to 100,000 or even 600,000 cubic meters per year for rivers draining the western Olympic Peninsula. Maximum discharge can exceed 1000 to 4000 cubic meters per second for steep and extensive watersheds such as Washington's Quinault River. Commensurate with the seasonal precipitation cycle, the flow of coastal rain forest rivers might best be characterized as "seasonally intensive": there is a high-flow period during winter to late spring, south of 48 degrees north latitude, and during summer and early fall north of this latitude (Figure 6.3). The pattern of river discharge, or "hydroperiod," depends on the drainage basin topography and elevation, the type and extent of natural vegetation, the structure of aquifers underlying the drainage basin, and the presence of dams.

River flow affects water temperature, pH, and conductivity among other characteristics (Kempe and Lammerz 1983); it also provides the energy and medium to transport dissolved nutrients and other inorganic and organic compounds from terrestrial ecosystems—sediments, from fine clay particles to boulders, organic matter as large as trees, and, all too often, human refuse. Commensurate with its geological youth and erodibility, the rivers of the coastal rain forest discharge over 130 million metric tons of suspended sediment into the northeastern Pacific Ocean (Lean et al. 1990)—a mass equivalent to more than 400,000 fully loaded Boeing 747s. Heavier (bedload) sands and fine gravels also move out of the rivers and estuaries and may provide much of the sediment that feeds the beaches along the coastline. Fine sediments settle out in marshes and mudflats at the periphery of estuaries, providing the "building material" that (in combination with organic matter generated by vegetation) maintains and raises marsh elevations and promotes marsh expansion across unvegetated mudflats. Sediment erosion is also a major natural source of inorganic nutrients, although human activities now supply increasing amounts of fertilizers, organic wastes, and other inorganic materials.

Oceanic Forcing

The land/sea ecotone pulses with lunar tides. Tidal ranges (from high to low tide) in the northeastern Pacific Ocean are approximately 10 feet (3 meters) but may increase to 16 feet (5 meters) at the head of deep fjords and inlets, such as those that surround Hecate Strait, and can reach over 25 feet (7.8 meters) in Skidegate Channel (Thomson 1981). The pulse itself, however, fluctuates. As the tidal wave in the northeastern Pacific Ocean propagates northward, it produces a mixture of diurnal tides (one cycle of a high and a low tide per day) and semidiurnal tides (two cycles per day) (Figure 7.6). High and low tides also vary in amplitude. High tides are progressively higher, and low tides progressively lower, over about a week's time; then the cycle reverses and the tidal range decreases (Figure 7.6). The strength of this two-week cycle alternates with the phases of the moon, producing higher ranges (spring tides) during full and new moons and lower ranges (neap tides) on half moons. This tidal pattern creates an abrupt ecotone: there are sharp shoreline gradients in physical stresses, including drying and heating, that strongly influence the distribution and abundance of intertidal organisms. The effect of drying and heating varies geographically, as the potential sunlight and temperature extremes are much higher in the southern latitudes of the coastal temperate rain forest than in the more temperate, fog-shrouded northern latitudes.

Tidal effects are further complicated by the irregular shoreline of the coastal rain forest. And nowhere are these complications more evident than in estuaries. Shallowing and narrow passages contort, magnify, and change the timing of the tidal wave. For example, the wave that is almost symmetrical with the lunar phases at Tatoosh Island, Washington, on the open coast is intensely modified as it progresses deep into a fjord like Hood Canal (Figure 7.6). Similarly, the currents induced by the ebb and flood of the tide are highly modified by features of the land/sea ecotone. The 3.7-km/h ebb or flood current or "tidal stream" at the broad entrance of the Strait of Juan de Fuca is very different from the roaring 9–11-km/h current at the shallow entrance of the Columbia River estuary, especially when the strong currents during spring ebbs combine with high river flow (Figure 7.6). That same energy can also be dissipated by the deep bathymetry of a fjord, as illustrated by the minor tidal stream in southern Hood Canal, or by the gravitational force of freshwater river flow, as illustrated by the truncated tidal currents at Cathlamet, 67 kilometers (42 miles) upstream of the Columbia River entrance, where the flooding tide will stop but not reverse river flow.

Ecological Gradients and Habitats

Abrupt and gradual physical gradients along the coast give rise to diverse biotic communities. The meeting and mixing of water, sediments, and energy from the land and the sea result in dynamic processes that affect not only the terrestrial/marine ecotone but adjacent ecosystems as well. Physical gradients range in scale from the tiny (such as the interface between a plant root and sediments, where nutrient exchange must occur) to the huge (salinity gradients across a large estuary). Some organisms inhabiting the ecotone are permanent residents. Although these often sort themselves out into habitats in distinct parts of the ecotone, even species living in one stable location must adapt to the substantial fluctuations in physical conditions they are likely to experience. Mudflat organisms such as clams and annelid worms, for instance, burrow deep to minimize the chance of being washed away with surface sediments, as well as to avoid the high temperatures and abrupt salinity changes at the mud surface. Other species are transients using the ecotone as a transition between major stanzas in their life history; these species must have physiological and behavioral adaptations for crossing the whole ecotone. At the interface of tidal fresh water and brackish water, for example, juvenile salmon must acclimate to seawater at a critical time during which they must also feed and grow sufficiently to avoid predators further down in the estuary and the open ocean.

Freshwater Tidal Habitats

Freshwater tidal habitats, the landward margin of the ecotone, comprise complex wetland sloughs, forests, and marshes that are saturated but infrequently covered by water except during floods and extremely high spring tides. The freshwater tidal zone can be quite broad in low-gradient estuaries where the tidal wave propagates far inland. Although salt water does not penetrate into

a.

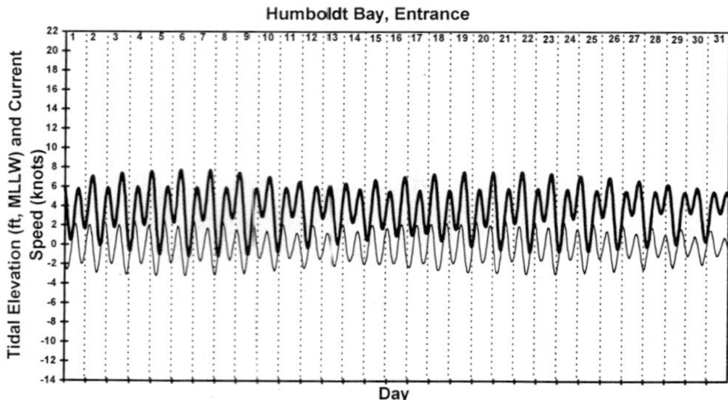
Humboldt Bay, Entrance
Tidal Elevation (ft, MLLW) and Current Speed (knots)
Day

b.

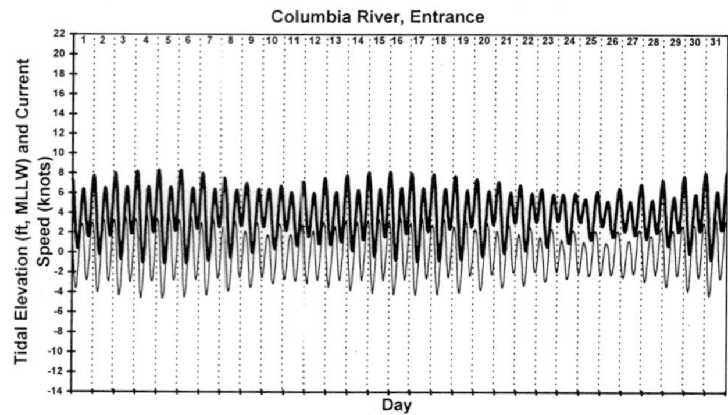
Columbia River, Entrance
Tidal Elevation (ft, MLLW) and Current Speed (knots)
Day

c.

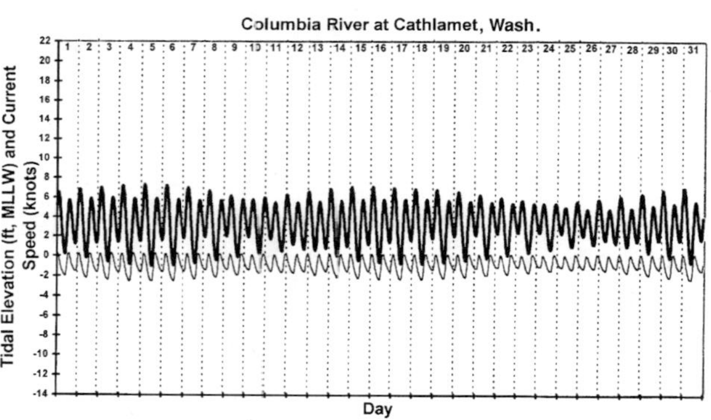
Columbia River at Cathlamet, Wash.
Tidal Elevation (ft, MLLW) and Current Speed (knots)
Day

d.

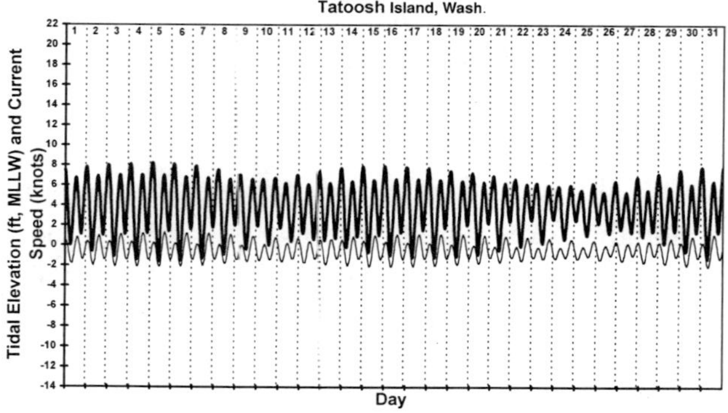
Tatoosh Island, Wash.
Tidal Elevation (ft, MLLW) and Current Speed (knots)
Day

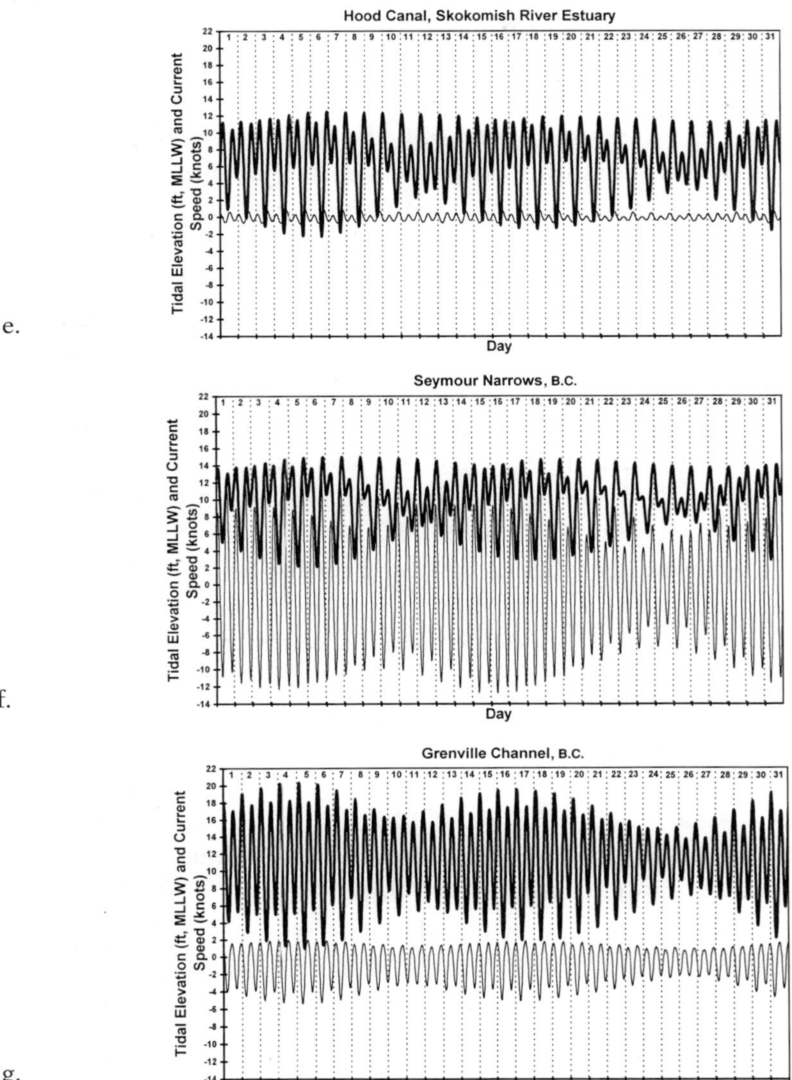

Figure 7.6. Comparison of tidal waves during May 1996 for different locations in the coastal temperate rain forest: (*a*) entrance of Humboldt Bay; (*b*) Columbia River entrance; (*c*) Columbia River at Cathlamet (67 kilometers upriver); (*d*) Tatoosh Island, Washington, at the entrance of the Strait of Juan de Fuca; (*e*) off the Skokomish River delta in southern Hood Canal, Washington. (*f*) Seymour Narrows in Seymour Inlet, mid-coast of British Columbia; and (*g*) Grenville Channel, north coast of British Columbia. The dark line at the top is tidal amplitude (in feet) relative to mean lower low water (MLLW); the thin line at the bottom is current speed (in knots); current direction is up-estuary (flooding) when positive and toward ocean (ebbing) when negative. Tidal amplitude and current are accentuated at the entrances to estuaries and deep in fjords like Hood Canal compared to open-ocean tidal ranges and speeds (at Tatoosh, for example) but decline as they progress upriver. The timing of peak current speed relative to slack flood or ebb also varies according to location. *Source:* Tide and current curves generated using Tides & Currents for Windows software published by Nautical Software, Inc. of Beaverton, Ore.

this realm, the water level does rise and fall with the tide. In some cases the tide actually reverses the river current, as it does for the first 55 kilometers of the Columbia River. Tidal fluctuation sometimes persists to Bonneville Dam more than 200 kilometers from the mouth. Because the river current slackens where it encounters the tidal wave, sediments and organic matter settle out and are often permanently trapped. Floods may seasonally move material accumulated in this manner in the river channel out into the estuary, but material that settles in quieter areas remains.

The forest of the freshwater tidal zone is dominated by Sitka spruce (*Picea sitchensis*), red alder (*Alnus rubra*), and (where flooding is very infrequent) western redcedar (*Thuja plicata*). Crabapple (*Pyrus fusca*), native salmonberry and blackberry (*Rubus* sp.), and currants (*Ribes* sp.) form dense "shrub-scrub" thickets where the forest canopy permits. The understory is most commonly carpeted by the sedge *Carex omnupta*, punctuated by water-parsley (*Oenanthe sarmentosa*) and skunk cabbage (*Lysichitum americnum*). Networks of blind sloughs with little freshwater drainage from adjacent uplands form an arterial system that saturates the whole basin and provides important corridors for fish, wildlife, and detritus. Because of their accessibility, tidal freshwater forests were among the first coastal environments to be harvested, colonized, and extensively altered by European settlers. Except for the coastal rain forest of western Vancouver Island, northern British Columbia, and Alaska, most of these habitats have been deforested, diked, and removed from the influence of rivers and tides. As a typical example, 77 percent of forested, shrub-scrub freshwater tidal wetlands existing in the Columbia River estuary around 1870 had been diked, filled, or dredged by 1980 (Plate 5; Thomas 1983). Today only vestiges of that habitat remain, some found along the Chehalis River in Washington (Figure 7.3).

Freshwater tidal habitats act as the ecotone's "prefilter." Water, along with the associated material and energy transmitted from the landscape, diffuses through the deltas. As a result, the intricate complexes of sloughs, marshes, swamps, and forests become repositories of organic matter of all sizes, especially large woody debris. As this material decomposes, it becomes more available to the estuarine food web. While tidal freshwater deltas may store water and organic material and regenerate nutrients during the wetter periods of the year, in drier seasons they release water, organic matter, and nutrients to the estuary.

Trees and logs in sloughs and creeks also provide extensive refuges for fish and, as well, substrate and food for invertebrate prey. Riparian vegetation is also used by beavers in freshwater tidal creeks to build ponds used by juvenile coho salmon (*Oncorhynchus kisutch*) in winter. Logging, grazing, road building, and other forms of development have severely curtailed the supply of this debris in many rain forest estuaries.

Tidal freshwater habitats appear to be especially important rearing and overwintering areas for coho and chinook salmon (*Oncorhynchus tshawytscha*) (Levings et al. 1995) and for anadromous trout (such as coastal cutthroat [*Oncorhynchus clarki*]). Moreover, sand carried into these habitats by rivers (as in the Fraser River watershed) provides a spawning substrate for eulachon (*Thaleichthys pacificus*) and a rearing habitat for juvenile starry flounder (*Platichthys stellatus*).

Over a longer time scale, the existence of this transition zone between marine and freshwater ecosystems probably influenced the recent evolutionary history of fish faunas of the coastal rain forest. During Pleistocene glaciation, the diversity of western freshwater fishes declined to those few groups that could find refuges and later recolonize (McPhail and Lindsey 1970). Many fish recolonized coastal rain forest watersheds from the sea. Although some freshwater fishes were able to recolonize from inland refugia, the continental divide presented a barrier impenetrable to many; marine fishes established themselves in fresh water, either as anadromous or resident species. To this day, the freshwater fish fauna of the many coastal islands of British Columbia and southeastern Alaska comprise only species that colonized from the sea. In addition to the salmonids, these include lamprey (*Lampetra* sp.), stickleback (*Gasterosteus aculeatus*), and sturgeon (*Acipenser* sp.).

Estuarine Habitats

The truly estuarine midsection of the ecotone is where the salt of the ocean is mixed to varying degrees with fresh water from the land (Figure 7.7). Mixing occurs horizontally—with increasingly salty water from the tidal freshwater realm to the coastal ocean—and to varying degrees vertically—with fresh water on the surface and the denser salt water near the bottom. In many estuaries, the vertical salinity gradient varies with the lunar tidal cycle: stratified during neap tides and mixed during spring tides. Such tidal variation affects the behavior and movement of fishes. In swift rivers, anadromous eulachon appear to swim upstream with the help of the tidal flow. Herring tend to spawn on the neap tides, the time of minimal tidal current in the 29-day tidal cycle (Hay 1990).

Many plants and animals tolerate only certain ranges of salinity and require specific frequencies and durations of coverage by the tide, soil porosity, organic content, and dissolved oxygen levels. Because these factors all change with the coursing of the tides across the land/sea ecotone, habitats blend in a true continuum. Lush Lyngbyei's sedge (*Carex lyngbyei*) and bullrush marshes that line the brackish, low-salinity stretches just downstream of the tidal freshwater zone are gradually replaced by true salt marshes of pickleweed (*Salicornia virginica*) and salt grass (*Distichlis spicata*) toward the ocean.

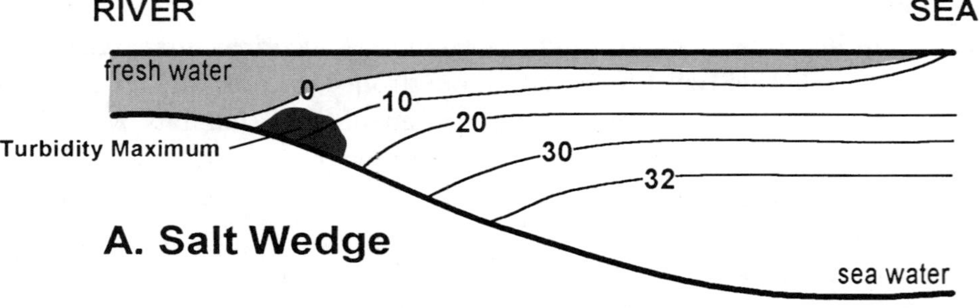

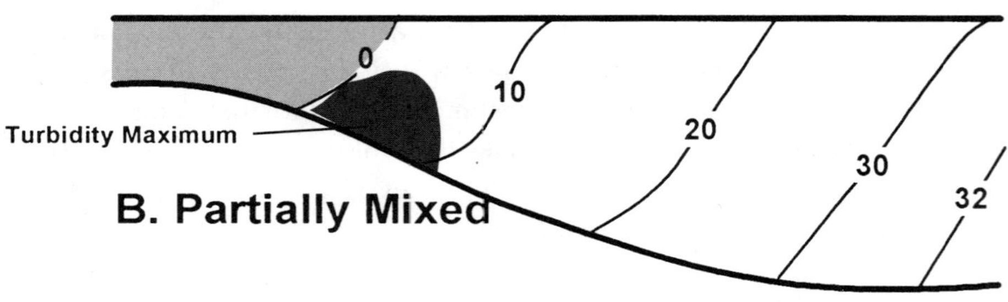

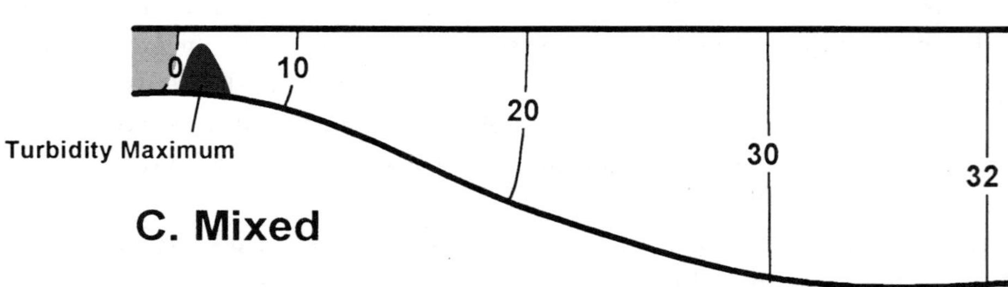

Figure 7.7. Schematic showing (*a*) salt wedge, (*b*) partial mixing, and (*c*) full mixing of salt with fresh water in estuaries. Turbidity maxima, where sediment and organic particles tend to be concentrated, are quite often associated with the upstream extent of salinity intrusion. Many estuaries along the coastal temperate rain forest land margin shift among these three types of circulation over tidal and seasonal cycles, with accompanying variability in turbidity maxima. *Source:* Modified from Postma (1980).

Sediment is trapped in many estuarine habitats, especially in marshes where the vegetation structure reduces the water's motion. These sediments enable ecological succession, since sand and mud transported into the estuary by the river provide a platform for colonizing riparian and marsh vegetation. In estuaries of muddy rivers with high runoff—for example, many of those at the heads of fjords in British Columbia and Alaska—marshes can expand over just a few years. In estuaries receiving comparatively little input of fresh water and sediments, however, such as those on the west side of Puget Sound and the Strait of Georgia, vegetation expansion is very slow and the transition between successive stages of estuarine communities may take decades. Riparian and floodplain forest areas at the undisturbed Stikine River estuary, for example, remained stable over a thirty-year period.

Sediment can be trapped even in turbulent main channels, where circulation features called "turbidity maxima" (areas of higher than normal concentrations of suspended particles) retain sediments in the landward-flowing bottom currents during flood tides (Figure 7.7). Turbidity maxima can be a particularly important feature in ecosystems with strong river outflow, such as the Columbia River estuary, because bottom-oriented zooplankton populations that dominate the estuary's overall food web can use them to keep from being carried out of the estuary and, moreover, can exploit the trapped particulate matter for food (Jones et al. 1990; Simenstad et al. 1990, 1992).

Sediments of estuarine marshes and mudflats also play a critical role in nutrient cycling. Above all, the extent of anoxia (low-oxygen conditions) below the sediment surface, as well as the populations of bacteria and other microbes there, regulate nutrient transformations. Water quality is degraded and ecosystem function can be disrupted when processes regulating nutrient cycling become unbalanced by significantly increased inputs of critical nutrients such as nitrogen or even overall organic matter. Estuaries are the the most fertilized ecosystems in the world by several orders of magnitude (LMER Coordinating Committee 1993). This is the result of natural input from watersheds, atmospheric deposition of nitrate, as well as wastewater discharge and leaching of fertilizers from intensively modified watersheds. Excessive nitrogen input into estuaries can increase the growth of algae (phytoplankton as well as seaweeds)—often far beyond their normal levels. Although few estuaries of the coastal rain forest show evidence of extreme nutrient enrichment (eutrophication), excess nitrogen input can severely affect the the land/sea ecotone through increased turbidity, degradation or loss of eelgrass beds and other submerged vegetation, and ultimately anoxia in the water column.

Transport of large organic debris into estuaries from their watersheds has historically contributed significantly to habitat structure, disturbance, and energy flow (Maser and Sedell 1994). Only 18 to 35 percent of the organic debris delivered to streams is actually transported downriver, but tree litter (leaves, needles, bark, cones), stems, root wads, large branches, and sometimes whole trees do make it into the estuary. This debris feeds estuarine and coastal ocean organisms such as gribbles (isopods; *Limnoria* spp.) and shipworms (*Bankia* and *Terredo*), and the larger components also add habitat complexity where they settle in estuarine wetlands. They also create significant disturbance when they drag across mudflats and marshes or batter against the exposed coastal intertidal zone.

Exposed Coast Habitats

Four types of shoreline habitat are formed by the interplay of coastal geology and the high-energy wave environment of the exposed coast: rocky headlands and seastacks (either highly exposed to wave action or protected behind islands); fine-grained, gradual, dissipative sand beaches; coarse-grained, steep, reflective gravel beaches or barriers; and long relatively flat benches of bedrock overlain by sediment, cobbles, and boulders (Dethier 1990). Each habitat type is characterized by a different set of plants, animals, and ecological processes.

The abiotic factors affecting these coastal communities reflect the interactions of the land and sea. First, substrate type (controlled by erosion and sediment delivery) is the primary determinant of intertidal community structure. Second, wave exposure affects communities on rocky shores directly, and organisms in soft-sediment habitats indirectly, by controlling the size and motion of particles. Third, sediment scouring is a significant determinant of rocky shore communities; even plants and animals adapted to high wave energies are removed by severe storms. Fourth, dilution of surface salinities by freshwater input from estuaries and by rivers that do not form enclosed estuaries (such as Washington's Hoh River) may affect coastal organisms, their planktonic larvae, or key planktonic prey of fish and seabirds. And fifth, coarse woody debris derived from rivers and from coastal landslides smashes organisms on rocky shores and causes sediment to build up on beaches. Drift logs on beaches temporarily protect beach and cliff sediments from waves; on sandy shores, they also provide a barrier to wind-transported sand (McKay and Terich 1992; Maser and Sedell 1994).

The biotic factors structuring coastal communities have been thoroughly studied along this shoreline, in some places for as long as thirty years. On rocky shores, community structure is often influenced or created by the organisms themselves; plants such as surfgrass (*Phyllospadix* spp.—related to eelgrass), kelps, and other algal beds, as well as animals such as mussels and large barnacles, cement themselves to the rock and create a three-dimensional structure that provides habitat for many other species. Sandy shores, unstable under constant wave action, generally lack such biostructures except for the temporary tubes of various worms. Competition on rocky shores is severe. Processes such as sand scouring, log smashing, and predation that create open space for settlement and growth are critical determinants of the types and diversity of life found on rocky shores (Paine 1966; Dayton 1975). Who wins and who loses in the competition for space depends in large part on the susceptibility of different organisms to wave action. Because disturbances are frequent and unpredictable, these ecotone habitats are characterized by a mosaic of communities in various stages of development.

Predation and herbivory are important on all shore types (except perhaps on gravel beaches where too few organisms may be available to support consumers). On rocky shores, predation on the highly competitive California mussel (*Mytilus californianus*) by seastars (*Pisaster ochraceus*) and dogwhelks (*Nucella* spp.) is a key determinant of community structure. Feeding by chitons (e.g., *Katharina tunicata*) and limpets (*Lottia* spp.) affects types and diversities of algae. The effects of predation on other types of shore communities, and the effects of terrestrial or subtidal consumers on rocky shore communities, are less well known. Humans and other terrestrial mammals may be important predators in some areas.

A final biotic process of key importance on shorelines is recruitment—the ability of plants and animals to get to new places on the shore, settle, and survive to join an adult population. Many invertebrates and algae have a planktonic dispersal phase ("propagule") during which they are carried more or less passively by longshore currents or (to their detriment) out to sea away from their natural habitat. Oceanographic processes—including upwelling, the relative strength of onshore and offshore flows, and the direction of currents—can thus affect recruitment by controlling the supply of propagules back to the shore. Links with the terrestrial realm may also affect shoreline organisms during this vulnerable stage. Despite increasing efforts to understand the coupling of oceanographic and nearshore processes, these interactions are poorly understood. Reducing the surface salinity of the Columbia River freshwater plume, for example, could affect propagule survival considerably to the north (in winter) or south (in summer) of the estuary itself. Sediment and wood transported down rivers has unknown but possibly significant effects on propagules as well.

Critical Foraging Habitats and Food Web Interactions

Freshwater tidal habitats, estuaries, and exposed shorelines are all highly productive places supporting large biomasses and diversities of consumers, including fish, birds, and small mammals valued by humans. These upper-level consumers also illustrate the links between terrestrial and marine ecosystems that characterize the coastal temperate rain forest. Although these linkages contribute to overall productivity, they also make the organisms in the land/sea ecotone susceptible to disruption by natural or human-caused events.

The principal "fuel" for most of the organisms in estuarine and freshwater tidal habitats is detritus—dead and decaying organic matter—made up primarily of plant material (leaves, needles, even trees) either exported from the watershed, generated within the estuary itself (macroalgae, eelgrass), or washed in from the open ocean (Figure 7.8). Both tides and rivers deposit organic debris in riparian and high intertidal zones, where it is fragmented by animals and finally decomposed by bacteria. Detrital decomposition is especially intensive in sloughs and protected marshes. Microscopic algae (diatoms) on the mud and sand flats and macroalgae (kelps and other seaweeds) colonizing hard substrates can also be important. In fjords, which have physical and ecological structures more like those of the open ocean, direct grazing on living plants is more common.

The invertebrate prey of estuarine fish and other predators either originate in the estuary or are transported from upriver. Marshes, riparian shrub-scrub, and wetland forests of tidal freshwater and estuarine habitats provide structural habitat for "fish-food" invertebrates and, in some situations, for the fish themselves. For instance, juvenile chinook salmon in open estuarine waters sometimes feed on drift insects such as aphids, ants, bark lice, spiders, and mites from upstream riparian habitats. As these drift organisms are commonly concentrated along tidal fronts, it is likely that juvenile salmon seek out these physical features to feed on the accumulated prey. Juvenile chinooks also consume midge larvae and adults, which are much more abundant in sedges than in sand or mud flats. The midge larvae not only feed on the plant detritus but may use the vegetation to climb out of the water when they metamorphose to adults.

Salmon take advantage of the seasonal dynamics of prey production, runoff, and spring tide patterns that generate opportune feeding conditions at the freshwater/seawater interface. Studies at the Carnation Creek estuary in British Columbia have illustrated that river and tidal currents trap and concentrate terrestrial, estuarine, and marine organisms in tidal fronts and riptides. Tschaplinski (1987) has found that diurnal rates of drift in the estuary in late June through early July exceed rates in the stream by factors of between 2 and 4. As a result, species that depend on these floating aggregations of living things (neuston)—species like juvenile coho salmon that are rearing in the brackish transition zone—tend to consume more prey in the "entrapment zones" of estuaries than in the rivers they just traversed.

Intertidal and shallow subtidal areas of the terrestrial/marine ecotone are vital feeding grounds for birds. The resident fish community in particular provides a relatively stable food base (compared to the seasonally migratory

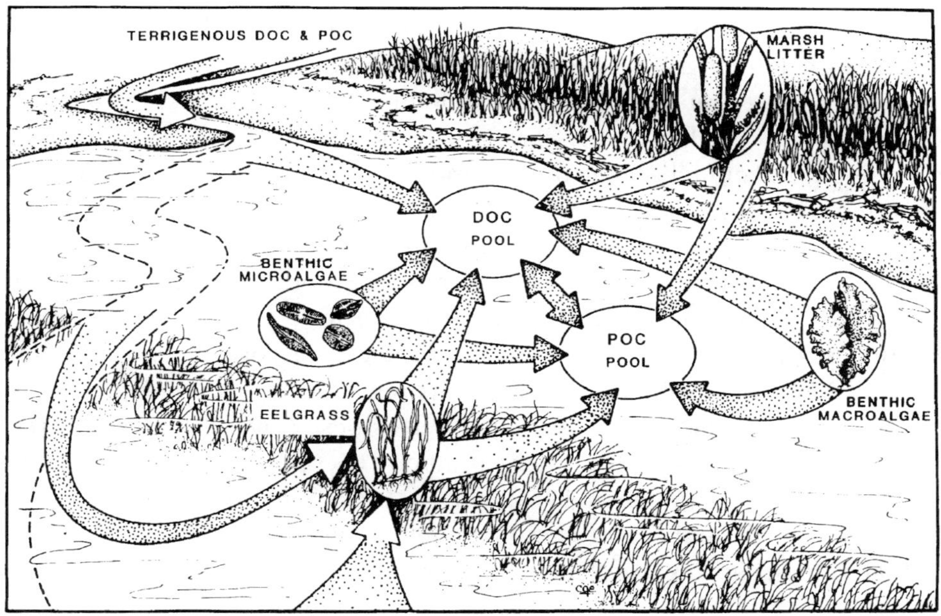

Figure 7.8. Sources of detritus to estuarine food webs. DOC, dissolved organic carbon; POC, particulate organic carbon. *Source:* Simenstad and Armstrong, unpublished.

fishes) for marine seabirds and other coastal predators (Hay et al. 1989, 1992). Key prey include sculpin (*Cottidae*), juvenile rockfish (*Sebastes* spp.), greenling (*Hexagrammus* spp.), and perch (*embiotocids*). Seabirds, such as gulls and alcids, feed in estuaries and in the open ocean but spend much of their lives at sea. Gulls feed extensively along the shoreline in both rocky and soft-sediment environments where they can significantly affect abundances of mussels and gooseneck barnacles (*Pollicipes polymerus*) on rocky shores (Irons et al. 1986; Marsh 1986; Wootton 1994), while most other seabirds feed further out to sea. Marbled murrelets forage at sea but nest exclusively in old-growth rain forests up to 80 kilometers from the ocean. Seabird colonies on offshore islands and seastacks accumulate significant amounts of nutrients—initially gathered as fish offshore and released nearshore as guano, which may increase phytoplankton production and alter intertidal communities under certain conditions (Bosman and Hockey 1986; Wootton 1991). The most abundant species that form colonies along the temperate rain forest coastline are Cassin's auklets, storm petrels, common murres, gulls, Caspian terns, and cormorants (Strickland and Chasan 1989).

Shorebirds feed mainly along intertidal and nearshore marine environments, both in estuaries and on the coast. Small flocks of oystercatchers, turnstones, tattlers, and sandpipers may significantly reduce populations of limpets (Frank 1982) and dogwhelks (Feare and Summers 1986) along rocky shores. Thousands of shorebirds roost and forage on sandy beaches during spring and fall migrations; common species include western sandpipers, sanderlings, and dunlin. Millions of shorebirds rely on the rich resources of coastal estuaries to fuel their migrations (Butler and Vermeer 1994). Western sandpipers and dunlin are the most numerous species, but over fifty species of shorebirds use the estuaries every year. Caspian terns, blue herons, gulls, cormorants, and Canada geese also nest in coastal estuaries, as do snowy plovers, bald eagles, and peregrine falcons. The impact of shorebird feeding on local invertebrate populations is poorly understood because of the difficulty of conducting experiments with these predators and because of the naturally patchy distribution of their worm and crustacean prey. By selecting larger prey, shorebirds can significantly reduce the overall densities or average sizes of certain invertebrate populations, especially in muddy (as opposed to sandy) habitats (Schneider 1978; Quammen 1984; Wilson 1989). The long-term and large-scale consequences of this predation are unknown.

Waterbirds, such as ducks and geese, are found nearshore on the coast and in estuaries. Many breed in freshwater habitats but overwinter and stage for migrations in coastal estuaries. Scoters, for example, feed on mussels and barnacles while they overwinter in the Strait of Georgia (Vermeer and Levings 1977). Black brant rely on eelgrass for overwintering in these estuaries—indeed, their annual arrival is sometimes celebrated by coastal communities. Hundreds of thousands of ducks of many species feed and rest in estuaries during fall migration. Truly terrestrial birds, such as crows and eagles, also forage extensively at the terrestrial/marine ecotone. Crows forage for clams in mudflats (Richardson and Verbeek 1986) and for buried worms in sandy areas, but their overall impact on prey populations has also not been studied.

Use of the land/sea margin by mammals is much less well known. A wide variety of mammals forage on rocky and sandy shores at least occasionally. Deer are often seen consuming algae in Olympic National Park. Mink take small fish from tidepools. River otters consume fish as well, mostly in the shallow subtidal zone, and sea otters come into the nearshore region from the open sea to consume urchins, bivalves, crabs, and other prey. Raccoons gather shorecrabs and fish from rocky areas and dig for bloodworms in sandy areas. Bears have been observed consuming beach grasses and scraping barnacles off the rock with their paws.

Other mammals are among the most disruptive exotic species that have been introduced into the coastal rain forest. Introduced rat species, shown elsewhere to have a significant impact on certain intertidal mollusks (Navarrete and Castilla 1993), are part of a food web (involving mussels and snowy owls) that concentrates heavy metals in the Fraser River estuary (Brown et al. 1977). Even feral horses and cattle have an impact on salt marshes in some areas (Furbish and Albano 1994). With the exception of the substantial effects of sea otters on shallow subtidal areas, the impacts of mammals on prey species at the land/sea margin are unknown.

Life History Transitions

Anadromous fishes—those that spawn in fresh water but migrate to salt water for a large segment of their lives—directly cross the terrestrial/marine ecotone during critical stages of their life histories or concentrate there. Because they traverse the ecotone and pass through multiple life-history steps in the process, anadromous salmon and smelts are most obviously characteristic of the bioregion, but many other species, such as green and white sturgeon (*Acipenser medirostris, A. transmontanus*), also move between fresh and salt water.

The transitions at both ends of the Pacific salmon life cycle are abrupt (Groot and Margolis 1991). Once they enter the tidal freshwater zone on their initial migration to the sea, juvenile salmon must be capable of adapting rapidly to an extreme change in physiological conditions (from preventing loss of salts to preventing loss of water from the body), water flow, habitat, food sources, and predators. Different salmon species have evolved different approaches to dealing with this variation (Table 7.1). While chum salmon (*O. keta*), several forms of chinook salmon, and a few sockeye (*O. nerka*) stocks spend up to several months in estuaries, pink salmon (*O. gorbuscha*) delay their ocean migration much less, and coho, other forms of sockeye, chinook, and steelhead (*O. mykiss*) almost seem to dash to the ocean (Simenstad et al. 1982; Healey 1982; Groot and Margolis 1991). The duration of residence in estuaries by these "outmigrating" juveniles appears to be strongly related to both their size and their ability to adapt rapidly to the physiological stresses of saline waters: smaller "fry" (such as "ocean type" chinook) tend to spend much longer in their transition through the ecotone than larger "smolts" (such as "stream type" chinook) that make a marked physiological transition as they migrate downriver. In fact, in some watersheds coho and steelhead fry spend considerable time rearing in tidal freshwater habitats during the winter, using that portion of the ecotone long before they actually migrate to the ocean in the spring and summer.

Patterns of habitat use during estuarine rearing also vary by species and, to some interdependent degree, by fish size: fish move into deeper, more open bodies of water as they grow larger and approach the open ocean. Fry of chum, pink, and chinook salmon are commonly found in very shallow waters, particularly in marshes, mudflats, and eelgrass meadows. Larger chinook, coho, and sockeye salmon are more prevalent in open waters of the main distributary channels and along tidal fronts. Species differences in diets while crossing the

Table 7.1. Different Life History Strategies of Anadromous Pacific Salmonids

Species	Freshwater Rearing (years)	Tidal Freshwater Rearing (months)	Estuary Rearing (months)	Ocean Rearing (years)	Estuarine Habitat	Estuarine Prey
Pink salmon (*Oncorhynchus gorbuscha*)	<0.02	<0.02	1–4.5	1	shallow water along beaches, eelgrass; ultimately open surface waters	epibenthic harpacticoid/pelagic calanoid copepods; epibenthic gammarid amphipods and cumaceans; pelagic larvaceans
Chum salmon (*Oncorhynchus keta*)	<0.02	<0.02	1.25–5.75	2–5	shallow water along beaches, eelgrass, mudflats; ultimately open surface waters	epibenthic harpacticoid/pelagic calanoid copepods; epibenthic isopods and cumaceans; pelagic decapod larvae; drift adult insects; pelagic amphipods; larvaceans
Coho salmon (*Oncorhynchus kisutch*)	<0.02–2	1–6?	1.25–3.75	0.3–1.3	sloughs and marshes, beaches; ultimately open surface waters	pelagic decapod larvae; epibenthic amphipods; pelagic euphausiids and calanoid copepods; fish (herring) larvae; drift adult insects

Steelhead (*Oncorhynchus mykiss*)	1–3	?	short?	1–7	main distributary channels, open waters	amphipods; fish
Sockeye salmon (*Oncorhynchus nerka*)	1–2	<0.02	3.0	2–3	main distributary channels, open surface waters	pelagic euphausiids; amphipods and decapod larvae; epibenthis amphipods and shrimp
Chinook salmon (*Oncorhynchus tshawytscha*)	"ocean type" <0.02–0.25; "stream type," 1–2	"ocean type," <0.02; "stream type," 1–6	1.5–7.25	1–5	"ocean type": salt marshes, gravel beaches mud/sand flats, ultimately open waters; "stream type": distributary channels and open water	epibenthic insect larvae and amphipods; drift adult insects; pelagic decapod larvae and amphipods; fish (herring, surf smelt) larvae; mysids

ecotone resemble an evolutionary "partitioning" of the available prey populations (Table 7.1). While in shallow-water habitats, chum, pink, and chinook salmon fry feed on small, epibenthic crustaceans and insects associated with bottom sediments and algae; chinook smolts feed extensively on amphipods, fish larvae, and drift insects, even when they are out in open fjords and inlets; coho and larger pink and chum salmon consume plankton such as calanoid copepods and crab larvae. Salmon feeding focuses on concentrations of prey—such as dense tube mats of amphipods on mud and sand flats, swarms of epibenthic harpacticoids in eelgrass meadows, or patches of zooplankton or neuston or drift (insects) aggregated along tidal fronts. Many of these prey originate in habitats in the land/sea margin itself, as in salt marshes and eelgrass meadows.

Smelts, Pacific herring (*Clupea pallas*), and Pacific sand lances (*Ammodytes hexapterus*) are commonly termed "forage fishes" because, being small and silvery and occurring in large, dense schools, they are exceedingly important to coastal food webs as prey of larger fish, birds, and marine mammals. The smelts (family Osmeridae) are small fishes distributed through circumpolar regions of the Northern Hemisphere. The Pacific coast of North America is rich in the numbers of smelt species, although the biomass of most of these species is low relative to the capelin of the Bering Sea and North Atlantic Ocean. Six genera are represented by ten species (Table 7.2); of these, seven are closely associated with the rain forest coast.

Most smelts are marine, but longfin (*Spirinchus thaleichthys*) and rainbow smelt (*Osmerus mordax*) and eulachon are anadromous; these spawn in rivers or lakes, and their larvae and young migrate or are transported to estuaries (where they may rear for at least a year) or into the coastal sea. The nonanadromous species are common, even concentrated, in shallow coastal waters or in estuaries; nonanadromous smelts and sand lances generally spawn on fine gravel to sand beaches, often in the intertidal zone or the open surf.

Smelts as a group may be particularly sensitive to climate variability and thus may be good indicators of broad climatic changes that affect the whole coastal rain forest region. For instance, in 1994 eulachon populations declined sharply in several rivers, including the Columbia, Fraser, and Klinaklini (Knight Inlet), even though the spawning times vary by more than three months among these rivers. The geographic distribution of eulachon resembles that of the coastal rain forest itself. In part, this similarity may be explained by the mutual dependence of the fish and forests on the heavy coastal precipitation. Populations of eulachon seem to occur predominantly in watersheds that drain the coastal mountains, where rivers usually have a pronounced spring freshet. Eulachon are not found, on the contrary, in the smaller rivers of the coastal islands. A general decline in eulachon across the

Table 7.2. Distribution of Smelts, Herring, and Sand Lances

Species[a]	Common Name	North Pacific Distribution
Smelts (Osmeridae):		
Mallotus villosus	capelin	Korea to Strait of Juan de Fuca
Thaleichthys pacificus	eulachon	Monterey Bay to Bering Sea
Spirinchus thaleichthys	longfin smelt	Monterey Bay to Prince William Sound
Spirinchus starksi	night smelt	Central California to SE Alaska
Hypomesus pretiosus	surf smelt	Long Beach to Prince William Sound
Hypomesus olidus	pond smelt	Gulf of Alaska to N. and Arctic
Hypomesus transpacificus	delta smelt	Sacramento–San Joaquin river system
Osmerus mordax dentex	rainbow smelt	Vancouver Is. to Arctic
Allomerus elongatus	whitebait smelt	San Francisco to Vancouver Is.
Herring (Culpeidae):		
Culpea harengus pallasi	Pacific herring	Korea to N. Baja
Sand lance (Ammodytidae):		
Ammodytes hexapterus	Pacific sand lance	Sea of Japan to S. California

[a]The marine areas adjacent to the coastal temperate rain forest support more smelt species than anywhere else in the world. Four of the species have distributions confined to areas adjacent to the coastal temperate rain forest.

region was accentuated by the drastic 1994 declines mentioned above; stocks in several other rivers (including the Skeena in British Columbia and the Taku in southeastern Alaska) may also have experienced this decline.

Pacific herring rely on the terrestrial/marine ecotone to varying extents throughout their geographic range (Korea to northern Baja California), although all spawn on shallow subtidal kelps and eelgrass at seemingly specific, obligatory sites. This site fidelity may be related to the adaptive significance of larval transport by longshore currents into coastal bays and estuaries with high food resources. Herring spawning areas comprise a substantial portion of some sections of the coast—for instance, about 3000 of the approximately 20,000 kilometers of tidal coastline in British Columbia have been used for spawning by herring during the last fifty years (Hay and Kronlund 1987). In any one year, herring spawn along 300 to 500 kilometers of coast. Spawning is sometimes strongly affected by upland influences. Throughout their southern distribution (from Neah Bay, Washington, to San Francisco Bay), herring almost always spawn in the protected waters of estuaries, perhaps seeking shelter from the exposed surf. In much of their northern range (from southern British Columbia to Prince William sound, Alaska), by contrast, herring spawn and rear larvae and juveniles mainly in the protected marine waters of the many inlets and fjords. Here they also are influenced by terrestrial effects, as most herring populations appear to spend their juvenile stages in nearshore habitats, often within a few hundred meters of shore.

Dungeness crab (*Cancer magister*) populations also depend on terrestrial/ marine links at the ecotone. Although adult crabs live offshore where they are captured commercially, and generally spawn outside estuaries, many larval crabs find their way into coastal estuaries to rear for their first year or two before returning to the ocean. Studies along the central and southwestern Washington coast (Grays Harbor and Willapa Bay) have indicated that crabs in the estuary grow significantly larger during their first year than those rearing offshore (Armstrong and Gunderson 1985; Gunderson et al. 1990). The survival of the crabs during the critical few months after they metamorphose into their bottom-dwelling form depends on refuge habitats (bivalve shells, wood litter, eelgrass, macroalgae) on intertidal flats.

Understanding the Consequences of Altering Ecotone Processes

Human activities alter the exchange of material between land and sea throughout the coastal rain forest, affecting many processes at the ecotone. Because both terrestrial and oceanic sources of material vary naturally, nutrient cycles and plant and animal populations along the rain forest coast tolerate a range of conditions. Sediment transport and deposition in estuaries and on ocean beaches, for example, depend on both the sediment source and the energy available to move it. Human land-use practices can increase sediment accretion in estuarine and coastal habitats beyond its natural range of variation. Conversely, diverting or depressing peak river flows can reduce sediment delivery and increase erosion.

Clearcut logging increases peak river flows and the amount of suspended sediment they carry. During rain-on-snow events, runoff from clearcuts is between 21 and 90 percent higher than from comparable naturally forested sites in the Oregon and Washington Cascade Range (Harr 1986; Berris and Harr 1987). Mass soil movements or "wasting events" are more frequent and more intense where slopes are disturbed by construction of logging roads (Sidle et al. 1985; Swanson et al. 1987; Naiman et al. 1992). Beschta (1978) has documented threefold increases in river-borne suspended sediments from the cumulative impacts of road building, logging, and subsequent slash burning associated with clearcuts (Figure 7.9). The cumulative effects of farming and urban development, although less dramatic, are similar.

Modifying river flows is likely to affect processes of the land/sea ecotone in even more fundamental ways. The Columbia River estuary represents an important example. Historically, freshwater runoff in the Columbia peaked in

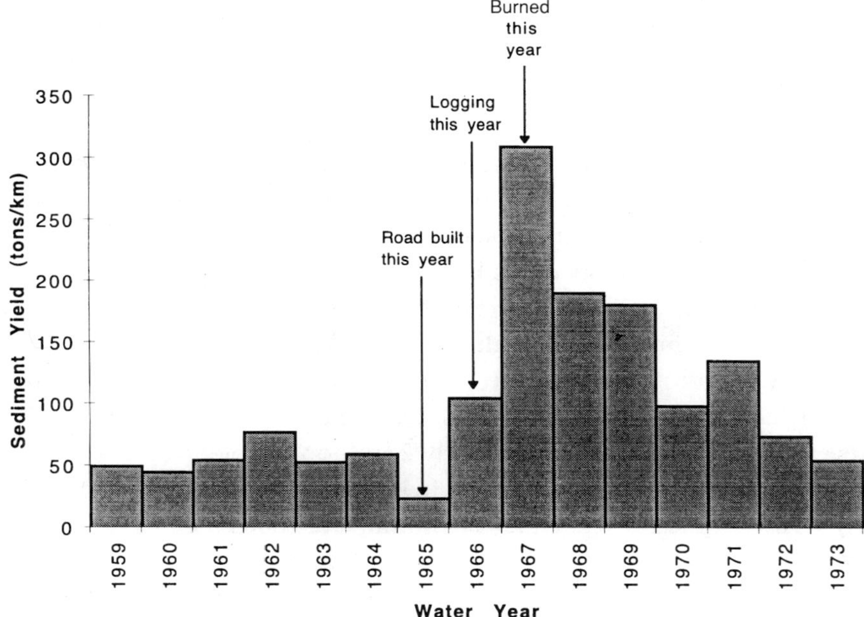

Figure 7.9. Increased annual suspended sediment yield from Needle Branch of the Alsea River watershed, Oregon, after road building and 82 percent clearcut logging in 1965–1966. *Source:* Modified from Beschta (1978) and Scrivener (1987).

the spring, as it does for most intact snowmelt rivers that drain high mountain ranges and interior basins. Today this water is stored behind major dams until the normally low-flow summer and early fall season; peak releases are allowed in winter, when demand for hydroelectric power is highest. Although these changes in the timing and amount of runoff have probably altered the freshwater tidal habitats of the upper Columbia River estuary, no baseline data exist for comparison and the responses by biological communities in the affected wetlands are unknown.

Losses of wetland and other habitats have altered the ecological integrity of estuaries in many parts of the coastal temperate rain forest. Urbanization and diking in the deltas and estuaries have significantly reduced marsh, riparian, and floodplain forest habitats, particularly in Washington and southwestern British Columbia. The Stikine, Fraser, and Columbia river watersheds (Figure 7.1) illustrate contrasts in ecological integrity along the terrestrial/marine ecotone.

The Stikine River and its estuary in southeastern Alaska offer the epitome of ecological integrity: there is little development in the estuary proper (del Moral and Watson 1978); there is no evidence of impacts from upland forest harvest; and the river is unimpeded by man-made structures. Brackish sedge marshes in the estuary expanded by 522 hectares between 1948 and 1979, while the upriver riparian communities remained stable (Beak Consultants Ltd. 1981). Juvenile salmon rear extensively in the freshwater tidal creeks. Coho salmon fry are particularly abundant and appear to depend on the riparian vegetation. The marshes alone on the Stikine annually provide about 800 metric tons of organic carbon to the estuarine ecosystem (Beak Consultants Ltd. 1981).

The Fraser River estuary in southwestern British Columbia, by contrast, has been extensively urbanized and diked. In part because freshwater spawning habitat in the upper watershed has been protected, the Fraser supports some of the largest salmon runs in the world. Most of the food consumed by young salmon when they migrate through the lower Fraser is produced in the riparian and wetland margins of the river. But the structure of this estuarine ecosystem has changed dramatically since the arrival of Europeans (North and Teversham 1984). At present, nearly 2 million people live next to the estuary. Almost all of the floodplain forest in the lower river has been cleared. On the North Arm of the Fraser, surveys from the late 1800s indicate nearly 3000 hectares of wetlands and floodplain forests but, as in the Columbia River (Plate 5), development of the delta has reduced the wetland area to some 109 hectares today—equivalent to a 96 percent reduction in the contribution of organic carbon to the estuary. In other parts of the estuary, human activities have eliminated marsh habitat (Thom 1997).

We do not fully understand the implications of these changes. We do know that restoration of estuarine wetland habitat in the Fraser will be difficult because potential recovery sites along the river's margins are constrained by development. About 6 hectares of marsh habitat have been restored since 1985, but this gain came at the expense of other viable habitat, such as mud and sand flats (Langer et al. 1994). Although management practices have slowed the loss of the habitat in the lower river, new challenges to understanding and managing vital freshwater habitats arise as urbanization spreads in the upper 80 kilometers of the estuary. Industrial and urban development have already eliminated numerous tidal freshwater creeks on the floodplain, sacrificing considerable juvenile coho rearing habitat. Further upriver, but still within tidal influence, native habitats such as cottonwood stands on gravel bars still persist. These and other riparian remnants must be identified and conserved before they, too, disappear.

Although the lower Columbia River and its estuary are not so urbanized as the Fraser River delta, dredging and control structures to maintain a navigation channel have modified the original flow patterns. Marshes along the estuary's shorelines remain capable of producing detritus and cover to support midges and other invertebrate fish food. Mud and sand flats and emergent marshes, however, have been lost from some areas (Plate 5). Projects to restore or establish marsh habitats on sand islands created from dredged material have been under way for a number of years, but it is too early to judge the long-term success of these endeavors. Sherwood et al. (1990) and Simenstad et al. (1992) have proposed that the detritus supplied to the Columbia River estuary from upstream is qualitatively different from the detritus produced by its estuarine wetlands (Figure 7.10). Detritus arising from phytoplankton produced in the reservoirs behind hydroelectric dams upstream now enters the estuarine ecosystem, where it potentially replaces detritus formerly supplied by wetland habitats that have been destroyed. The consequences of this difference in detrital source for the function of the Columbia River estuary are not known.

The same processes that promote settling and trapping of sediments and organic matter can also result in the accumulation of toxic compounds that drain from developed landscapes—especially compounds such as pesticides, heavy metals, PCBs, and organochlorides that tend to be associated with organic particles. For this reason, inputs of pollutants from heavily altered watersheds like the Columbia River may directly affect estuarine organisms quite apart from the consequences of direct modification of the estuary itself. Toxicants accumulated in the estuary enter food-web pathways and contaminate fish and cause reproductive failures of top-level predators such as eagles and ospreys.

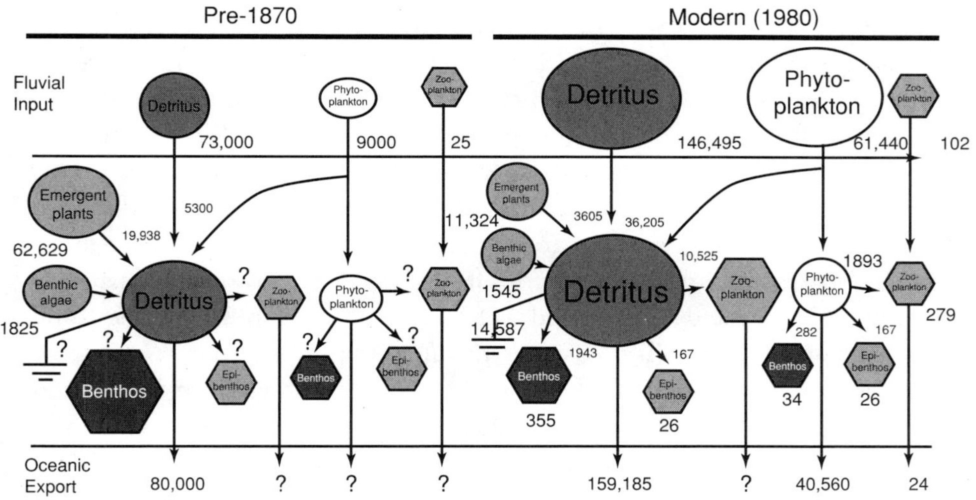

Figure 7.10. Postulated historical shifts (pre-1870 vs. 1980) in organic carbon sources (10^6 metric tons of carbon per year) and pathways for the detritus-based food web in the Columbia River estuary. *Source:* Adapted from Sherwood et al. (1990) and Simenstad et al. (1993).

Adapting Management to Natural Variability

The mosaic of habitats at the terrestrial/marine ecotone of the coastal rain forest depends to some degree upon chronic levels of natural disturbance. The rain forest coast is structured by pulses from freshwater floods, winter storms, and other unpredictable events. Floods erode and accrete sediment, relocate channels, extract and deposit large woody debris, and push fresh water far into the ocean. In the process, fine sediments, nutrients, and large woody debris are deposited in marshes; coastal barrier spits are breached by the ocean; and whole new habitats become available for colonization by plants and animals. When natural floods are suppressed, however, habitats stabilize, their diversity declines, and sediments accumulate in estuaries.

Natural processes in the terrestrial/marine ecotone characteristically operate at scales in space and time that differ sharply from the localized, short-term focus of most resource and habitat management efforts. Management of the social, economic, and cultural totem of the coastal rain forest—Pacific salmon—illustrates the need to consider the large-scale and long-term processes that influence the stocks available for commercial, recreational, or subsistence harvest. The resilience of salmon stocks depends on the cumulative and interactive effects of harvest, habitat degradation (from logging, hydroelectric power generation, irrigation withdrawals), habitat loss (of estuarine wetlands and riparian stream corridors), and natural climate cycles that alter freshwater conditions (stream discharge) and marine conditions (upwelling, salinity, temperature). A prime example of these interactive effects is the Carnation Creek watershed in British Columbia. Holtby and Scrivener (1989) describe how severe habitat alteration and adverse climatic conditions have magnified the effect of high harvest levels on the variability in coho and chum salmon abundance. The two species differ in their responses: chum salmon are more susceptible to habitat alteration but less influenced by climatic variability; coho can better tolerate habitat alteration but prove more vulnerable to variability in ocean conditions. Salmon, however, constitute only a few links in the dynamic food web of the land/sea interface; the same sources of natural variability cascade through the entire web. We cannot manage salmon, and anticipate the effects of our harvest, without taking into account natural variability and the tendency of habitat alteration to magnify impacts.

Vulnerability to Establishment of Exotic Species

The geological youth of the coast and the frequent disturbances experienced along shorelines continually generate new habitats that may be particularly vulnerable to invasion by exotic species. Additional disturbance or stress due to human activities may further increase the vulnerability of these habitats to the establishment and spread of invasive species. And where human activities have pushed native species toward extinction, exotic species may be especially likely to take hold (Vermeij 1991).

The settlement and development of estuaries, along with increases in ship traffic around the world in the past two centuries, have created many opportunities for exotic species to invade niches created by disturbance and maintained by human activity. Organisms piggybacking on oysters or other species favored for aquaculture have been introduced to many estuaries in the coastal rain forest in the ballast water of ships. Their impacts are mostly unmonitored. Carleton and Geller (1993) found 367 exotic taxa in ballast water of ships docked in Coos Bay, Oregon, but have not sought to identify those that have actually established populations in the bay or evaluate the ecological consequences of their arrival. An introduced Japanese clam, *Corbicula manilensis*, now dominates the tidal freshwater and brackish estuarine benthic community of the Columbia River estuary. A pelagic copepod, *Pseudodiaptomus inopinus*, introduced into the Columbia River estuary during the last decade is now one of the more common zooplankton species (Cordell et al. 1992). It appears to be a common prey of the American shad (*Alosa sapidissima*), populations of which have increased dramatically over the last century since its intentional introduction to the West Coast.

Deliberate introductions of the Eastern oyster (*Crassostrea virginica*) and Japanese oyster (*C. gigas*) decades ago were responsible for inadvertantly introducing at least twenty-five other species of plants and macroinvertebrates to Willapa Bay, Washington (Hedgpeth and Obrebski 1981). Zipperer (1996) has recently estimated that 40 to 81 percent of the total density, and 28 to 39 percent of all species, of benthic macroinvertebrates found in that bay may be exotics. One introduced species of particular concern is the eastern Atlantic smooth cordgrass (*Spartina alterniflora*), which arrived in the bay a century ago and is spreading very rapidly and accelerating the buildup of mudflats. Another troubling arrival, Japanese eelgrass (*Zostera japonica*), is rapidly colonizing mudflats from British Columbia to California. Predicting the ecological responses to these introductions, and devising appropriate management approaches, is not simple.

Many introduced species—for example, striped bass (*Roccus saxatilis*) and American shad—have become accepted as valued components of coastal ecosystems. Many exotic species have detrimental effects on populations of native species through competition for space and food; some, such as *Spartina* spp., may alter whole communities. Yet even spartina may have beneficial effects, including creation of habitat or food for native fish and wildlife. Despite its different growth form and distribution (intertidal elevation) in estuaries, Japanese eelgrass is accorded the same habitat value by some regulatory agencies in Washington State as the valued native eelgrass, *Zostera marina*. Fish in Willapa Bay feed on nonnative benthic crustaceans. Nonnative American shad in the Columbia River estuary support a valued recreational fishery and have become a significant component in the diet of many larger predators.

Measures used to control exotic species, such as aerial herbicide and pesticide applications, can pose hazards to nontarget species, further complicating management efforts. In disturbance-prone settings like estuaries, invasions by exotic species may be amenable to control only by reducing the mode and frequency of introductions—a problem that may require international agreements on ballast water disposal and the transport and release of living material into coastal waters.

New Approaches to Management

The complex interactions among climate, freshwater runoff, and the transport of sediments, nutrients, organic matter, and organisms along the temperate rain forest shoreline determine the production of valued natural resources. Yet present-day resource management tends to ignore how these factors vary in space and time and, moreover, how human actions affect important processes at the land/sea ecotone.

Our ignorance of how the land/sea interface functions, and the variability of the complex interactions that affect it, pose a challenge to the management of the coastal rain forest region. Our base of knowledge about specific organisms (especially salmon) is reasonably sound, but our appreciation of their dependence on large-scale and long-term processes is limited. We often focus on small questions—such as the discrete life stages of Pacific salmon rather than their entire life cycle. Both scientists and managers must expand their viewpoints to the appropriate scales of space and time that influence the processes and resources of the terrestrial/marine ecotone. We need to approach the land/seascape holistically, perhaps by using geographical information system (GIS) databases and models, which can build and test various management scenarios. Complex ecosystem processes are best understood using interdisciplinary teams of scientists that integrate, for example, climatology, hydrology, geomorphology, geochemistry, terrrestrial and aquatic botany, fisheries biology, and ecology.

The variability of natural systems demands that management efforts be treated as experiments requiring careful monitoring to evaluate ecosystem responses and permit timely adjustments in management programs. Where intensive management seems necessary—as in cases of rampant disease, pests, and exotic species—interdisciplinary science should be applied to understand the causes of undesirable ecosystem responses and to direct actions toward the source rather than the symptoms. Solutions may often require a more integrated and long-term view of natural ecosystem cycles and fluctuations, (such as large-scale shifts in climate or ocean conditions and disturbance regimes, as well as modification of human activities that alter fundamental processes of the coastal rain forest ecotone.

Coexisting with Uncertainty

Coastal temperate rain forest ecosystems—and human use and abuse of them—do not terminate at the forest margin: watersheds and the coastal zone are inextricably linked. Dynamic flows of fresh water, nutrients, sediments, large woody debris, fine particulate organic matter, and even organisms of the estuary and coastal zone depend on interactions between landscape structure and the climatic processes. Both the landscape and the climate are subject to long-term natural change and short-term human manipulation. Living resources, especially anadromous fish, convey energy in the form of organic matter and nutrients through this ecotone, essentially recycling from the land to sea and vice versa, but filtered to some degree by the ecotone. Moreover, the tidally influenced reaches of floodplains and estuarine shorelines provide fish with critical transition, rearing, and spawning habitats. The structure of coastal areas—such as the dominant rocky intertidal zone or seagrass and kelp bed communities—is also both directly and indirectly influenced by land/sea ecotone interactions through far-reaching physicochemical (salinity, sedimentation) and ecological (disturbance) influences. A considerably broader understanding of the terrestrial/marine ecotone, especially its tidal freshwater habitats, is urgently needed to give resource management in the coastal rain forest a more complete ecological foundation.

Both economic and social forces must be taken into account in an era when we are beginning to realize that human interests actually do depend on a harmonious relationship with nature (Norton 1988; Kessler et al. 1992). Manipulating energy and altering ecosystem processes to maximize individual resources is ultimately counterproductive to sustaining the diversity, complexity, and resiliency of entire ecosystems, which provide important commodities, uses, and values to society (Kessler et al. 1992). Sustaining economic and social services and goods that depend on ecosystem-scale processes, such as those that sustain the land/sea ecotone of the coastal rain forest, demands landscape-scale management based on scientific knowledge.

It remains to be seen whether we will learn to coexist with uncertainty and evolve as a society without jeopardizing the large-scale and long-term ecosystem processes that sustain the resources and environment. Our success or failure in this effort will depend to a large degree on our appreciation of the dynamic nature of the rain forest coast and the consequences of our intentional or unintentional—the outcome is the same—intervention in the fluxes of material, energy, and life forms across the land/sea interface. Human actions, such as altering hydrological and other critical processes, rearranging the landscape, imposing unnaturally high or low levels of disturbance, and introducing exotic species, will ultimately favor less desirable resources and ecosystem states that will be difficult to reverse. Alternatively, restoring degraded ecosystems is a monumental task for which we have few successful models—in part because we have focused on rehabilitating patches rather than renewing processes. Evaluating the ecological and economic costs of management decisions implicitly includes understanding the exhaustive costs of repairing the consequences of bad decisions. The coastal temperate rain forest bioregion can only benefit from a more comprehensive understanding of landscape processes and taking an adaptive view of sustainable ecosystem management. Shaping our activities to accommodate large-scale and long-term processes that permit ecosystems to sustain themselves should be the key to better management.

Acknowledgments

The authors are grateful to Dan Bottom for his insightful review of an early draft of this chapter.

References

Armstrong, D. A., and D. R. Gunderson. 1985. "The role of estuaries in Dungeness crab early life history: A case study in Grays Harbor, Washington." In B. Melteff, ed., *Proceedings of the Symposium on Dungeness Crab Biology and Management.* Lowell Wakefield Fisheries Symposia Series, no. 3. Alaska Sea Grant Report 85-3.

Atwater, B. F. 1987. "Evidence for great holocene earthquakes along the outer coast of Washington State." *Science* 236:942–936.

Barnes, C. A., A. C. Duxbury, and B. A. Morse. 1972. "Circulation and selected properties of the Columbia River effluent at sea." In A. T. Pruter and D. L. Alverson, eds., *The Columbia River Estuary and Adjacent Ocean Waters.* Seattle: University of Washington Press.

Beak Consultants Ltd. 1981. *Preliminary Analyses of the Potential Impact of Hydroelectric Development of the Stikine River System on Biological Resources of the Stikine River Estuary.* Report prepared for British Columbia Hydro and Power Authority, Vancouver, B.C.

Berris, S. N., and R. D. Harr. 1987. "Comparative snow accumulation and melt during rainfall in forested and clear-cut plots in western Cascades of Oregon." *Water Res. Res.* 23:135–142.

Beschta, R. L. 1978. "Long-term patterns of sediment production following road construction and logging in the Oregon coast range." *Water Res. Res.* 14:1011–1016.

Bosman, A.L., and P.A.R. Hockey. 1986. "Seabird guano as a determinant of rocky intertidal community structure." *Mar. Ecol. Prog. Ser.* 32:247–257.

Brown, D. A., C. A. Bawden, K. W. Chatel, and T. R. Parsons. 1977. "The Wildlife community of Iona Island jetty, Vancouver, B.C., and heavy metal pollution effects." *Envir. Cons.* 4:213–216.

Butler, R. W., and K. Vermeer, eds. 1994. *The Abundance and Distribution of Estuarine Birds in the Strait of Georgia, British Columbia.* Occasional Paper 83, Canadian Wildlife Service, Ottawa, Ontario.

Carefoot, T. 1977. *Pacific Seashores: A Guide to Intertidal Ecology.* Seattle: University of Washington Press.

Carleton, J. T., and J. B. Geller. 1993. "Ecological roulette: The global transport of nonindigenous marine organisms." *Science* 261:78–82.

Chelton, D. B., and R. E. Davis. 1982. "Monthly mean sea-level variability along the west coast of North America." *J. Phys. Oceanog.* 12:757–784.

Cordell, J. R., C. A. Simenstad, and C. A. Morgan. 1992. "The Asian calanoid copepod *Pseudodiaptomus inopinus* in Pacific Northwest rivers—biology of an invasive zooplankter." *Northwest Env. J.* 8:164–165.

Dayton, P. K. 1975. "Experimental evaluation of ecolgical dominance in a rocky intertidal algal community." *Ecol. Mono.* 45:137–159.

del Moral, R., and A. F. Watson. 1978. "Vegetation on the Stikine Flats, southeast Alaska." *Northwest Sci.* 52:137–150.

Dethier, M. N. 1990. *A Marine and Estuarine Habitat Classification System for Washington State.* Olympia: Natural Heritage Program, Washington Dept. of Natural Resources.

———. 1993. "Results from the 1993 monitoring season, Olympic National Park coastal inventory study." Unpublished report to the National Park Service.

Duxton, P. K. 1975. "Experimental studies of algal canopy interactions in a sea otter-dominated kelp community of Amchitka Ksland, Alaska." *Fisheries Bull.* 73(2):230–237.

Feare, C. J., and R. W. Summers. 1986. "Birds as predators on rocky shores." In P.G. Moore and R. Seed, eds., *The Ecology of Rocky Coasts.* New York: Columbia University Press.

Frank, P. W. 1982. "Effects of winter feeding on limpets by black oystercatchers, *Haemotopus bachmani*." *Ecology* 63:1352–1362.

Furbish, C. E., and M. Albano. 1994. "Selective herbivory and plant community structure in a mid-Atlantic salt marsh." *Ecology* 75:1015–1022.

Groot, C., and L. Margolis, eds. 1991. *Pacific Salmon Life Histories.* Vancouver: University of British Columbia Press.

Gunderson, D. R., D. D. Armstrong, Y.-B. Shi, and R. A. McConnaughey. 1990. "Patterns of estuarine use by English sole (*Parophrys vetulus*) and Dungeness crab (*Cancer magister*)." *Estuaries* 13:59–71.

Harr, R. D., B. A. Coffin, and T. W. Cundy. 1989. "Effects of timber harvest on rain-on-snow runoff in the transient snow zone of the Washington Cascades." Interim final report submitted to Timber, Fish and Wildlife, Sediment, Hydrology, and Mass Wasting Steering Committee for Project 18 (rain-on-snow). Portland: Pacific Northwest Forest and Range Experiment Station, USDA Forest Service.

Hart, J. L. 1973. *Pacific Fishes of Canada.* Bulletin 180. Ottawa: Fisheries Research Board.

Hay, D. E. 1990. "Tidal includence on spawning time of Pacific herring (*Clupea harengus pallasi*)." *Can. J. Fish. Aquat. Sci.* 47:2390–2401.

Hay, D. E., and A. R. Kronlund. 1987. "Factors affecting the distribution, abundance, and measurement of Pacific herring (*Clupea harengus pallasi*) spawn." *Can. J. Fish. Aquat. Sci.* 44:1181–1194.

Hay, D. E., M. J. Healey, L. Richards, and J. R. Marliave. 1989. "Prey fishes in Georgia Strait." In C. Vermier and R. Butler, eds., *The Ecology and Status of Marine and Shorebirds in the Strait of Georgia, British Columbia.* Ottawa, Ontario: Canadian Wildlife Service.

Hay, D. E., M. J. Healey, D. M. Ware, and N. J. Wilimovsky. 1992. "Distribution, abundance, and habitat of prey fish on the west coast of Vancouver Island." In K. Vermeer, R. W. Butler, and K. H. Morgan, eds., *The Ecology, Status and Conservation of Marine and Shoreline Birds on the West Coast of Vancouver Island*. Occasional Paper 75. Ottawa, Ontario: Canadian Wildlife Service.

Healey, M. C. 1982. "Juvenile Pacific salmon in estuaries: The life support system." In V. S. Kennedy, ed., *Estuarine Comparisons.* New York: Academic Press.

Hedgpeth, J. W., and S. Obrebski. 1981. *Willapa Bay: A Historical Perspective and a Rationale for Research.* FWS/OBS-81/03. Washington, D.C.: Office of Biological Services, U.S. Fish and Wildlife Service.

Holland, M. M., comp. 1988. "SCOPE/MAB technical consultations on landscape boundaries: Report of a SCOPE/MAB workshop on ecotones." *Biol. Intern.*, Special Issue, 17:47–106.

Holtby, L. B., and J. C. Scrivener. 1989. "Observed and simulated effects of climatic variability, clear-cut logging and fishing on the numbers of chum salmon (*Oncorhynchus keta*) and coho salmon (*O. kisutch*) returning to Carnation Creek, British Columbia." In C. D. Levings, L. B. Holtby, and M. A. Henderson, eds., *Proceedings of the National Workshop on Effects of Habitat Alteration on Salmonid Stocks.* Canadian Special Publication in Fisheries and Aquatic Sciences 105, Ottawa.

Irons, D. B., R. G. Anthony, and J. A. Estes. 1986. "Foraging strategies of glaucous-winged gulls in a rocky intertidal community." *Ecology* 67:1460–1474.

Jackson, P. L., and A. J. Kimerling, eds. 1993. *Atlas of the Pacific Northwest.* 8th ed. Corvallis: Oregon State University Press.

Jay, D. A., and C. A. Simenstad. 1994. "Downstream effects of water withdrawal in a small, West Coast river basin: Erosion and deposition on the Skokomish River delta." *Estuaries* 17:702–715.

Jones, K. K., C. A. Simenstad, D. L. Higley, and D. L. Bottom. 1990. "Community structure, distribution, and standing stock of benthos, epibenthos, and plankton in the Columbia River Estuary." *Prog. in Oceanog.* 25:211–241.

Kempe, S., and U. Lammerz. 1983. "Statistical interpretation of hydrochemical data from major world rivers." *SCOPE/UNEP Sonderband Heft* (Hamburg, Germany) 55.

Kennelly, S. J., and A. J. Underwood. 1993. "Geographic consistencies of effects of experimental physical disturbances on understorey species in sublittoral kelp forests in central New South Wales." *J. Exp. Mar. Biol. Ecol.* 168:35–58.

Kessler, W. B., H. Salwasser, C. W. Cartwright, Jr., and J. A. Caplan. 1992. "New perspectives for sustainable natural resources management." *Ecol. App.* 2:221–225.

Langer, O. E., R. U. Kistritz, and C. D. Levings. 1994. "Evaluation of the no net loss compensation strategy used to conserve Fraser River estuary fish habitats." In *Proceedings of the 1994 Submerged Lands Management Conference,* New Westminster, B.C., October 4–6.

Lajoie, K. R. 1986. "Coastal tectonics." In R. E. Wallace, panel chairman, *Active Tectonics.* Washington, D.C.: National Academy Press.

Lean, G., D. Hinrichsen, and A. Markham. 1990. *Atlas of the Environment.* New York: Prentice-Hall.

Levings, C. D., D. E. Boyle, and T. R. Whitehouse. 1995. "Distribution and feeding of juvenile Pacific salmon in freshwater tidal creeks, lower Fraser River, British Columbia." *Fish. Mgmt. and Ecol.* 2:299–308.

Lichatowich, J. A., and L. E. Mobrand. 1995. *Analysis of Chinook Salmon in the Columbia River from an Ecosystem Perspective.* Portland: Bonneville Power Administration.

LMER Coordinating Committee. 1993. "Understanding changes in coastal environments: The LMER Program." *Eos: Trans. Amr. Geophys. Union* 73:481–485.

Marsh, C. P. 1986. "Rocky intertidal community organization: The impact of avian predators on mussel recruitment." *Ecology* 67:771–786.

Maser, C., and J. R. Sedell. 1994. *From the Forest to the Sea: The Ecology of Wood in Streams, Rivers, Estuaries and Oceans.* Delray Beach, Fla.: St. Lucie Press.

McKay, P. J., and T. A. Terich. 1992. "Gravel barrier morphology: Olympic National Park, Washington state, U.S.A." *J. Coast. Res.* 8:813–829.

McPhail, J. D., and C. C. Lindsey. 1970. "Freshwater fishes of northwestern Canada and Alaska." *Fish. Res. Bd. Canada Bull.* 173.

Miller, R. R. 1958. "Origin and affinities of the freshwater fish fauna of western North America." In C. L. Hubbs, ed., *Zoogeography.* Washington, D.C.: American Association for the Advancement of Science.

Mitch, W. J., and J. G. Gosselink. 1986. *Wetlands.* New York: Van Nostrand Reinhold.

Moyle, P. B. 1993. *Fish: An Enthusiast's Guide.* Berkeley: University of California Press.

Naiman, R. J., T. J. Beechie, L. E. Benda, D. R. Berg, P. A. Bisson, L. H. MacDonald, M. D. O'Connor, P. L. Olson, and E. A. Steele. 1992. "Fundamental elements of ecologically healthy watersheds in the Pacific Northwest coastal ecoregion." In R. J. Naiman, ed., *Watershed Management: Balancing Sustainability and Environmental Change.* New York: Springer-Verlag.

Navarrete, S. A., and J. C. Castilla. 1993. "Predation by Norway rats in the intertidal zone of central Chile." *Mar. Ecol. Prog. Ser.* 92:187–199.

North, M. E. A., and J. M. Teversham. 1984. "The vegetation of the floodplains of the lower Fraser, Serpentine, and Nicomeki Rivers, 1850–1890." *Syesis* 17:47–66.

Norton, B. G. 1988. "The constancy of Leopold's land ethic." *Cons. Biol.* 2:93–102.

Norton, T. A. 1978. "The factors influencing the distribution of *Saccorhiza polyschides* in the region of Lough Ine." *J. Mar. Biol. Assoc. U.K.* 58:527–536.

Paine, R. T. 1966. "Food web complexity and species diversity." *Am. Nat.* 100:65–75.

Postma, H. 1980. "Sediment transport and sedimentation." In E. Olausson and I. Cato, eds., *Chemistry and Biogeochemistry of Estuaries.* Chichester, U.K.: Wiley.

Quammen, M. L. 1984. "Predation by shorebirds, fish, and crabs on invertebrates in intertidal mudflats: An experimental test." *Ecology* 65:529–537.

Richardson, H., and N.A.M. Verbeek. 1986. "Diet selection and optimization by northwestern crows feeding on Japanese littleneck clams." *Ecology* 67:1219–1226.

Richey, J. E., J. I. Hedges, A. H. Devol, P. D. Quay, R. Victoria, L. Martinelli, and B. R. Forsberg. 1990. "Biogeochemistry of carbon in the Amazon River." *Limnol. and Oceanog.* 35:352–371.

Risser, P. G. 1990. "The ecological importance of land-water ecotones." In R. J. Naiman and Henri Décamps, eds., *The Ecology and Management of Aquatic-Terrestrial Ecotones.* UNESCO Man and the Biosphere Series, vol. 4. Carnforth, U.K.: Parthenon.

Schneider, D. 1978. "Equalisation of prey numbers by migratory shorebirds." *Nature* 271:353–354.

Scrivener, J. C. 1987. "Changes in composition of the streambed between 1973 and 1985 and the impacts on salmonids in Carnation Creek." In T. W. Chamberlin, ed., *Proceedings of a Workshop: Applying 15 Years of Carnation Creek Results.* Nanaimo, B.C.: Pacific Biological Station.

Shaffer, J. A., and D. S. Parks. 1994. "Seasonal variations in and observations of landslide impacts on the algal composition of Puget Sound nearshore kelp forest." *Botanica Marina* 37:315–323.

Sherwood, C. R., D. A. Jay, R. B. Harvey, P. Hamilton, and C. A. Simenstad. 1990. "Historical changes in the Columbia River estuary." *Prog. in Oceanog.* 25:299–357.

Sidle, R. C., A. J. Pierce, and C. L. O'Loughlin. 1985. "Hillslope stability and land use." Water Resources Monograph 11. Washington, D.C.: American Geophysical Union.

Simenstad, C. A., and K. L. Fresh. 1996. "The estuarine imperative: A synthesis of early marine life history of Pacific salmon in the Pacific Northwest." Unpublished manuscript.

Simenstad, C. A., K. L. Fresh, and E. O. Salo. 1982. "The role of Puget Sound and Washington coastal estuaries in the life history of Pacific salmon: An unappreciated function." In V. S. Kennedy, ed., *Estuarine Comparisons.* New York: Academic Press.

Simenstad, C. A., D. A. Jay, and C. R. Sherwood. 1992. "Impacts of watershed management on land-margin ecosystems: The Columbia River estuary as a case study." In R. J. Naiman, ed., *Watershed Management: Balancing Sustainability and Environmental Change.* New York: Springer-Verlag.

Simenstad, C. A., C. D. McIntire, and L. F. Small. 1990. "Consumption processes and food web structure in the Columbia River estuary." *Prog. in Oceanog.* 25:271–298.

Simenstad, C. A., B. S. Miller, C. F. Nyblade, K. Thornburgh, and L. J. Bledsoe. 1979. "Food web relationships of northern Puget Sound and the Strait of Juan de Fuca: A synthesis of the available knowledge." Research Report EPA-600/7-79-259. Washington, D.C.: U.S. Environmental Protection Agency.

Strickland, R., and D. J. Chasan. 1989. *Coastal Washington: A Synthesis of Information.* Seattle: Washington Sea Grant.

Swanson, F. J., L. E. Benda, S. H. Duncan, G. E. Grant, W. F. Megahan, L. M. Reid, and R. R. Ziemer. 1987. "Mass failures and other processes of sediment production in Pacific Northwest forest landscapes." In E. O. Salo and T. W. Cundy, eds., *Streamside Management: Forestry and Fishery Interactions.* Contribution 57. Seattle: Institute of Forest Resources, University of Washington.

Thom, R. M. 1987. "The biological importance of Pacific Northwest estuaries." *Northwest Env. J.* 3:21–42.

Thomas, D.W. 1983. *Changes in Columbia River Estuary Habitat Types over the Past Century.* Astoria, Ore.: Columbia River Data Development Program.

Thomson, R. E. 1981. *Oceanography of the British Columbia Coast.* Canadian Special Publication in Fisheries and Aquatic Sciences 56, Ottawa.

Tschaplinski, P. J. 1987. "The use of estuaries as rearing habitats by juvenile coho salmon." In T. W. Chamberlin, ed., *Proceedings of a Workshop: Applying 15 Years of Carnation Creek Results.* Nanaimo, B.C.: Pacific Biological Station.

Vermeer, K., and C. D. Levings, 1977. "Populations, biomass, and food habits of ducks on the Fraser River intertidal area, British Columbia. *Wildfowl* 28:49–60.

Vermeij, G. 1991. "When biotas meet: Understanding biotic exchange." *Science* 253:1099–1103.

Wilson, W. H., Jr. 1989. "Predation and the mediation of intraspecific composition in an infaunal community in the Bay of Fundy." *J. Exp. Mar. Biol. Ecol.* 132:221–246.

Wissmar, R. C. 1986. "Carbon, nitrogen, and phosphorus cycling in Pacific Northwest wetlands." In R. Strickland, ed., *Wetland Functions, Rehabilitation, and Creation in the Pacific Northwest: The State of Our Understanding.* Publication 86-14. Olympia: Washington Department of Ecology.

Wootton, J. T. 1991. "Direct and indirect effects of nutrients on intertidal community structure: Variable consequences of seabird guano." *J. Exp. Mar. Biol. Ecol.* 151: 139–153.

———. 1994. "Predicting direct and indirect effects: An integrated approach using experiments and path analysis." *Ecology* 75:151–165.

Zipperer, V. T. 1996. "Ecological effects of the introduced cordgrass, *Spartina alterniflora,* on the benthic community structure of Willapa Bay, Washington." M.S. thesis, Department of Fisheries, University of Washington, Seattle.

The Kennedy Lake Technical Working Group

IAN GILL, PETER K. SCHOONMAKER, AND KIM HYATT

Kennedy Lake is a U-shaped low-elevation lake, 65 square kilometers in extent, on the west coast of Vancouver Island. A traditional fishing area of the Tla-o-qui-aht Nation, the lake supplied aboriginal fishers with an abundance of sockeye salmon for thousands of years. For almost a century, it supported a commercial fishery; until 1930, its sockeye supplied a cannery at the mouth of the Kennedy River.

At the height of its productivity, Kennedy Lake might have produced as many as 250,000 sockeye salmon in a season. On average, about 85,000 fish returned each year. It was a key center of activity and a source of great bounty on the west coast of Vancouver Island.

In 1994, an estimated 3000 fish struggled back to their spawning beds in the Kennedy Lake system. Warning bells had first sounded in the early 1960s, when the sockeye population crashed. The commercial fishery was discontinued after 1971, but the stocks failed to recover. In 1982, the Tla-o-qui-aht First Nation ceased its aboriginal food fishery.

In 1991, the Tla-o-qui-aht people invited various stakeholders to work with them to restore the natural productivity of the Kennedy Lake system. They cautioned that they were not interested in a quick fix, a simple boost in the numbers of fish in the lake. The Tla-o-qui-aht had a much grander ambition: to restore a superb natural system that for thousands of years had produced salmon for food, trade, and ceremony.

Months of planning led to the formation of the Kennedy Lake Technical Working Group. Its first meeting was held in Tofino in January 1992 at Tin Wis. The group includes First Nations, Canada's federal Department of Fisheries and Oceans, the provincial Ministry of Forests, the forest products company MacMillan Bloedel, the conservation organization Ecotrust, the private environmental consulting company ESSA Technologies, and others. All sides have committed to a sustained, cooperative approach to solving the mysteries behind salmon population declines in the Kennedy watershed.

At the first meeting, Tla-o-qui-aht elders spoke compellingly of the importance of salmon to their communities, both as a source of food and an object of ceremony. They also described the integral role the sockeye fishery played in the cultural training of their young people. Over countless generations, the young people were taught how and where to fish. They were told stories and, through stories and fishing practice, were imbued with a respect for the sockeye. In Tla-o-qui-aht communities, valuable cultural information was transmitted in this way. Thus when the Tla-o-qui-aht voluntarily shut down their food fishery in 1982, they not only lost an important source of sustenance; they also closed one of their most valued schools.

The working group decided to try to integrate traditional and scientific knowledge of the Kennedy system into an understanding of its natural productivity, why it collapsed, and how it can be restored. The group has spent three years assembling historical information and promoting research on current conditions. Researchers have analyzed historical and contemporary data. They have conducted fieldwork and counts to determine current levels of spawner abundance. They have mapped key habitat areas and sought to pinpoint the factors now affecting the sockeye population. They have held workshops to share and analyze data and assign ongoing tasks. They have created a computer simulation model to test assumptions and hypotheses. A vivid picture of the history of natural and human impacts on the Kennedy Lake system has emerged from this work. The group has sharpened its assessment of what currently ails the system and has framed a hypothesis concerning what must be done to rejuvenate it.

Aboriginal people have been fishing in the Kennedy Lake system for uncounted generations. Under traditional management, each family had a distinct territory in which to fish. Members of one family recall catching "big canoe loads" of fish at the mouth of the Clayoquot River near Cold Creek. Clayoquot Arm was considered the best part of the lake for fishing, with large harvests taken every fall. Until about fifty years ago, a village with perhaps 500 residents at the mouth of the Clayoquot River was involved in canning operations.

By the turn of the century, non-Native boats were catching their own "big loads" of fish. By the late 1930s, a severe decline in the sockeye population had occurred—probably caused by overfishing during a period of low stock productivity. The onset of World War II put a temporary end to commercial gillnet fishing, but paradoxically overall fishing pressures may have increased during the war. Commercial fishing resumed in 1946 and continued until 1970. During this time there was a lack of effective management control, a problem exacerbated during the 1960s by large increases in fishing intensity resulting from radical advances in fishing technology (such as power blocks and drum seines). As catches grew in size, the ability of the stock to maintain itself near its former levels was destroyed. Although information on catches during that period is spotty, the data suggest a familiar pattern: an ever-escalating sockeye harvest, climaxed by population collapse.

Overfishing a naturally variable stock was likely the major cause of this decline. Secondary causes included logging, predation by harbor seals, and competition and predation of juvenile salmon by three-spined sticklebacks. Despite the discontinuance of net and seine fishing, and the small proportion of Kennedy Lake sockeye taken by the western Vancouver Island troll fishery, these stocks have remained depressed for over twenty years.

Overfishing and seal predation are probably not in fact responsible for the present failure of Kennedy Lake sockeye to recover. Juvenile recruitment and ocean conditions may be the limiting factors. The most likely cause of poor juvenile survival in Kennedy Lake appears to be competition and predation by sticklebacks. One hypothesis is that presmolt juvenile sockeye have fallen into a "predation-competition pit" in which the species that eat them (or consume their prey) are so much more numerous that the sockeye population cannot rebound. If this is true, the Kennedy Lake system will remain in a stable state with relatively few sockeye and relatively many sticklebacks, until sockeye can swamp their stickleback competitors and predators.

It appears to the technical working group that restoration of sockeye abundance will depend on detailed understanding of the interactions between these two species, perhaps in preparation for manipulations of both populations when ocean conditions are particularly favorable for sockeye. This latter requirement will mandate continued monitoring of sockeye escapement and juvenile survival. But if sockeye are to be restored to their former abundance, the working group will need a more accurate estimate of that historical abundance and its variation through time.

While continuing its monitoring efforts, the technical working group is also taking the next steps toward restoring the Kennedy Lake ecosystem through two related initiatives. First, the group plans to elucidate sockeye–stickleback interactions through a three-step process that includes modeling, observation, and experimentation. The modeling component of this effort is already well under way. This model will be augmented with more detail on the life history of sticklebacks and specific interactions with sockeye in the Kennedy Lake system. The results from running this model will be compared with historical data in order to improve the model and suggest experiments using enclosures containing varying numbers of the two species. Second, sockeye and stickleback population fluctuations through time will be documented by examining a variety of sources: cannery and hatchery records, interviews with net, troll, and native fishers, escapement counts, and analysis of acoustic trawl surveys of juvenile fish.

The results of this research should help the Kennedy Lake Technical Working Group to understand population, community, and ecosystem dynamics in the lake and determine manipulations that might nudge the system toward a state resembling its historical condition. Only then will the people of the Tla-o-qui-aht First Nation be able to use the lake as they have for millennia: as a source of food, trade, and ceremony—and, above all, as a sophisticated school to pass their rich traditional knowledge on to successive generations.

Ian Gill is executive director of Ecotrust Canada, based in Vancouver, B.C. Peter K. Schoonmaker is vice president of Interrain Pacific in Portland, Oregon. Kim Hyatt is a research scientist with the Department of Fisheries and Oceans in Nanaimo, B.C.

□ □ □ □ □ □ □ □ □ □ □ □ □ □

Institute for Sustainable Forestry

SETH ZUCKERMAN

The Institute for Sustainable Forestry grew out of the fertile, logged-over soils of southern Humboldt County, on California's North Coast. Homesteaders like Jan and Peggy Iris moved there in the 1970s and bought land after its rich Douglas-fir and redwood had been harvested to feed the post–World War II building boom. The logging left behind trees too small to process economically in those days, as well as hardwoods—tan oak, madrone, bay laurel, and others—for which there was no market.

After the homesteaders built their houses and planted their gardens and orchards, they realized that the inevitable wildfires which burned through the forests where they had settled posed a great risk to their lives and property. Although the newcomers organized a volunteer fire company, Jan and Peggy recognized that a lot could be done to prevent fires from reaching dangerous intensities. The second-growth forests were dense and brushy; branches extended from ground level all the way into the canopy, creating a "fire ladder" that would carry flames into the treetops where they would spread destructively.

The Irises began to thin the young forest and prune the lower branches, creating stands that were more open and inviting to people and also less vulnerable to conflagration. At first the wood they cut in the process went solely for firewood. But as time went on, the beauty of the hardwoods struck them; it seemed like a waste for potentially magnificent lumber to go up in smoke. After Jan took a course in hardwood processing through the University of California, the Irises were ready to develop a small-scale logging and milling system for their region. They named their enterprise "Wild Iris Forestry."

Jan and Peggy's first customers were their neighbors. Using roads built for the original logging thirty years before, they towed a portable sawmill into the woods. Operating with truck winches, pulleys, and rigging, they were able to snake logs out of the forest without driving onto undisturbed ground and compacting the soil. They pulled the logs to a landing to be milled. Since they could get by with pickup trucks instead of full-size log trucks, it was possible to maintain narrower roads that disrupt drainage patterns less and generally cause less erosion.

The Iris's method of choosing which trees to cut was noteworthy, too. Instead of basing that decision on the mixture of wood they wanted to harvest, they made a choice based on their vision for the forest that would be left after they finished harvesting. This often meant leaving the straightest, biggest trees to keep growing; taking hardwoods that were slowing the growth of the scarcer Douglas-fir; and thinking ahead to clear a space to fall larger trees when their turn eventually came.

They considered aesthetics, too, leaving some curved madrones to add variety to the look of the forest, and cutting stumps flush with the ground and rubbing the tops with dirt so they would not stand out visually. The hardwood lumber had to be air-dried for a long time—a year for every inch of thickness to the board—and then kiln-dried. Flooring became one of Wild Iris's main markets, so they milled boards to tongue-and-groove and put out a product comparable in quality to the more familiar eastern hardwoods.

But for Wild Iris, the effort was worth it. Their philosophy was to increase the value of the wood as much as possible before it left the region—to maximize the number of people who could earn their livelihood from each tree cut. They would sooner sell their wood to a cabinetmaker in a nearby town than ship it to a wood boutique in the San Francisco Bay area, even if it meant making less money. In time, word of the Iris's work spread. More of the neighbors—some of whom started out skeptical of any logging after seeing what previous loggers had done to the landscape—became interested in having the Wild Iris treatment.

People began to call from outside the region to learn about the Irises and tour the lands they had logged. It proved difficult to carry on a small business while also propagating information about good forestry. After a speaking appearance at a 1990 restoration forestry conference outside Eugene, Oregon, they decided to form the nonprofit Institute for Sustainable Forestry to carry on educational work while they pursued their own business.

A few months later, the institute was born in a meeting of a few dozen forest activists. This brainstorming session—held in the new kitchen of the Iris's hillside home, replete with wood from their gentle logging operations—resulted in a declaration of principles that have helped the institute generalize from the work that Jan and Peggy began. These "Ten Elements of Sustainability" demand forest practices that mesh with the natural processes at work in the forest; they call for the protection of water, old growth, and all resident species; they pay respect to other issues, as well, such as workers' rights, the integrity of cultural sites, and the scale of timber operations.

The Institute for Sustainable Forestry uses the Ten Elements as the basis of a program to certify wood and other forest products as ecologically harvested. Modeled after organic produce certification, the Pacific Certified Ecological Forest Products program is aimed at giving consumers a way to support beneficial forestry practices. The institute is coordinating its efforts with the Forest Stewardship Network (a web of local certifying groups) to create a more widely recognized national trademark. Overall, the aim is to help local activists and practitioners of good forestry determine what practices can be certified as ecologically sound in their regions.

Wild Iris suffered a tragic setback in 1990 and 1991 when Jan fell ill with cancer and ultimately lost his life to the disease. (Jan continues on the institute's advisory board as its "guardian angel.") The institute is raising capital to expand Wild Iris and create a processing facility that would buy certified wood from small operators. Others at the institute work on disseminating information about good forestry and help landowners develop long-term management plans for their forests.

The institute also conducts research on the availability of hardwood timber and the demand for the lumber that can be made from it. In a rural region whose residents hope to make part of their living by working the woods, forest practices that benefit the forest and the people in it are of great value. While others find the voice to say "No!" to bad logging, the Institute for Sustainable Forestry is creating something to which we can say "Yes!"

Seth Zuckerman is a restoration consultant and member of the board of advisers of the Institute for Sustainable Forestry in California's Mattole River valley.

□ □ □ □ □ □ □ □ □ □ □ □ □ □

Clatsop EDC's Salmon Program

JIM HILL

In 1975, citizens of Clatsop County in northwestern Oregon formed the Clatsop Economic Development Council (CEDC) in an effort to improve a badly depressed local economy. One of the first committees the CEDC organized focused on the fishing industry and sought opportunities to boost the fishing economy.

With the commercial salmon industry on the lower Columbia River in gradual decline even then, the Fisheries Committee began investigating salmon enhancement. Initial efforts involved renovating a 2-acre farm pond to make it suitable for rearing and releasing salmon smolts. Approximately $78,000 including local contributions, in-kind donations of fish food, and a governor's grant through the federal Comprehensive Employment and Training Act (CETA) was sufficient to pay for CEDC's first release of 50,000 coho smolts in 1977.

The success of the salmon rearing efforts with the first pond prompted construction of a second earthen rearing pond on the same property. The property owners were excited to be part of a salmon enhancement effort that would contribute to the economic vitality of a local industry. The second pond was completed in 1979, all with donated equipment and labor. CETA funds paid personnel to care for the fish, and with help from local cash contributions and donated fish food, the 1979 release from both ponds totaled about 1.5 million fall chinook smolts.

A third rearing pond was constructed in 1981 on land belonging to the Crown Zellerbach Corporation (a large timberland owner) with support from local contributions, CETA, Oregon's Department of Fish and Wildlife (ODFW), the Pacific Northwest Regional Council (PNRC), and the Port of Astoria. Crown Zellerbach donated use of their land and provided all the heavy equipment and labor for construction of the third pond. Local Job Corps students constructed a diversion dam funded by the PNRC grant to fill the pond. In 1981, about 3.5 million salmon smolts were released from all three rearing ponds.

In 1985, CEDC began investigating other sites for salmon rearing. With no more freshwater sites available in the county, the CEDC Fisheries Project initiated a small-scale net-pen rearing program in the Youngs Bay estuary near

Astoria. In 1986, about 30,000 fall chinook were released from a single pen. With grant funding from a variety of sources including the Bonneville Power Administration, the Clatsop County Economic Development Department, and the Port of Astoria, the net-pen program developed into a tremendous success. By 1994, over 3 million salmon smolts were released from the pens.

The salmon smolts released by the Fisheries Project augment the numbers of salmon produced by the ODFW in Youngs Bay. Together, Fisheries Project and ODFW releases grew from 1.2 million in 1977 to more than 5 million in 1993. Fall salmon harvests in Youngs Bay averaged 350,000 pounds from 1979 to 1993, and the ex-vessel value surpassed $1.5 million in 1988 (a figure which has since declined due to ocean conditions that have depressed salmon returns off the Pacific Northwest).

Though grants in support of CEDC's Fisheries Project have been generous, the year-in and year-out operation and maintenance funds are always difficult to raise. Beginning in 1981, local fishers participating in the Youngs Bay fishery agreed to a voluntary assessment. Since then, fishers have voluntarily assessed themselves 5 percent of their poundage value, and the processors who purchase their fish have matched the fishers' contributions. This self-imposed 10 percent "tax" on the value of the fishery is directed to the Fisheries Project. This program has enjoyed nearly 100 percent local participation, and it continues to be a critical piece of the funding puzzle.

The accelerating decline of salmon fishing in the Columbia mainstem has finally prompted a regionwide focus on the approach CEDC has developed over nearly twenty years. The Youngs Bay terminal fishery—so-called because mature fish return to the net pen "terminal" where they were raised—provides a protected coastal site from which it is possible to harvest salmon that do not mix with the depressed and endangered salmon stocks of the Columbia. In recent years this fishery has grown in importance. It contributes significantly, too, to the lower Columbia economy: every dollar from the fishery ripples through the communities and multiplies its economic impact.

Terminal fisheries like the one in Youngs Bay allow managers to reduce or eliminate mixed-stock harvests to protect weak or declining wild stocks, yet still harvest hatchery fish for regional economic gain. As CEDC expands the terminal fishery concept to other lower Columbia sites, the effort is revitalizing a tradition and lifestyle that are an integral part of the lower Columbia Basin and the Pacific Northwest. Increased survival from terminal areas like Youngs Bay not only offers more efficient and cost-effective use of hatchery fish but offers the opportunity to make the best use of the public money available for fisheries. The modest $78,000 invested in salmon enhancement by residents, the state, and the fishing community nearly twenty years ago has paid off handsomely in Clatsop County.

Jim Hill is director of the Clatsop County Economic Development Council's fisheries project, based in Astoria, Oregon.

□ □ □ □ □ □ □ □ □ □ □ □ □ □

The Pacific Forest Trust

Constance Best

Three of every four acres of forest in the coastal Pacific Northwest of the United States are owned privately. The owners of this forestland must evaluate the returns they receive on their forest resources and consider if there may be other "higher and better uses" for the land, such as residential development. A changing economy, population growth, and related demographic shifts—the average age of a nonindustrial private forestland owner is now sixty-five years—create attractive non-forest-based economic opportunities for landowners. Converting forest to residential lots sacrifices most other forest values, such as habitat, watershed, or the protection of biodiversity. The world's most productive forest is being fragmented, simplified, and simply lost as a result.

The Pacific Forest Trust (PFT) was created in 1993 by a group of forestland owners, conservationists, and foresters to address these threats to the integrity of the Pacific Northwest's forests. Based in Boonville on California's redwood coast, the Pacific Forest Trust is dedicated to the restoration, enhancement, and preservation of the region's private forestlands. The trust's primary strategy is to work *with* landowners, using their enlightened self-interest to address the long-term challenge of sustaining forests and the economic and ecological wealth they generate. Solutions emerge when landowners become willing partners in the conservation of the coastal temperate rain forest.

When the economic and ecological benefits of long-term forest stewardship are made clear and compelling to the landowners whose decisions determine the forest's future, forest resources will be sustained—and will pay for themselves as well. Providing the information, demonstration, and incentives that landowners need to make that transition to long-term stewardship is the essence of the PFT.

PFT seeks to remove the many barriers that inhibit landowners' commitment to long-term forest stewardship. A central dilemma is that people must manage trees with lifetimes significantly longer than their own. While a 75-year-old man or woman is considered elderly, most major commercial timber species in the Pacific Northwest are just reaching maturity at 80 to 120 years, and many have the potential to live hundreds of years more. Many forest functions, such as habitat, are attained only as forests age. Redwoods, central to the forest economy of coastal northern California, happily reach ages of 1500 to 2000 years.

This presents a considerable economic challenge, given the time value of money (a dollar today is considered more valuable than a dollar twenty years from now) and landowners' need for periodic income. With so many private forestland owners near or past retirement age, the transfer of forest assets intact from one generation to the next—essential to restoring and maintaining older forests—faces the impact of estate taxes. The high-quality wood from older trees commands a premium, and the payment of estate taxes often drives premature or excessive timber harvest or the sale of forestland for residential development.

Another significant barrier to forest stewardship is that timber harvest still pays the bills for the forest as a whole. Natural forests yield a complex array of nontimber products, from mushrooms to water to carbon storage, but the markets in which these products can be traded are either poorly developed or nonexistent. Most private landowners naturally tend to focus their management efforts on timber. The Pacific Forest Trust is working to stimulate markets for nontimber forest products and educate landowners about the economic value of these resources. Managing a forest for its natural diversity of products can offer landowners a hedge against the inevitable risks in their forest enterprise. Management for multiple tree species, other botanicals and edibles, recreation, fish and wildlife, water production, and even the storage of atmospheric carbon has the potential to enhance economic returns—while encouraging the protection and maintenance of a complex, older forest.

The Pacific Forest Trust has launched two initiatives that begin to address this array of pressures and opportunities: broadening the use of conservation easements on forestland and encouraging the management of forests to promote carbon storage and slow the pace of global warming.

Conservation easements (deeded restrictions to a property's title granted to a nonprofit land trust or conservancy or to a government agency) have typically been employed to restrict development on natural lands through creation of "forever wild" areas. On agricultural lands, they have been used to prohibit nonfarm development and maintain open space. Lands encumbered by such easements remain private property and can be used, sold, or transferred like any other property, subject to the enduring terms of the easement. The use of conservation easements on forestlands in the Pacific Northwest has been limited; their use on *commercial* forestlands has been nonexistent.

PFT is pioneering the design and application of conservation easements to working forestlands, using easements to ensure that the land remains permanently in forest uses, and guiding long-term forest management to protect the ecological character and functions of the land. Working with foresters, biologists, landscape ecologists, land use planners, attorneys, and appraisers, the Pacific Forest Trust is developing a new generation of practical, management-oriented conservation easements.

The value of conservation easements donated by landowners can generally be deducted from income and estate taxes. These tax benefits, often quite significant, can help defray other land management costs, providing compensation for conservation of nontimber resources. In effect, conservation easements can help "monetize" forest values for which no market yet exists, such as habitat or water quality.

Nonindustrial forestland owners often cite estate taxes as a primary barrier to growing older trees: because older trees are more valuable, they create a higher tax liability that compels forestland heirs to harvest excessively or subdivide and sell. The next decade will see millions of acres of forestland transferred from one generation of owners to the next. This transfer could generate a regionwide "forest disturbance" as forests are cleared or developed to pay tax bills. Through workshops and publications, PFT is educating landowners about how conservation easements can maintain forest assets intact through the estate process.

A conservation easement can strip off such speculative values as real estate development from a forest property, allowing the Internal Revenue Service to appraise the land exclusively for its forest uses. Without such a limitation, the IRS must appraise the property at its liquidation value, taking into consideration complete timber harvest and subdivision resale potential. A conservation easement, combined with other estate planning tools, can dramatically reduce the value of the forest property within an estate. By reining in the estate tax bill, the forest and its many values may be transferred intact to the landowner's heirs.

If essential forest services that we now receive for free could be monetized, forest landowners could be compensated for keeping their land in trees and stewarding it well. The public receives, but does not pay for, a variety of benefits from private forestlands including habitat for fish and wildlife, clean water, and carbon storage. PFT is working to develop a "market" for one of these, carbon storage.

Carbon dioxide makes up 77 percent of human-caused greenhouse gases, the primary cause of global warming. Forests of the Pacific Northwest are the most effective carbon "sink" in the world, turning more atmospheric carbon into biomass than any other forest type when they are managed well. Managed poorly, as is often the case with industrial forestry, they can become a net source of carbon to the atmosphere.

PFT has developed an approach to maximize carbon storage through alternative forest management of the major commercial tree species in this region. Working with private forestland owners, PFT's Forests Forever Fund program markets the carbon storage capacity of private forestland to carbon producers (such as industries and public utilities) that are seeking to offset the impact of their releases of carbon dioxide.

Working with Dr. Mark Harmon of Oregon State University, PFT has developed a sophisticated computer model that simulates forest carbon storage under different management regimes. Other such computer models are either global in scope (and therefore not readily applicable to the regional landscape), or they focus on the metabolism of individual trees, or they represent landscapes using a single species, single age, and unvarying condition not found in nature. By contrast, PFT's model represents the landscape and stand level at which landowners actually manage their trees, and it incorporates a mix of tree species and ages.

Analyses of stewardship forest management using PFT's model show that 140 to 170 percent more carbon can be stored in Douglas-fir forests managed for longer rotations—80 to 120 years—and periodic thinning than those managed on the 40- to 60-year rotations typical of industrial management. In another instance, 200 percent more carbon could be stored in a redwood forest under stewardship forestry than under the management prescribed by the California Forest Practices Act.

The Forests Forever Fund operates by linking forest landowners and carbon producers to achieve their respective goals. Producers offset their carbon releases at a competitive price; forestland owners receive compensation for the opportunity cost they incur by deferring harvest and altering management practices to enhance forest diversity. As a result, forests are restored and permanently maintained, yielding a variety of public and private benefits. PFT accomplishes this by acquiring conservation easements designed to create permanent carbon sinks on private forestland. The easement acquisitions are funded by carbon dioxide producers, who can receive credit for the carbon stored under the terms of the easement. PFT's computer model projects and monitors the carbon storage results, verified periodically by field testing.

PFT is preparing a pilot project on a redwood site in Mendocino County, California. If successful, it is likely to leverage significant amounts of private-sector capital to offset forestland owners' costs of long-term forest stewardship. By bringing this forest function into the marketplace, forestland owners can receive compensation for restoring and maintaining old-growth forests, bringing complexity and diversity back to forest structure on private land, and providing fish and wildlife habitat, water production, and other nontimber resources in perpetuity.

Constance Best is president of the Pacific Forest Trust, based in Boonville, California.

□ □ □ □ □ □ □ □ □ □ □ □ □ □

The Skeena Watershed Committee

EVELYN PINKERTON

The Skeena Watershed Committee is a multiparty planning body established in 1992 to address fish management issues of mutual concern to local First Nations, commercial and sport fishers, and the federal and provincial agencies charged with fisheries management responsibility on the Skeena River in northern British Columbia.

The Skeena watershed encompasses some 32,000 square kilometers and includes over 150 tributaries, most of which support six species of Pacific salmonids. Sockeye is the most important species for commercial purposes in the Skeena, with historical catches exceeding 2 million fish in peak years.

The Skeena and its tributaries flow through the traditional territories of Native peoples who now live in fifteen communities and have organized themselves into three political groupings: the Tsimshian Tribal Council, the Nat'oot'en Band, and the Gitksan-Wet'suwet'en Government (recently split into two separate governments). Many of the Native communities are on the coast; some Native residents have participated in the commercial gillnet and seine fisheries since the early part of this century. The rest of the communities are on the river and fish communally owned sites with gillnets, dipnets, or gaffs. Many have always claimed and practiced the right to sell fish, although their right to do so was disputed by the government until 1992.

First Nations' rights to a priority allocation of fish were affirmed in 1990 in the Supreme Court of Canada's *R.* v. *Sparrow* decision and other cases. These decisions and changes in public opinion led the national government in Ottawa to adopt an Aboriginal Fisheries Strategy in 1992. The strategy involves interim one-year agreements between the federal Department of Fisheries and Oceans (DFO) and selected First Nations, specifying harvest quotas and, in some cases, the right to sell fish.

Three of the most important fisheries conflicts in British Columbia are prominent in the Skeena watershed. The first is biological: should fisheries be based on maximum sustainable yield of the most abundant or enhanced stocks, or should management attempt to protect a greater diversity of stocks? The second concerns First Nations' rights to fish—both the allocation of fish between First Nations and non-Native commercial fishers and the sharing of management authority among First Nations and other agencies and interests. The third conflict is between the interests of sportfishers, who seek steelhead, coho, and other salmon upriver, and commercial fishers at the river mouth who unintentionally catch these species while targeting more abundant sockeye.

Artificial spawning channels completed by 1971 on tributaries to Babine Lake near the Skeena headwaters greatly increased the abundance of two particular sockeye stocks. While these "enhanced" stocks could tolerate a harvest rate as high as 80 percent, other stocks of sockeye, coho, and steelhead with which they shared the river could only tolerate much lower harvest rates.

Poor data concerning the status of smaller stocks in mixed-stock fisheries complicated management on the Skeena. And the agencies authorized to make management decisions usually relied on advisory bodies with only a limited capacity to address these problems. A watershed orientation offered a structure in which all the fishers could perceive that conservation and sustainable management of many stocks was in their long-term best interest.

In 1991, the First Nations communities of the Skeena formed the Skeena Fisheries Commission to coordinate their activities and agreements with the government under the Aboriginal Fisheries Strategy. The commission forms the First Nation component of the Skeena Watershed Committee. Gitksan communities living 200 kilometers upriver from Prince Rupert were party to one of the early agreements pursued under the strategy.

Under the strategy, the DFO agreed that the Gitksan could harvest some 50,000 enhanced sockeye for commercial sale if they were caught with selective gear that excluded other species. This "Excess Salmon to Spawning Requirements" (ESSR) fishery targeted the two enhanced Babine Lake sockeye stocks, using beach seine nets from which steelhead and coho salmon were released live. The Gitksan ESSR fishery thus targeted sockeye salmon that would otherwise be wasted, while conserving the scarcer natural-spawning stocks. The Gitksan developed cooperative enforcement, stock analysis, and sale monitoring programs with the DFO.

The Skeena Watershed Committee created a new forum in which the Gitksan and other First Nations could participate with commercial and sportfishing interests in a watershed-based, multiparty, comanagement arrangement. The most important component, initially, was a harvest plan for the commercial fishery. The DFO supported the initiative financially. Although the department's traditional client group has been the commercial fishery, it has responded to the mixed-stock fishery problem when other groups demanded this, and it fit into new policies. The DFO's growing concern regarding mixed-stock problems coincided with the decision to implement the experimental ESSR commercial fishery on the Skeena under the Aboriginal Fisheries Strategy. By early 1992, all the parties fishing on the Skeena and federal and provincial governments reached a memorandum of understanding that laid the groundwork for comanagement.

After two years of limited success, the DFO (which had sponsored, convened, and chaired the Skeena Watershed Committee up to that point) relinquished control of the process by hiring a professional facilitator. This was an important step that moved the committee out of the realm of the traditional advisory committee. Government cannot be both the sponsor and the convenor of a process, retaining all the power, or the process will be perceived, correctly, as just another way to impose the government's agenda.

The Skeena Watershed Committee has already made a contribution to resolving fisheries management conflicts on the Skeena, but the commercial fishery has largely resisted using selective strategies or gear to reduce pressure on vulnerable stocks. Eventually, the fear of losing much of the sockeye run to the upriver First Nations (who are using selective strategies in the ESSR fishery and also considering them for their subsistence food fishery) may motivate them to seek other ways to target sockeye during the steelhead and early coho run. The Skeena Watershed Committee is likely to be the forum in which these issues are resolved.

One element of successful conflict resolution in multiparty comanagement settings is that common goals are set by the parties involved and incorporated into an explicit agreement among them. The Skeena Watershed Committee has created an important forum in which First Nations and sportfishers have had a chance to further their shared goals for stock diversity.

Evelyn Pinkerton is a maritime anthropologist and a research associate at the School of Community and Regional Planning at the University of British Columbia in Vancouver. This profile was adapted from an article published in *Environments* 23(2):51–68 (University of Waterloo, Ontario), 1996.

□ □ □ □ □ □ □ □ □ □ □ □ □ □

Ecoforestry: An Approach to Ecologically Responsible Forest Use

Mike Barnes and Twila Jacobsen

The Ecoforestry Institute, a nonprofit, educational organization with branches in the United States and Canada, seeks to educate the public and a new generation of ecoforesters by offering education and training programs, establishing demonstration forests where ecoforestry is practiced and taught, and by certifying "ecoforest products" which come from forests that are sustained and restored while products are harvested on a long-term sustainable basis.

Over the past three years, the Ecoforestry Institute/US and the Ecoforestry Institute Society/Canada have promoted a dialogue with leading ecological scientists, philosophers, practitioners of "sustainable forestry," and indigenous forest dwellers across the Cascadia Bioregion about the principles and practices of ecoforestry. Ecoforestry is rooted in an indigenous caretaker land ethic, in a deep-ecological critique of mechanistic philosophy and science, and in the practices of those who have a track record of sustaining and restoring multispecies, multiaged forests while harvesting forest products.

The well-being of the entire ecosystem is the basis of the well-being and quality of life for each of us, an insight common among indigenous peoples but rare in modern industrial culture. Ecoforesters seek to reestablish a reciprocal relationship of caretaking and use of our surrounding forested landscape by becoming native to place.

Based largely on the European agricultural model, industrial forestry converts natural forests into tree plantations on a fifty-year clearcutting rotation in the forests of the Pacific Northwest. The forest products companies, which own most of the private forestlands, seek to maximize short-term timber yields and corporate profits. Yet they still tend to discount claims that their silvicultural practices lead directly to the degradation of terrestrial and aquatic habitat and biodiversity.

Ecoforestry, in contrast, is about caring for forests. Caring for forests does not mean managing or controlling them. Forest ecosystems are far too complex for humans to understand, much less "manage." What must be managed is not the forest, but rather our human interventions in the forest, in order to assess our impacts and minimize harmful disturbances in the short and long term. Ecoforestry, therefore, attempts to understand and work with the dynamic patterns and processes that characterize forests—the diversity and complexities of interacting communities of beings, each playing a vital role in a fully functioning forest ecosystem.

Based on the principle that "nature knows best" how to evolve complex, healthy forests, ecoforestry seeks to maintain maximum photosynthetic gain and native diversity while harvesting a volume of forest products well within the forest's annual growth increment. Ecoforesters focus initially, not on what to remove from the forest, but rather on what to *leave* in order to maintain and restore a fully functioning forest. By maintaining a multispecies, multiaged forest with a sufficient volume of coarse woody debris and standing snags, the structural and genetic diversity of the forest is allowed to develop through natural patterns and processes. By harvesting less than the annual growth, the remaining trees are augmented by additional sunlight, water, and space. In this way, the total biomass and solar conversion process of a forest continue to increase.

Ecoforesters start with a forest assessment to understand the history and the present conditions of the forest at both the landscape and stand levels. An inventory of all forest species of wildlife, trees, and plant communities begins this work. Based on the inventory, old-growth forests, secondary forests, biological corridors, riparian areas, meadows, and so on are demarcated on a map. Assisted by the map, management decisions can protect existing wildlands and habitat of threatened and endangered species, as well as designate areas for harvesting various types of products, including timber, medicinals, and special forest products. Monitoring systems are established prior to harvest in order to study changes in the forest caused by human impacts and to provide data to adapt forest management practices in the future.

The Ecoforestry Institutes are establishing demonstration forests throughout the bioregion. In British Columbia, two demonstration forests provide tours and education programs: Merv Wilkinson's Wildwood Forest, in Ladysmith, and Bob and Nancy McMinn's Kindwood Forest thirty minutes north of Victoria.

In the United States, the Ecoforestry Institute/US has established its headquarters and first demonstration forest at Mountain Grove, a 420-acre forest located forty-five minutes south of Roseburg in southern Douglas County, Oregon. With support from the World Wildlife Fund, a forest assessment including a survey of native plants and grasses is in preparation. Soon monitoring systems and long-term research plots will be designed and implemented. The Mountain Grove Demonstration Forest began offering public tours in the fall of 1995.

A team of leading ecological scientists sponsored by the Pacific Certification Council will meet at Mountain Grove to review the PCC's criteria and standards for the certification of "ecological forest products." The same team will help the Ecoforestry Institute to design long-term research plots at Mountain Grove to validate these criteria and standards on the ground.

Mountain Grove also serves as a teaching forest. The second year of an Ecoforesters' Certification Program began in July 1995. The program has three components: a ten-day ecoforestry intensive course; a fifteen-day apprenticeship training program with master ecoforesters in both British Columbia and Oregon; and a six-month distance learning course with weekly readings and videos. Since ecoforestry is an adaptive discipline that deepens as practice continues, we are developing continuing education courses and workshops to supplement the certification program.

The Ecoforestry Institutes in the United States and Canada participate in a global network of forest dwellers, forest workers, and forest-interdependent communities committed to protecting and caring for forests around the world. Part of a global ecological movement, the institutes seek to shift the paradigm in the human/forest relationship to one in which humans work with nature to protect wild, old-growth, and unfragmented forests, to sustain fully functioning second-growth forests from which we harvest products, and to restore degraded forests, streams, and entire watersheds.

Mike Barnes and Twila Jacobsen are codirectors of the Ecoforestry Institute/US in Glendale, Oregon.

□ □ □ □ □ □ □ □ □ □ □ □ □ □

Heritage Stocks: A "Good News" Strategy to Save Salmon

GUIDO R. RAHR III

Over 100 years ago, Livingston Stone startled scientists and fishermen gathered at the twenty-first annual meeting of the American Fisheries Society with an eloquent plea to establish national salmon parks. "Provide some refuge for the salmon and provide it quickly, before complications arise which may make it impracticable," Stone exhorted the society. "If we procrastinate and put off our rescuing mission too long, it may be too late to do any good. After the rivers are ruined and the salmon gone they cannot be reclaimed."

Stone's words were prescient. No salmon parks were established then; none have been established since in the Pacific Northwest; and Pacific salmon have declined to a level even Stone would have had difficulty imagining. In 1991, fisheries biologists Willa Nehlsen, Jack Williams, and James Lichatowich identified 214 stocks of native salmonids in California, Oregon, Idaho, and Washington facing moderate to high risks of extinction. More recent research has revealed 100 additional stocks facing extinction in the Pacific Northwest. Possibly never in the evolutionary history of salmonids have so many rivers in their natural range been hostile to the survival and reproduction of these native wild fish.

Neither laws and policies nor money have been able to stem the decline so far. In many cases it has taken the United States Endangered Species Act to move federal and state agencies to protect dwindling native salmon stocks, but the ESA offers protection only after target populations have reached perilously low levels. The most dramatic case is the Snake River sockeye, which declined from an average of 3000 fish in the 1950s to one fish in 1994, three years after it was listed as "endangered." And attempting the recovery of small salmon populations is phenomenally expensive. According to the government's General Accounting Office, federal agencies and private organizations spent more than $1 billion on salmon in the Columbia River Basin between 1981 and 1991. More than half of this was spent on hatchery production, but programs to replace declining native salmon runs with hatchery fish further endanger native stocks.

Is there another way? Is there still a chance to heed Stone's impassioned plea of a century ago? Ironically, much more is known about salmon and steelhead stocks that are in decline than about those that are still healthy. Only in the last few years have comprehensive inventories of native stocks been compiled. In 1994, the first survey of healthy native stocks of salmon and steelhead in the Pacific Northwest identified ninety-nine stocks at least one-third as abundant as they would be in the absence of human pressures (and twenty stocks at least two-thirds as abundant). Yet no one has outlined a strategy to safeguard the vitality of these stocks and the rivers that sustain them.

Such an approach makes good sense. Healthy stocks are not only important in themselves: wild, locally adapted stocks that persist at healthy levels can serve as "source populations" for future efforts to recolonize adjacent habitats where native stocks have gone extinct, once the factors responsible for their extinction have been removed. In just this way, salmon returned to rivers cut off from them for millennia during the Ice Ages.

Oregon Trout, a regional fish conservation organization, proposes to identify and recognize healthy native salmon and steelhead stocks as "heritage stocks" and to designate the rivers where they live as "Wild Fish Heritage Rivers." Every five years, stocks would be reassessed and those that declined below the "healthy" threshold would be removed from the program. Twenty-one fall chinook, seven winter steelhead, and six summer steelhead stocks are rated "healthy" in Oregon today. But the rating is no guarantee of the future. The spring chinook of the North Fork of the John Day River, the last native spring chinook stock rated "healthy" in the entire Columbia Basin, dropped from 4000 fish in 1994 to fewer than 400 fish in 1995.

Designating and publicizing heritage stocks would build local awareness and support for recovery and management of native fish. The lasting health of most, if not all, of the region's native salmon lies in the hands of communities—the people who share the watersheds where fish live. If local communities are aware and take pride in the fact that they share their watershed with one of the last healthy native salmon stocks in the country, support for wild fish and their habitats will grow where it matters most.

Wild salmon in the Pacific Northwest offer a regional analogy to the seeds of native and heritage fruits and vegetables that have catalyzed grassroots movements in other regions of North America and the world. An exceptional resource of biodiversity, neglected until it is in serious jeopardy, offers people an opportunity to connect with the living fabric of the world through meaningful conservation close to home.

The "heritage stocks" idea proposed by Oregon Trout—like the hands-on conservation exemplified by Kent Whealy's Seed Savers Exchange and what heirloom seed company Seeds of Change founder Kenny Ausubel calls "backyard biodiversity"—offers a "good news" strategy to involve people not simply in protecting biodiversity (in the form of native stocks of wild fish) but in living with it, actively enhancing it, and coming to depend upon it as a resource that builds and enriches the human community. It is an idea that Livingston Stone would applaud.

Guido R. Rahr III directs Oregon Trout's healthy-stocks program, based in Portland, Oregon.

□ □ □ □ □ □ □ □ □ □ □ □ □ □

The Prince William Sound Science Center

NANCY BIRD

Alaska's Prince William Sound is a diverse and productive array of ecosystems located at the northern boundary of the coastal temperate rain forest. The region is endowed with spectacular scenery and a wealth of natural resources. Until recently, the waters of the sound provided abundant harvests of herring, five species of salmon, and three species of crab plus shrimp and halibut. Mining operations, limited timber harvests, and fox farms diversified the economic base of commercial fishing that began in the 1880s and remains the region's chief industry. In the mid-1970s, another major industry entered the sound: the storage and transport of crude oil from Alaska's North Slope.

The Prince William Sound Science Center was organized in 1989 through the efforts of fishers and scientists living in Cordova, a fishing community of about 3000 residents located on the eastern side of the sound. Cordova offers easy access both to the maritime environment and to the Copper River Delta, the largest contiguous wetland on the Pacific Coast of North America and a critical rest-stop for hundreds of thousands of migrating shorebirds and waterfowl. The center's founders were concerned about increasing demands on the sound's wealth of fish, timber, and mineral resources. They particularly recognized the need for an independent local source of credible scientific information. The center was designed to promote a base of knowledge that might help communities around the sound to avoid another boom and bust cycle and build sustainable economies on the natural diversity and abundance of species in the sound.

The devastating 1989 *Exxon Valdez* oil spill on Bligh Reef catalyzed the formal establishment of the Science Center and changed its course of development. While the spill offered new funding opportunities, it also complicated the development of the center's research program. Research shifted from a planned focus on natural baselines to the distinction between natural and human-caused changes in Prince William Sound.

Prince William Sound was the area hit hardest by the oil spill. In the years immediately following the spill, the returns of commercially harvested pink salmon and herring were in the range of preseason predictions. But in 1992, a dramatic decline in both populations began; it continued into 1995. Is this decline connected to the oil spill? Scientific studies, to date, indicate a probable but still uncertain connection. Meanwhile, residents dependent on fisheries and other resources in the region continue to be concerned over the long-term impacts of the oil spill and want to understand the changes in the environment.

Frustrated by the lack of studies designed to answer why the sound's ecosystem was no longer producing its former abundance of herring and pink salmon, and in turn the species dependent on them, fishers literally put their boats and lives on the line. For three days in August 1993, they blockaded all oil tanker traffic to and from the Valdez oil terminal. This dramatic action catalyzed creation of a fisheries research planning group composed of scientists, fishers, local resource managers, environmentalists, and other interested residents in the region.

The Science Center was still a fledgling organization at that time, but its small research staff included people with expertise in fisheries ecology and oceanography. With an initial ecosystem study plan in hand, they took a leading role in the Planning Group. This unique group, composed of scientists and nonscientists, spent the next two and a half months writing a research plan focused on the unanswered questions about the biological and physical processes of Prince William Sound.

The resulting plan, called Sound Ecosystem Assessment (SEA), received high marks from academic peer reviewers in December 1993. The interdisciplinary program is designed to understand the underlying physical and biological factors that control populations of key marine animals. In early 1994, thirteen research projects based on the SEA plan were under way, funded by the *Exxon Valdez* oil spill settlement, a multimillion-dollar fund administered by six state and federal agency representatives.

The Science Center's SEA research projects have an annual budget of about

$1.5 million and have nearly tripled the center's staff. Funding for SEA is expected to continue, on a declining scale, for another three to five years.

For basic operations, the center depended on private foundation grants, a loan from the City of Cordova, and a few small research grants. The competition for funding among the academic and government organizations with which the center works is intense, and the center is at a comparative disadvantage. At the same time, the center must have the active support of residents in the region—support that is based on good communication and dissemination of research results. Communication and education activities have also become major programs for the Science Center.

Since 1990, the center's small education staff has developed a model partnership program called Science of the Sound. This program offers hands-on science experience and understanding of the local environment to residents around the sound, including an outreach program that transports educators from Cordova to outlying villages. The latest expansion of the Science of the Sound program is "From the Forest to the Sea," a summer science camp for children nine to fifteen years old. The camp is a cooperative effort of the Science Center, the USDA Forest Service, the 4-H Club, and the University of Alaska Cooperative Extension Service.

It is all too common for residents of remote regions to be the last to hear the latest scientific information about resources and the ecosystems surrounding them. While the Science Center is not an advocacy organization, a key component of its mission is to distribute the best available information to residents of the greater Prince William Sound region so that they will be able to participate in resource management issues effectively.

The Science Center is a unique entity: not-for-profit, independent, strongly academic yet not affiliated with a university, and not an advocacy group. It seeks to disseminate widely the results of its work to the general public, as well as to the scientific community. Its vision of community-based science, though not without its limitations, is alive and well in Prince William Sound.

Nancy Bird is vice president of the Prince William Sound Science Center in Cordova, Alaska.

8. Pacific Salmon: Life Histories, Diversity, Productivity

WILLA NEHLSEN AND JAMES A. LICHATOWICH

Pacific salmon begin and end their lives in rivers at the margins of the North Pacific Ocean, in both Asia and North America, from approximately 32 to 70 degrees north latitude (Meehan and Bjornn 1991). In North America, salmon spawn in rivers from southern California to northern Alaska, a range that covers even more than the 22 degrees of latitude that encompass the coastal rain forest (Plates 6 and 7). Seven species of Pacific salmon—pink (*Oncorhynchus gorbuscha*), chum (*O. keta*), sockeye (*O. nerka*), chinook (*O. tshawytscha*), coho (*O. kisutch*), steelhead (*O. mykiss*), and sea-run cutthroat trout (*O. clarki clarki*)—live in the waters of the coastal temperate rain forest and adjacent landscapes of the Pacific Northwest. Nearly all of the streams, rivers, and estuaries in the bioregion, as well as the northeast Pacific Ocean, are (or were) part of the salmon's extended ecosystem.

The freshwater life histories of these migratory fishes occur within large and small watersheds. A small watershed may show comparatively little variation in its geology, soil, potential natural vegetation, and overall pattern of land use. A larger watershed is likely to vary a great deal in these and other attributes, so in the course of their migration, salmon may encounter several ecosystem types. Some Columbia River salmon populations, for example, spawn and rear in the more arid interior ecosystems east of the coastal rain forest, as far east as Idaho; these juveniles then migrate through the rain forest zone on their way to sea. The significance of the coastal temperate rain forest to salmonids, then, embraces instances in which the fish spend a limited portion of their freshwater lives there, as well as instances in which salmon complete their entire life histories in rain forest watersheds.

Regardless of the time they spend there, Pacific salmon are a key component of coastal temperate rain forest ecosystems. They bring nutrients from the marine ecosystem into terrestrial ecosystems: returning salmon substantially enrich carbon and nitrogen cycles in the vicinity of spawning areas (Bilby et al. 1996). And the salmon's ecological influence extends well beyond the streambank: at least twenty-two forest-dwelling mammals and birds feed on salmon carcasses (Cederholm et al. 1989).

Salmon are key components of human cultures as well. Pacific Northwest Indian tribes have received physical and spiritual sustenance from salmon for at least ten millennia, and salmon have been a dominant part of their economies for at least 2500 years (Chapter 10; Matson and Coupland 1995). For the last 150 years, salmon have been an economic, cultural, and recreational mainstay of non-Native communities too. Today salmon are harvested in commercial, tribal, and recreational fisheries. Income from commercial salmon harvests and from recreational fishers is critical to most coastal and many inland communities. Until recently, Pacific salmon fisheries generated more than $1 billion per year in personal income and supported more than 60,000 jobs in the region (Oregon Rivers Council 1992). However, since 1993, commercial salmon fishing has been restricted to minimal levels in California, Oregon, and Washington to protect salmon populations listed under the Endangered Species Act of 1973, significantly reducing the contribution that salmon fisheries make to the local economies.

Pacific salmon are generally anadromous—they spend a portion of their lives at sea and return to fresh water to spawn (Figure 8.1). They begin their lives as newly hatched fry in headwater streams, in mainstem rivers, or in the mixing zone between coastal estuaries and freshwater streams, depending on the species. With some species, the young spend only a few days in fresh water; with others, freshwater residence can last two years or more. During this period, they feed and try to avoid predators. Their survival depends on adequate cover to avoid predators and on the quality of the substrate and riparian vegetation, which are primary sources of food. Survival is diminished by poor water quality—such as excessive siltation or lethally high water temperatures—or by floods that wash juveniles with inadequate refuge habitat out to sea.

After rearing in fresh water, juvenile salmon migrate to the sea, most during the spring and summer. Once they reach the coast, juveniles of some species (especially chum) feed for a few days or weeks in estuaries before going to sea. The various species of Pacific salmon remain in the ocean for one to five years. After growing and maturing in the ocean, adult salmon return to the streams

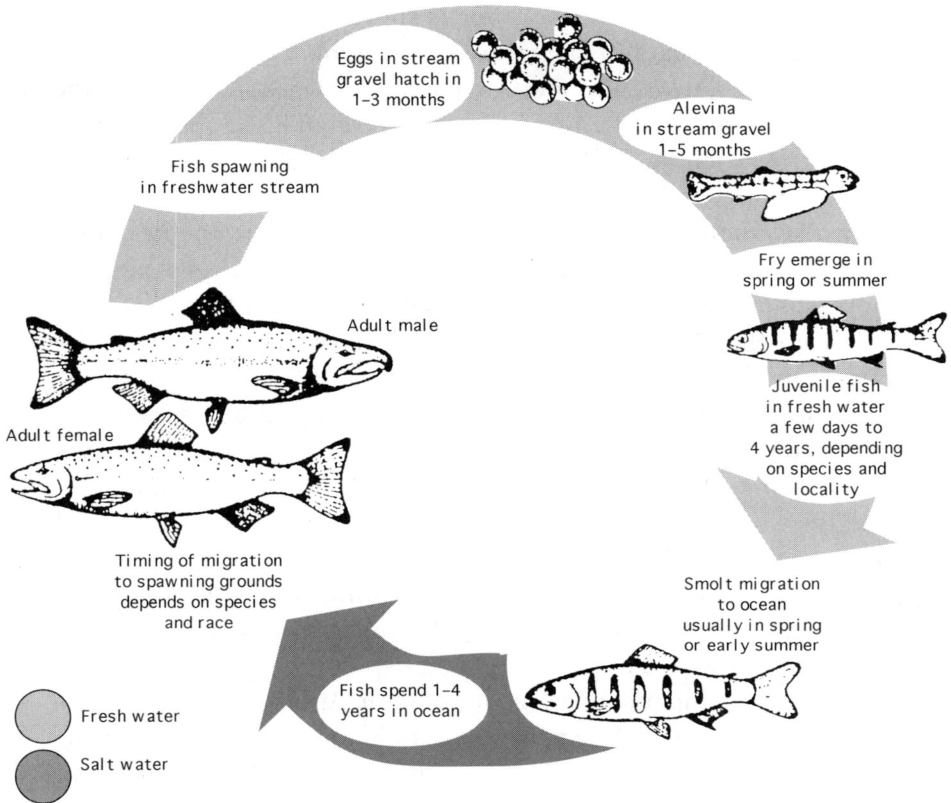

Figure 8.1. Typical life cycle of anadromous salmonids. *Source:* Meehan and Bjornn (1991). Reprinted with permission of the American Fisheries Society.

of their birth to spawn. Of the seven Pacific salmon, only steelhead and sea-run cutthroat trout survive spawning.

Each salmon species has a life history pattern possessing distinctive features such as age structure, the timing and distribution of spawning within a particular watershed, and the length of juvenile residence in fresh water. Matsen (1994:48–51) provides one of the better sketches of the anadromous Pacific salmon:

> Each species, and in some cases each race or run within a species, carries a schedule for fresh- and saltwater migration in its genetic code. Chinook salmon, adapted to long or steep rivers, build strength and size during as many as five years at sea after a year in fresh water. A seiner off Alaska caught a 145-pounder, though surely bigger chinooks have lived and live now.
>
> The coho's anadromous rhythm is similar to the chinook's, its

> closest relative, though its sea time is shorter, usually only eighteen months or two years. If salmon were automobiles, cohos would be the sports cars, fast, agile, and compact. The biggest ones, late in the season when all the fattening is done, can weigh fifteen pounds, and anglers say catching a coho on a fly rod is sublime.
>
> The annual cycles of the sockeye are the most varied of the tribe, from a few months to three years in fresh water, and from one to four years in the sea. Their life spans are so varied because they depend on more ecological combinations, patterns, and sizes of lakes, streams, and rivers than any other salmon. Some biologists spend their entire careers on the complicated sockeye. . . .
>
> Chums are the blue-collar salmon, dependable, nothing fancy, the most widely distributed of all species, ranging from Korea around the Pacific Rim to Monterey Bay, California. Second only to chinooks in size, the workerlike chums usually return in two waves, summer and late fall, when they are the last of the salmon to reach their home rivers. They leave their streams within months of spawning, and return with a territorial precision notable even among salmon after two to five years at sea.
>
> Pinks, by comparison, are zoom salmon. After just a few months in fresh water and a single winter at sea, they return in great swarms of three- to six-pound fish. Throughout their range from the Sea of Japan to the Sacramento River, "humpies" spawn in alternating big and small years, and we know why. Because of their short life cycles, each year's pink salmon never mix with another's, so runs in odd- and even-numbered years have become genetically isolated and radically different in size.
>
> Until 1992, steelhead were a part of the *Salmo* tribe, but taxonomists reclassified the species as *O. mykiss*. They remain, however, the same object of passion of sports anglers, revered for their exuberance and, increasingly, for the difficulty of catching one on rod and reel. [Along with sea-run cutthroat trout,] they are the only member of the *Oncorhynchus* tribe that spawns more than once.

Sea-run cutthroat trout are perhaps the salmon species most closely tied to the rain forest coast. Rarely more than 14 inches long as adults, these small cousins ply coastal streams and nearshore waters, spawning yearly in their native streams. For detailed treatment of the life histories of all the salmonids, the interested reader may refer to Meehan and Bjornn (1991) and relevant chapters of Groot and Margolis (1991). The variations in how salmon undergo the transition from fresh water to salt water are described in Chapter 7.

The characteristic life histories provide central themes upon which populations within each species have developed profuse variation. Life histories of salmon populations reflect adaptation to local habitats (see Nehlsen et al. 1991 for a discussion) and can be important determinants of overall productivity. Life history diversity, a key element of the biodiversity within a species, is expressed fully in watersheds with complex habitat structure that are well connected from spawning areas to estuary. Fish with different life histories use the available habitat in different ways and make different contributions to overall productivity.

Coastal temperate rain forest ecosystems support a considerable diversity of aquatic habitats (Chapter 6). Variations in the amount and timing of river discharge and water temperature with latitude are particularly important for salmon. At both the species and population levels, these factors constrain the life history possibilities open to salmon.

Life History at the Species Level

At the landscape scale, the distribution of salmon species and the diversity of their life histories reflect latitudinal variation in the environment's physical characteristics (Schalk 1977). In the northern part of the range (above about 60 degrees north latitude), the predominant species are pink and chum salmon, whose young spend only brief periods in fresh water. The reason is simple: unstable riparian environments with wide fluctuations in discharge, freezing temperatures, and limited food supply resulting from cold water temperatures do not favor a long period of juvenile rearing. Even the northern ocean provides a more supportive rearing environment. Sockeye salmon, also found in the northern latitudes, rear in lakes, which are more stable than the riverine environments. In the southern part of the range, however, increasing stability in flow levels and food availability favor coho and chinook salmon, species with longer juvenile freshwater residence. All Pacific salmon species are found at the intermediate latitudes between northern Oregon and southern Alaska.

Geographic trends in the commercial catch of salmon over nearly six decades (1920–1977) demonstrate the north-to-south variation in the distribution of salmon species (Figure 8.2) (Fredin 1980). Over that period, the majority of pink and sockeye salmon were caught in Alaska; fewer were taken in British Columbia and Washington waters; negligible numbers were caught in waters off Oregon and California. Conversely, chinook and coho accounted for most salmon caught in Washington, Oregon, and California. Variations in the distribution of salmon species have shaped the cultures and economies of both Native peoples and non-Native fishers.

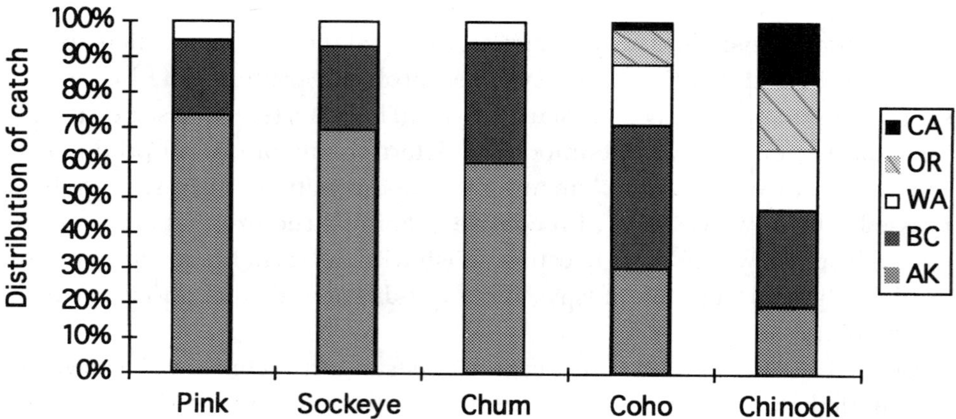

Figure 8.2. Percentage of commercial harvest of pink, sockeye, chum, coho, and chinook salmon caught in commercial fisheries in Alaska, British Columbia, Washington, Oregon, and California: 1920–1977. *Source:* Fredin (1980).

Take chinook salmon, for example. Chinooks possess a striking and well-documented diversity of juvenile life histories. There are generally two patterns of juvenile freshwater residence (Healey 1991). "Ocean-type" fish usually migrate to sea within six months of emergence. "Stream-type" fish migrate to sea in the spring of their second year or later. Stream-type life histories are found in rivers north of 56 degrees north latitude and in populations that spawn in the upper reaches of rivers that penetrate long distances inland, such as the Fraser and Columbia rivers. Between 56 degrees north latitude and the Columbia River, both life history patterns are present. South of the Columbia River, the ocean-type life history predominates (Figure 8.3). This variation in chinook life history is believed to be a response to growth opportunity (temperature and day length) and distance from the sea (Taylor 1990).

Within a single watershed, however, the distinction between stream-type and ocean-type life histories blurs. Under healthy habitat conditions, a population of juvenile chinook shows several variations in the stream and ocean type of life histories; these patterns result from a combination of genetic and environmental factors. (For a review of the literature see Lichatowich and Mobrand 1995.)

The ocean type of life history is an important component of a population's productivity. Subyearling migrants tend to move downstream into lower river reaches for a summer rearing period, during which they take advantage of more stable environments and warmer temperatures to grow before entering the sea. This pattern has been observed in Oregon coastal rivers, the Columbia River Basin, and in the Nanaimo River. In many areas, lower river reaches that once supported growing juveniles during the summer (following the ocean type of life history) are now uninhabitable because habitat has been destroyed

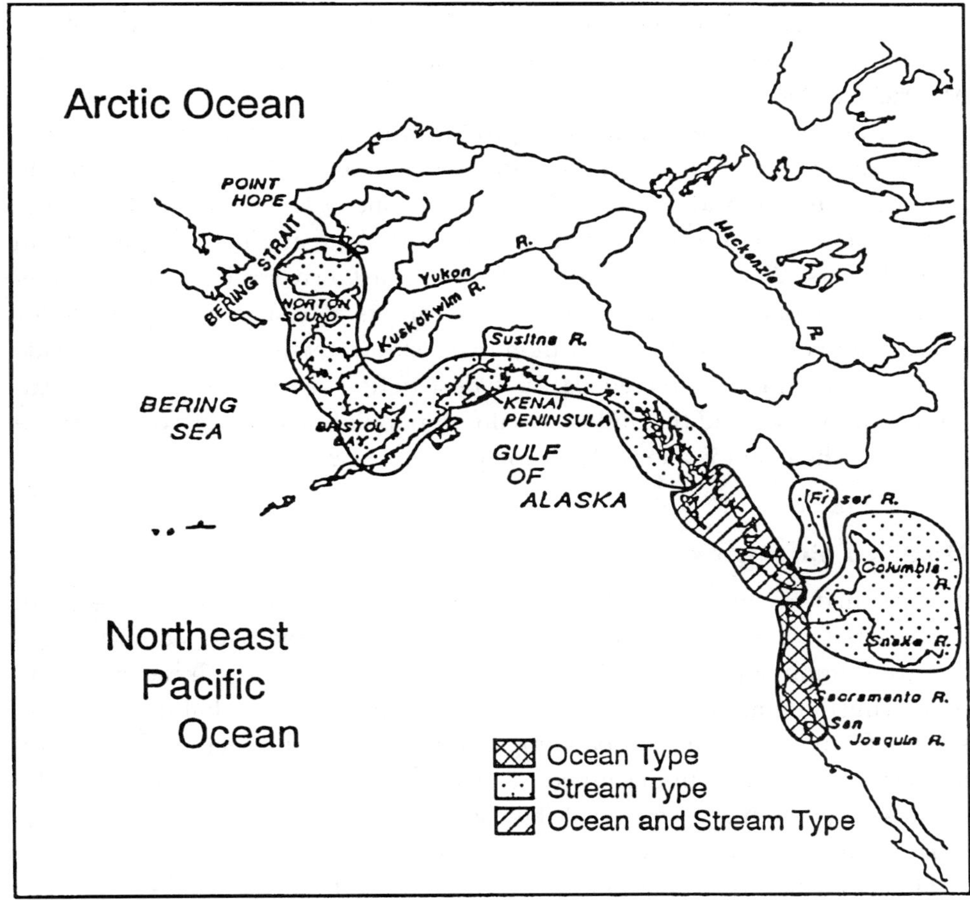

Figure 8.3. Distribution of stream-type and ocean-type life histories in chinook salmon. *Source:* Taylor (1990).

by farming, irrigation, grazing, and timber harvest. This degradation has allowed water temperatures to rise above lethal thresholds and reduced habitat complexity. Lower river areas were degraded earlier and more severely than headwaters: not only did development concentrate there first, but the effects of degradation throughout the watershed accumulate in the lower mainstems.

During the late 1800s and early 1900s, degradation in the Columbia River Basin reduced once-continuous habitats to discontinuous fragments in the streams flowing from the steppe and shrub-steppe biome east of the Cascade Mountains. Fragmentation occurred because areas of poor habitat (caused by human activity) created a barrier to summer migration between upper watershed areas and lower river areas. Because ocean-type juveniles could not undertake the summer migration to the sea, this barrier eliminated the ocean type of life history. This loss of the habitat connectivity that once allowed salmon to migrate between mainstems and subbasins is a major cause of lost productivity, especially in spring and summer chinook races (Lichatowich and Mobrand 1995).

Programs that seek to restore the productivity of migratory species such as Pacific salmon must pay close attention to the status of the habitat and conditions of connectivity and diversity within watersheds. Habitat degradation and loss of connectivity often fail to attract sufficient attention, especially in coastal rivers. Yet the lower reaches of many coastal rivers show signs of deterioration, including elevated water temperatures. Construction of roads for forestry and other purposes, common in coastal watersheds, also isolates side channels from mainstem reaches. Such destruction of connectivity affects watershed processes and the ability of salmon populations to carry out their life histories. The result is a loss of salmon productivity.

Habitat Complexity

A watershed fragmented by land use activities that destroy the connectivity and diversity of salmon habitat is likely to lose its capacity to support a productive salmon population (Lichatowich et al. 1995). When salmon spawning or rearing habitat is destroyed—or damage to migratory habitat prevents salmon from reaching good spawning or rearing habitat—overall productivity of the population is undermined. The land use activity that has had the greatest impact on ecosystems of the coastal rain forest is, of course, logging.

Logging poses a variety of threats to salmon habitat. (For a review see Hicks et al. 1991.) Removing vegetation from streamside areas, for example, increases stream temperatures, promotes streambank erosion, and diminishes the supply of large woody debris. Large wood in streams provides stable habitat structure, cover, and critical pool habitat for rearing salmon. Logging on hillslopes, and building forest roads to reach hillside cutting areas, alters watershed hydrology, increases peak flows, increases surface erosion and the incidence of landslides, and increases sediment to streams, filling pools and smothering incubating eggs. All of these factors simplify stream channels, reduce habitat complexity, and fragment salmon habitats. The reduced complexity of stream habitat has been one of the most pervasive consequences of past forest practices, one that is likely responsible for significant changes in fish communities (Bisson et al. 1992).

Status and Trends

Pacific salmon populations are declining throughout much of their North American range. The cumulative degradation of freshwater habitats—caused by dams, water withdrawals, logging, mining, grazing, farming, urbanization, and other development activities—has been a major factor in this decline, compounded by what is believed to be a cyclical downturn in ocean feeding conditions. Overfishing also has played a significant role. Finally, hatchery management practices that promote overfishing in mixed-stock fisheries, hybridization between native and hatchery stocks, and ecological and behavioral interactions between native and hatchery fish have been detrimental to native salmon populations (White et al. 1995).

Nehlsen (in press) has reviewed the current status of North American salmon and notes a north-to-south trend in the status of salmon populations: from relatively healthy in the north to very depleted in the south (Table 8.1). The Wilderness Society's (1993) map of the distributions of at-risk native Pacific salmon populations in Washington, Oregon, Idaho, and California further clarifies the poor status of salmon in the middle and southern portions of the range. The Wilderness Society concludes that Pacific salmon are either extinct or at risk of extinction across the vast majority of their natural range in the survey area. In over half of the total freshwater habitat of Pacific salmon in the four-state area, 50 to 100 percent of the native salmon species are extinct or at risk of extinction. In over 30 percent of the habitat, all Pacific salmon species are extinct. In only 6 percent of the freshwater habitat in the survey area are no Pacific salmon species known to be declining.

Ocean productivity cycles also play a major role in this north-to-south trend. Beamish and Bouillon (1993) have found that commercial catches of pink, sockeye, and chum salmon correlate with climatic conditions over the North Pacific Ocean. Catches of these northern species increased until 1989 (the last year of record in their 1993 study) and continue to be high. Conversely, Lawson

Table 8.1. General Trends in Salmon Population Status in Alaska, British Columbia, Washington, Oregon, Idaho, and California

Salmon Species	AK	BC	WA	OR	ID	CA
Pink	↑	↑	↑			
Sockeye	↑	↑	↑↓	↓	↓	
Chum	↑	↑↓	↑↓	↓		↓
Coho	↑	↓	↑↓	↓		↓
Chinook	↑	↑↓	↑↓	↓	↓	↓
Steelhead	↓	↓	↓	↓	↓	↓

Note: Down arrows (↓) indicate generally declining trends; up arrows (↑) indicate generally increasing trends. Double arrows (↑↓) indicate mixed trends.

Source: Nehlsen (in press)

(1993) and Lichatowich (in press) summarize literature indicating that low ocean productivity has depressed Oregon coastal coho populations. Francis and Sibley (1991) have observed a reciprocal relationship between pink salmon catch in the Gulf of Alaska and coho salmon catch in the Washington–Oregon–California region: when the catch of pink salmon is high, coho salmon harvests are low and vice versa. In brief, the contrasting status of Pacific salmon in the northern and southern portions of their range is due to northern species (pink, chum, and sockeye) enjoying a period of high ocean productivity while southern species (coho, chinook) are in a period of low ocean productivity.

Pacific salmon populations have always fluctuated in response to changes in climate that affect productivity. In recent decades, however, the peaks have been lower and the troughs deeper as a result of habitat destruction, intensive commercial harvest, and the negative effects of hatcheries (Figure 8.4). Salmon populations in Washington, Oregon, and California are currently responding to a deep trough in natural productivity and may recover considerably as ocean productivity improves. But this is no justification for complacency. Restoration activities must be pursued now, while public concern is high, to build a basis for strong future recovery; otherwise, the next trough in ocean productivity will bring salmon populations even closer to extinction. At the same time, the northern areas in Alaska and British Columbia that have healthy salmon stocks should act now to prevent further habitat degradation and further population declines when unfavorable ocean conditions return, as they surely will, in the future. Pacific salmon depend on a chain of ecosystems, including upland tributaries, mainstem migration corridors, lower-elevation floodplains, estuaries, and the ocean. Each species has a unique set of life history strategies and habitat needs. This diversity reflects the plethora of ecoregions embedded in the Pacific Northwest landscape, including the coastal temperate rain forest. These life history patterns govern each species' spawning and rearing distribution and the timing of its migrations. They influence fishing strategies, as well, and govern each species' or stock's vulnerability to habitat destruction.

The productivity of a salmon population depends on the maintenance of a complex and connected habitat template in which these diverse life histories can be expressed. Salmon populations in Washington, Oregon, Idaho, and California are generally declining—due in part to habitat fragmentation and the loss of biodiversity. The losses of biodiversity and productivity, owing largely to human activities, must be superimposed on a natural pattern of cyclical changes in ocean productivity that affect salmon.

Restoration of depleted populations depends on restoring habitat com-

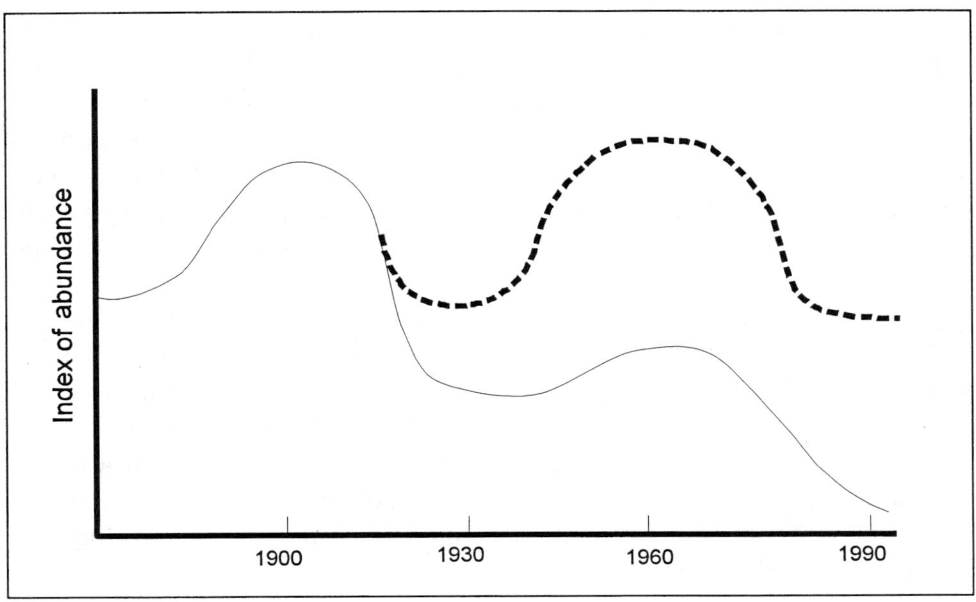

Figure 8.4. Hypothetical representation of salmon abundance in Oregon, Washington, Idaho, and California over the last 150 years. The solid line illustrates the response of salmon to natural fluctuations in climate and productivity. The dashed line represents the probable production without intensive harvest, habitat destruction, and the negative effects of hatcheries.

plexity and connectivity so that the diversity of life histories can be fully expressed. Freshwater habitat in Washington, Oregon, Idaho, and California should be restored now—to take advantage of a likely improvement in ocean rearing conditions and avoid further population declines during the next climatic trough. Freshwater habitat in Alaska and British Columbia should also be protected—to avoid a major loss of productivity and significant population declines when ocean conditions in northern waters deteriorate.

Salmon and Spirit

For many who consider the coastal temperate rain forest their native or adopted home, wild salmon speak most resonantly of the birthplace of spirit. They are tangible; they are visible, if only in the imagination. They are charismatic: pottery, clothing, and jewelry carry their likenesses. Even those uninterested in nature grant the importance of salmon to the cultural and ecological fabric of the Pacific Northwest. Yet the salmon are slipping away.

Though they are perhaps the most compelling link between people and place in this region, salmon are not an easy study. Some things about salmon we do understand or come close to understanding. Others lie within reach and require only more time, more cultural and scientific advancement, perhaps more humility. But much remains that we cannot easily understand, will have to stretch ourselves to grasp, and may still fail to comprehend. Then we will have to be content with articulating the questions or, failing that, struggling with a sense of mystery.

Today we stand on the verge of losing salmon and losing the future opportunities they present us. And yet, ironically, the need to save them presents an opportunity. If we are to save them, we must better understand ourselves and how our past ways have led us to where we are now. We must change how we view salmon and ourselves. Without question we in western North America will be changed by the salmon crisis—we already have been—either through the effort we devote to saving them or through our failure to do so, living with the loss, and explaining it to future generations.

Salmon are not predictable parts of mechanistic processes whose lives or whose habitats we can manipulate for predictable results. Salmon lead wild, uncertain, brief lives, yet their generations are governed by long time spans and vast spatial scales. They challenge us to think about things that are not natural or comfortable for us—such as uncertainty, chaos and unpredictability, and what today's actions might mean centuries from now. They challenge us to deal with nature as it is, with its complexities and mysteries, not as we wish it to be.

Can we conceive of a bioregion in which the natural world receives the priority required to save salmon? Can we be creative, even optimistic, about how this priority would change our cities, farms, and forests? Are we willing to make these changes to keep salmon and nature in our lives? Can we become the kind of people who can coexist with salmon and the natural world?

What is the alternative? Most residents of the northwest coast of North America are not native to this coastal rain forest. Finding a connection to the landscape and understanding ourselves in its context, therefore, is ever more difficult. Salmon remain one thread connecting people and landscape. Through time and seasons, through moving rivers, they follow life histories that evolved in ancestral habitats. Their existence allows people, even transplants from other places, to touch 10,000 years of the region's history. Salmon weave us into the landscape in a way that nothing else can.

Ultimately, failing to save the salmon means losing part of ourselves. Salmon connect us to something beyond us, and without limit. In stretching to know them, we may know ourselves better, and create a better future.

Acknowledgments

The authors wish to thank Thomas P. Quinn for his comments on an early draft of this chapter. The final section was adapted from "Without Them, Who Are We?" by Willa Nehlsen, which appeared in *Illahee* (Winter 1994).

References

Beamish, R. J., and D. R. Bouillon. 1993. "Pacific salmon production trends in relation to climate." *Can. J. Fish. Aqua. Sci.* 50:1002–1016.

Bilby, R. E., B. R. Fransen, and P. A. Bisson. 1996. "Incorporation of nitrogen and carbon from spawning coho salmon into the trophic system of small streams: Evidence from stable isotopes." *Can. J. Fish. Aqua. Sci.* 53(1):164–173.

Bisson, P. A., T. P. Quinn, G. H. Reeves, and S. V. Gregory. 1992. "Best management practices, cumulative effects, and long-term trends in fish abundance in Pacific Northwest river systems." In R. J. Naiman, ed., *Watershed Management: Balancing Sustainability and Environmental Change*. New York: Springer-Verlag.

Cederholm, C. J., D. B. Houston, D. L. Cole, and W. J. Scarlett. 1989. "Fate of coho salmon (*Oncorhynchus kisutch*) carcasses in spawning streams." *Can. J. Fish. Aqua. Sci.* 46(8):1347–1355.

Francis, R. C., and T. H. Sibley. 1991. "Climate change and fisheries: What are the real issues?" *Northwest Env. J.* 7:295–307.

Fredin, R. A. 1980. "Trends in North Pacific salmon fisheries." In W. J. McNeil and D. C. Himsworth, eds., *Salmonid Ecosystems of the North Pacific Ocean*. Corvallis: Oregon State University Press.

Groot, C., and L. Margolis, eds. 1991. *Pacific Salmon Life Histories*. Vancouver: University of British Columbia Press.

Healey, M. C. 1991. "Life history of chinook salmon (*Oncorhynchus tshawytscha*)." In C. Groot and L. Margolis, eds., *Pacific Salmon Life Histories*. Vancouver: University of British Columbia Press.

Hicks, B. J., J. D. Hall, P. A. Bisson, and J. R. Sedell. 1991. "Responses of salmonids to habitat changes." In W. R. Meehan, ed., *Influences of Forest and Rangeland Management on Salmonid Fishes and Their Habitats,* Special Publication 19. Bethesda, Md.: American Fisheries Society.

Lawson, P. W. 1993. "Cycles in ocean productivity, trends in habitat quality, and the restoration of salmon runs in Oregon." *Fisheries* 18(8):6–10.

Lichatowich, J. A. In press. "Evaluating the performance of salmon management institutions: The importance of performance measures, temporal scale and productivity cycles." In *Proceedings of Pacific Salmon and Their Ecosystems*. Seattle: University of Washington Center for Streamside Studies, College of Forest Resources, and College of Ocean and Fisheries Sciences.

Lichatowich, J. A., and L. E. Mobrand. 1995. *Analysis of Chinook Salmon in the Columbia River from an Ecosystem Perspective*. Portland, Ore.: Bonneville Power Administration.

Lichatowich, J. A., L. Mobrand, L. Lestelle, and T. Vogel. 1995. "An approach to the diagnosis and treatment of depleted Pacific salmon populations in Pacific Northwest watersheds." *Fisheries* 20(1):10–18.

Matsen, B. 1994. "From the tower: The evolution and biology of Pacific salmon." In N. Fobes, T. Jay, and B. Matsen, *Reaching Home: Pacific Salmon, Pacific People.* Seattle: Alaska Northwest Books.

Matson, R. G., and G. Coupland. 1995. *The Prehistory of the Northwest Coast.* San Diego: Academic Press.

Meehan, W. R., and T. C. Bjornn. 1991. "Salmonid distributions and life histories." In W. R. Meehan, ed., *Influences of Forest and Rangeland Management on Salmonid Fishes and Their Habitats.* Special Publication 19. Bethesda, Md.: American Fisheries Society.

Nehlsen, W. In press. "Pacific salmon status and trends—a coastwide perspective." In *Proceedings of Pacific Salmon and Their Ecosystems.* Seattle: University of Washington Center for Streamside Studies, College of Forest Resources, and College of Ocean and Fisheries Sciences.

Nehlsen, W., J. E. Williams, and J. A. Lichatowich. 1991. "Pacific salmon at the crossroads: Stocks at risk from California, Oregon, Idaho and Washington." *Fisheries* 16(2):4–21.

Oregon [Pacific] Rivers Council. 1992. "The economic imperative of protecting riverine habitat in the Pacific Northwest." Research Report 5. Eugene, Ore.

Schalk, R. F. 1977. "The structure of an anadromous fish resource." In L. R. Binford, ed., *For Theory Building in Archaeology: Essays on Faunal Remains, Aquatic Resources, Spatial Analysis, and Systemic Modeling*. New York: Academic Press.

Taylor, E. B. 1990. "Environmental correlates of life-history variation in juvenile chinook salmon, *Oncorhynchus tshawytscha* (Walbaum)." *J. Fish Biol.* 37:1–17.

White, R. J., J. R. Karr, and W. Nehlsen. 1995. "Better roles for fish stocking in aquatic resource management." *Am. Fish. Soc. Symp.* 15:527–547.

Wilderness Society. 1993. *Pacific Salmon and Federal Lands: A Regional Analysis.* Washington, D.C.: Wilderness Society.

9. Environmental History

Richard J. Hebda and Cathy Whitlock

□ □

Home to some of the world's largest and oldest trees, the coastal temperate rain forest of the North Pacific coast of North America conjures up a picture of an ancient and primeval superbeing. Images of tree limbs festooned with beard-like lichens, an understory choked by impenetrable thickets, and a landscape home to ancient First Peoples convey a sense of solidity and stability stretching back countless millennia. Might this be a false impression? Throughout much of the range of today's rain forest, glaciers ground their way from the mountains to the ocean only 14,000 to 15,000 years ago, a blink of geological time by any measure.

The coastal forests as we see them today are really snapshots in time of a long and changing series of ecosystems shaped by many processes ranging from global climate change to local human disturbance. How old is the coastal temperate forest? This is not simply a question of the age of the trees but a question of the age of the ecosystems that comprise this vegetation type. How did this remarkable assemblage of life forms arise?

Of great importance is the fact that the history of each ecosystem is part and parcel of its biodiversity. The history of the coastal rain forest is the reason why it is where it is and looks the way it does. Biodiversity is more than numbers, it is uniqueness; ecological uniqueness is a product of history. By understanding history, we gain valuable insight into how ecosystems function over ages, measured not in years and decades, but in centuries and millennia. From this understanding we may learn how to interact with the forest in a sustainable manner. A long-term perspective is especially important in the context of sustaining the forest for generations, particularly with global climatic change on the horizon.

In this chapter we describe the origin and history of the coastal rain forest in its many guises. Our objective is to convey the sense of a dynamic and complex entity with roots reaching far back in time, but one that has evolved its recent face not so long ago. We begin with a discussion of ecosystems and time, an important starting point for any consideration of ecosystem history. We describe briefly the methods involved in the study and reconstruction of ancient landscapes and forests. We conclude with a discussion of the implications of knowledge gained from studying ecosystem history for the bioregion's resources in the future.

Ecosystems and Time

"Ecosystem" is a familiar concept encompassing the living and physical components of a landscape at whatever spatial scale is of interest. An element of *distinctness* is implied. The ecosystem has features that distinguish it from adjacent ecosystems, implying boundaries in space.

Ecologists also recognize the dimension of time as part of the concept. The idea of a "climax" ecosystem, for example, implies that the biotic and physical components have interacted over an interval of time and assembled a web of life and land that has stability. This stable configuration can be disrupted by fire, windstorms, and disease, but eventually the climatic climax returns after a process called succession. These concepts imply that a coherent ecosystem, of which the coastal temperate rain forest climax is an example, will somehow reassemble itself no matter what the disruption. After all, this is what ecologists have observed, more or less, in the century or so they have studied ecosystems.

The perspective of a few decades or centuries has produced a false impression of ecosystem stability. Ecosystems come and go with passing millennia and sometimes even more quickly. Evolution and factors affecting biogeographic distributions alter the nature of the biotic pool available to a future ecosystem. Furthermore, the living components of an ecosystem respond individually as a region's physical circumstances, mainly its climate, change. Even if the same species reassemble into an ecosystem superficially similar to the original one, the new ecosystem cannot be exactly the same. The component species, especially plants, have evolved during the course of their history. Ecosystems do not migrate as coherent units: individual species do so by changing their ranges. Paleoecological studies clearly demonstrate that ecosystems have a finite existence in a place during an interval of time.

Since the time-dimensional ecosystem combines elements of life (*bios*), physical setting (*geos*), and time (*chronos*), let us call it a "biogeochron." The name derives from Krajina's (1965) term for regional ecosystem units in British Columbia. He called these units "biogeoclimatic zones." The biogeochron embodies similar concepts of the living and physical components (substrate and climate) of the landscape, but this notion includes the climatic factor as part of *geos*. The biogeochron can be applied at any spatial scale (from the forest stand to the continent) and is not only a regional concept like Krajina's biogeoclimatic zone. The coastal temperate rain forest, as we see it functioning today, can be viewed as a vast ecosystem or complex of ecosystems. The coastal temperate rain forest phase on the Pacific coast of North America is a biogeochron. As you will see in the following pages, although this rain forest seems an ancient entity, it is the result of changes to preceding biogeochrons and it, like them, will have a transient history. Biogeochrons never repeat exactly. Rather, they are assemblages of creatures uniquely interacting for a time under the set of physical circumstances characteristic of a particular place.

Why Do Ecosystems Change?

Ecosystems such as the coastal rain forest are influenced by environmental controls that operate on different space and time scales. Over the last 2.5 million years, variations in latitudinal and seasonal distribution of solar radiation have driven a succession of glaciations and interglaciations (Imbrie et al. 1992). In the last 750,000 years, for example, the earth has shifted from glacial to interglacial mode every 100,000 years. The warm interglacial periods last about 10,000 years. The Holocene, extending from 10,000 years ago to the present, with high summer radiation in the Northern Hemisphere, is the most recent interglacial period. Like former interglaciations, it is characterized by minimal ice cover, warm oceans, and widespread forest, temperate grassland, and warm desert.

On time scales of thousands of years, such as the period from the last glacial maximum to the present, variations in the size of the North American ice sheets have influenced environmental change far beyond the ice-sheet margins. During the last 20,000 years, changes in the size of the Laurentide and Cordilleran ice sheets and variations in the seasonal cycle of solar radiation have led to major environmental changes throughout the coastal rain forest region. That the vegetational adjustments occurred more or less at the same time emphasizes the magnitude of the climatic changes. General-circulation computer models provide insight into climatic changes over the last 20,000 years and their possible effect on the Pacific Northwest (Thompson et al. 1993). These models simulate the climate of particular times, taking into account variations in the seasonal cycle of solar radiation, ice-sheet size, sea surface temperatures, land surface characteristics, and atmospheric composition (Kutzbach et al. 1993). Comparison of the model simulations with paleoenvironmental data helps us to understand the large-scale causes of regional climate change and points to circumstances where model and data disagree and require further study. Thus these models help to explain paleoenvironmental change as it is recorded in the fossil record.

These variations in the large-scale controls of climate have a direct bearing on the climatic history of northwestern North America. The size and height of the ice sheets, for example, may have determined the path of the jet stream across the continent and thus the latitude of prevailing storm tracks. The amplitude of the seasonal cycle of solar radiation affected the strength of the eastern Pacific subtropical high-pressure system and the intensity of summer drought (Thompson et al. 1993; Barnosky et al. 1987).

Superimposed on these regional variations are changes in environment and climate that influence biotic distribution on time scales of centuries or less and at subregional spatial scales. Short-term climatic oscillations, such as the Little Ice Age (ca. 1650–1890 A.D.), have altered the frequency of natural fires and shaped plant communities in many Northwest forests (Hemstrom and Franklin 1982; Morrison and Swanson 1990; Graumlich and Brubaker 1986). Substrate characteristics also play a role in local vegetation composition. For example, prairie and oak woodland in the Willamette Valley, the central Puget lowland, southeast Vancouver Island, and the northeast Olympic Peninsula occur under summer drought conditions on coarse-textured, outwash-derived soils (Franklin and Dyrness 1988).

Studying Forest History

Ancient forests and landscapes are revealed and reconstructed through the discipline of paleoecology. This field of study combines techniques and concepts of biology and geology with a little intuitive interpretation to see what ancient ecosystems looked like. Basically the paleoecologist is a time traveler who studies geological exposures, lake and bog cores, and their fossil contents. Unlike a time traveler, though, the paleoecologist infers an incomplete image of the past on the basis of the bits and pieces preserved in sediments.

The characteristics of sediment layers deposited over millennia provide insight into physical events. The sediments also capture and preserve biological remains (Figures 9.1 and 9.2). The age of the sediment layers is established by radiocarbon dating and other dating techniques. In this chapter we relate our account in terms of "years ago," but we actually mean "radiocarbon years before 1950," the starting date for the radiocarbon decay clock. Today some ages are being converted to actual or calendar years before the present ("Cal

Figure 9.1. Ancient logs dredged up from sediments of Heal Lake, Vancouver Island. (Photo courtesy of R. Hebda.)

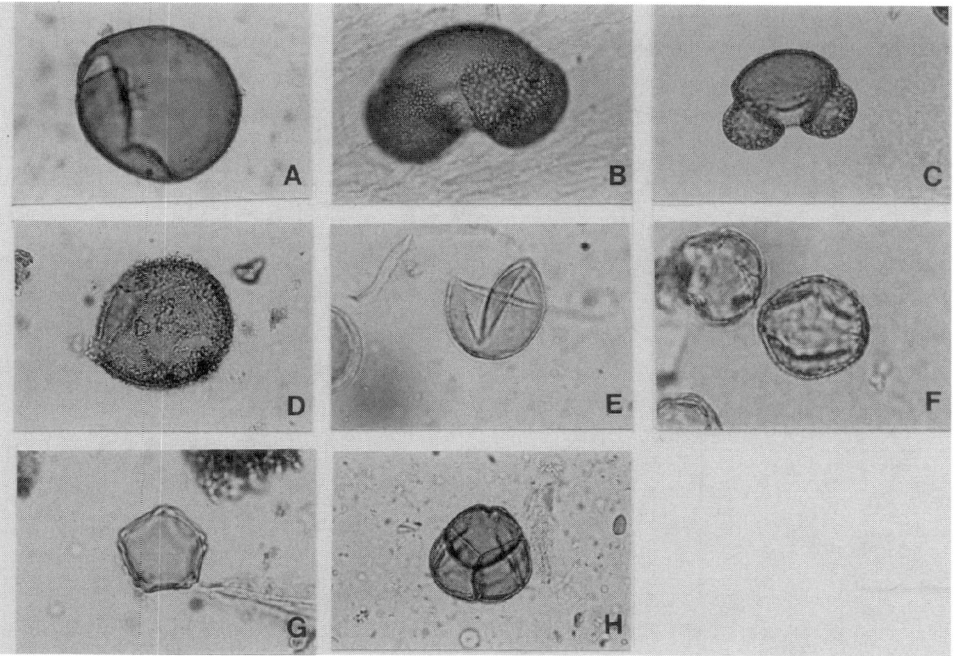

Figure 9.2. Common pollen types of the coastal rain forest: (*a*) Douglas-fir; (*b*) Sitka spruce; (*c*) lodgepole pine; (*d*) western hemlock; (*e*) western redcedar; (*f*) Garry oak; (*g*) red alder; (*h*) salal. Parts *a, b, c, d,* and *h* magnified about 300 times; parts *e, f,* and *g* magnified about 750 times. (Photo courtesy of R. Hebda.)

Yr BP") by using radiocarbon-dated tree-ring chronologies (Stuiver and Reimer 1993). Many pollen records, however, have not been converted in this manner, so we continue to use the "radiocarbon years before 1950" time scale.

Pollen, spores, seeds, and conifer needles are the main biological remains in the study of the history of the coastal rain forest. Of these, the study of pollen and spores, called palynology or pollen analysis, yields most of the information used to reconstruct ancient forests (Moore et al. 1991) (Figure 9.3). Different plant taxa (genera and species) produce distinctive pollen and spore types (Figure 9.2). These are dispersed by various agents, especially wind and water, and then deposited in the sediments accumulating in wetlands, lakes, and even the ocean. The dominant conifers of the coastal rain forest produce billions of wind-dispersed pollen grains, thus giving us a clear picture of their changing distribution.

The distinctive and remarkably tough outer coat of pollen and spores is constructed of sporopollenin, a waxy substance that withstands the onslaught of time. A few cubic centimeters of suitable sediment harbor hundreds of thousands, if not millions, of pollen grains and spores. These plant microfossils reflect in a direct way the composition of the vegetation that covered the region near the study site at the time the sediment was deposited.

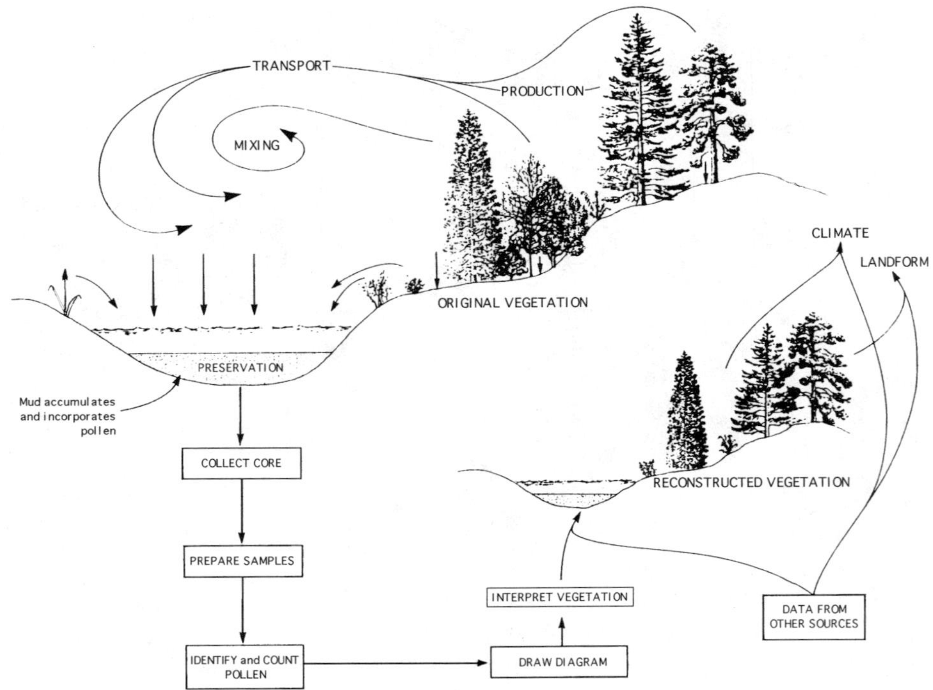

Figure 9.3. Diagram outlining the basis of pollen analysis.

Palynologists extract the pollen and spores through a series of physical and chemical treatments, mount the residues on microscopic slides, and identify and count the constituent remains at a magnification of 400 to 1000× (Moore et al. 1991). The results are usually presented as percentage data that are ordered into a "pollen" diagram. The pollen diagram can then be divided into stratigraphic assemblage zones that separate the most distinctive intervals.

The key to interpreting these zones is an understanding of each species' ecological characteristics. Information from other components of the sediment record, such as insect remains and chemical characteristics, helps to verify and fill out details of ancient environments (Berglund 1986). The reconstructed vegetation helps us to derive past climate, infer physical features of the landscape, and even detect human disturbance of ecosystems. Several pollen diagrams from sites in different settings provide the basis for a regional forest history.

The Coastal Temperate Rain Forest Area

For this account we include the strip of mainly coastal coniferous forest stretching from Kodiak Island, Alaska, southward through British Columbia, Washington, and Oregon (see Chapter 4; Franklin and Dyrness 1988; Meidinger and Pojar 1991; Viereck et al. 1992) (Figure 9.4). The region has three climatic gradients that influence forest composition: north to south; east to west; and sea level to mountaintop.

First, there is a gradual northward cooling with latitude, though this gradient is much modified by the enormous mass of the adjacent Pacific Ocean. Because of this trend, many southern coastal species such as grand fir (*Abies grandis*) and Garry oak or Oregon white oak (*Quercus garryana*) extend only as far north as southern British Columbia. In the north, tree species begin to reach the limit of their ranges (Heusser 1985). Second, climate becomes increasingly continental eastward from the coast as the moderating effect of the ocean declines. Valleys that stretch toward the interior have much hotter summers and cooler winters than the coast. There is also a rain shadow on eastward slopes of mountains. Third, climate becomes cooler and wetter with elevation, in part resembling the south–north gradient.

The current vegetation of the rain forest zone varies according to these three climatic gradients. In the south, moist oceanic sectors support western hemlock (*Tsuga heterophylla*), western redcedar (*Thuja plicata*), and Pacific silver fir (*Abies amabilis*) forests. In the north, Sitka spruce (*Picea sitchensis*), mountain hemlock (*Tsuga mertensiana*), western hemlock, and yellow-cedar (*Chamaecyparis nootkatensis*) are prominent, along with lodgepole pine (*Pinus contorta*) in bog environments. With increasing elevation, cold-intolerant species, such

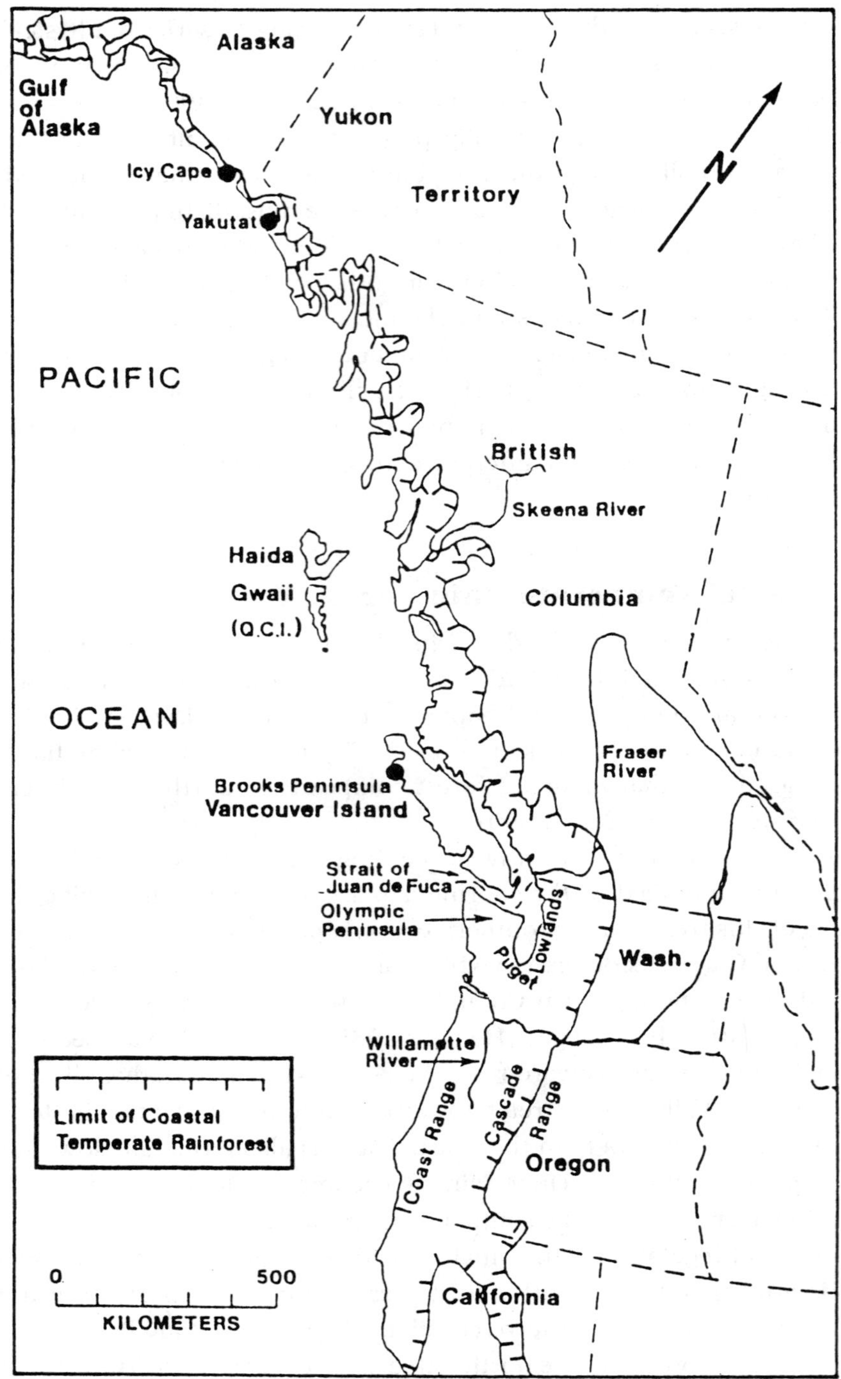

Figure 9.4. Selected locations mentioned in Chapter 9.

as western hemlock, drop out of the forests and are replaced by mountain hemlock and yellow-cedar, thus forming coastal subalpine forests. Further south, the coastal forest is dominated by Sitka spruce, western hemlock, and western redcedar; Douglas-fir is an important species in the early stages of forest development. The coastal rain forest follows a narrow strip in the drier Oregon climate. Port Orford cedar (*Chamaecyparis lawsoniana*) and redwood (*Sequoia sempervirens*) grow commonly in southwestern Oregon and California. Montane slopes of the Cascade Range support forests of Pacific silver fir, noble fir (*Abies procera*), and at the highest elevations below the timberline mountain hemlock, subalpine fir (*Abies lasiocarpa*), Engelmann spruce (*Picea engelmannii*), and whitebark pine (*Pinus albicaulis*). In the southern Cascade Range, sugar pine (*Pinus lambertiana*), California red fir (*Abies magnifica*), and at lower elevations white fir (*A. concolor*) reach their northern limits.

Drier, continental areas in the middle and southern coastal rain forest region support Douglas-fir forests with red alder (*Alnus rubra*) as well as some of the species just cited. Southern interior rain forests consist of western hemlock, western redcedar, and Douglas-fir on moist sites, as well as grand fir and western white fir. Within the southern section the driest areas of the rain forest, including southeast Vancouver Island, the San Juan Islands, the southern Puget lowland, and the Willamette Valley, support deciduous forest stands, mainly of Garry oak, and even nonforested assemblages (Franklin and Dyrness 1988).

Preglacial Forests

Although rain forest conifer and deciduous tree genera evolved in the late Cretaceous and early Tertiary periods (Stewart and Rothwell 1993), the first vegetation resembling the coastal rain forest appears during interglacial intervals recorded in Pleistocene deposits (Heusser 1977; Heusser and Heusser 1981; Alley and Hicock 1986). At sites in the Puget lowland and south Vancouver Island, ancient pollen assemblages resemble those of modern moist (western hemlock) and dry (Douglas-fir) phases of the coastal temperate rain forest. These assemblages date back to the last interglacial period, called the Sangamon Interglaciation, more than 100,000 years ago, and earlier interglaciations. These assemblages hint that coastal temperate conifer biogeochrons are characteristic of Pleistocene interglacial climates on the northwest coast of North America.

With the growth of ice sheets during the Wisconsin Glaciation 80,000 years ago, major changes occurred in the vegetation of the Pacific coast. One of the features of this glaciation was an alternation between periods of glacial advance and retreat. As far south as Washington State, full glacial intervals were characterized by nonforested or parkland assemblages of plants apparently dominated by grasses, sedges and members of the aster family (Asteraceae), and scattered conifers, such as mountain hemlock and lodgepole pine (Figure 9.5). Nonglacial landscapes supported forests or parkland in which hemlocks and spruce predominated (Heusser 1977).

The pollen from the last Olympia Nonglacial Interval, about 60,000 to 30,000 years ago, gives us a glimpse of coastal forests before the most severe part of the last glaciation. During the Olympia interval much of the rain forest region supported cool moist conifer forests (Heusser 1977; Armstrong et al. 1985; Mathewes 1989) (Figure 9.5). Western hemlock predominated with spruce, mountain hemlock, and fir on the moist west coast of the Olympic Peninsula. Further north, and in the more continental Fraser River valley, mountain hemlock and spruce held sway and western hemlock was much less abundant than today. Douglas-fir, found mainly in the driest parts of the region today, was absent over much of its range, presumably because the climate was too cool. Western redcedar and yellow-cedar, a diagnostic twosome of moderate and cool forms of today's coastal rain forest, respectively, were apparently a rare component of Olympia nonglacial forests.

The central Coast Range of Oregon supported an open forest of western white pine, western hemlock, and fir before 25,000 years ago and included mountain hemlock, red alder and Sitka alder, and possibly yellow-cedar. The assemblage has no modern analog, but the combination of taxa suggests that conditions were cooler and wetter than today (Worona and Whitlock 1995).

The Great Freeze

Between 25,000 and 14,000 years ago, the earth passed through the height of the last glaciation and the continental ice sheets, notably the Laurentide ice sheet, reached maximum extent. This period is referred to as the Fraser Glaciation in the Pacific Northwest. Using a full-sized Laurentide ice sheet in the general-circulation computer model, simulations suggest that the climate of northwestern North America was affected in three ways (Kutzbach et al. 1993; Thompson et al. 1993; Whitlock 1992). First, the mass of cold Laurentide ice cooled northern midlatitudes. Second, the jet stream veered south of its present path; as a result, winter storms tracked south of today's path, and the southern coastal rain forest region was robbed of winter moisture during the full-glacial period. Third, clockwise (anticyclonic) circulation over the Laurentide ice sheet resulted in stronger easterlies along the southern margin of the ice and southeasterlies in coastal British Columbia and Alaska, bringing dry cold, continental air to the coast.

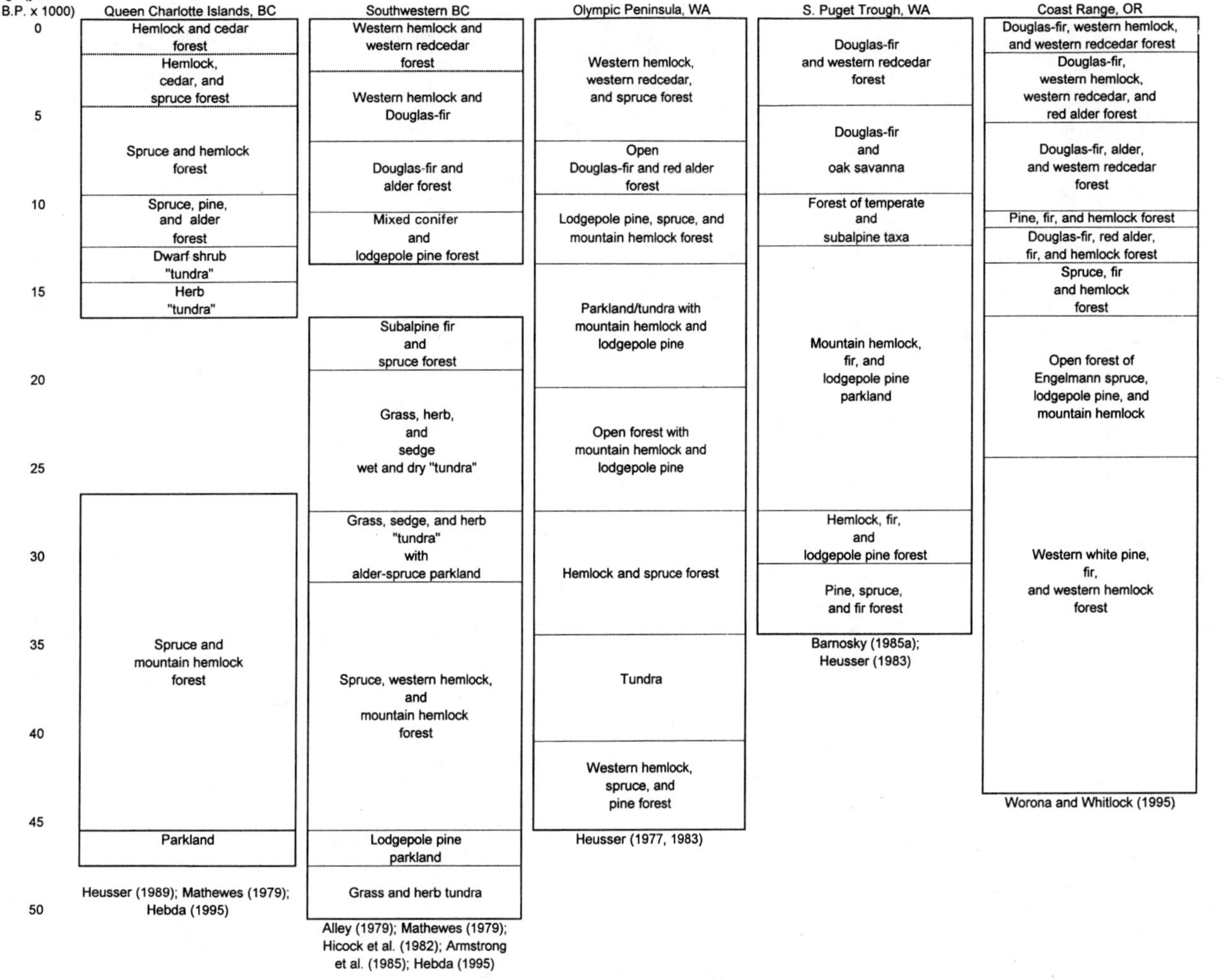

Figure 9.5. Summary of vegetation histories of selected regions of the coast of northwestern North America for the last 50,000 years.

The climatic features derived from computer models are consistent with paleoenvironmental reconstructions based on fossil assemblages. Nonforested and parkland communities were widespread during glacial time in the southern coastal rain forest region, and the conifer rain forest assemblages of the preceding nonglacial interval largely vanished (Heusser 1983; Hicock et al. 1982; Whitlock 1992). From 20,000 to 16,000 years ago, alpine glaciation was extensive, but the southern limit of the Cordilleran ice sheet was largely confined to British Columbia and little glacial ice reached the lowlands of Washington (Waitt and Thorson 1983; Clague 1989). Apparently the lack of winter precipitation resulting from the southward position of the jet stream inhibited the Cordilleran ice sheet from expanding southward. Perhaps because of the later northward shift in the jet stream, Cordilleran ice reached its southernmost limits 15,000 to 14,000 years ago, when ice lobes extended south into the Puget lowland and the Strait of Juan de Fuca (Waitt and Thorson 1983; Hicock et al. 1982). When the Cordilleran ice sheet was at its maximum, alpine glaciers in the Cascade Range and Olympic Mountains were actually smaller than before, and large areas of the Cascade Range were already ice-free (Waitt and Thorson 1983).

From 20,000 to 16,000 years ago, ice-free regions of western Washington and Oregon experienced a wider range of temperatures along a latitudinal gradient than at present. In western Washington, the vegetation consisted of tundra and subalpine parkland (Figures 9.6 and 9.7). Farther south, open forest of Engelmann spruce, lodgepole pine, fir, and mountain hemlock grew in central Oregon. Apparently the climate was more continental than today, because high-elevation inland species such as Engelmann spruce and subalpine fir extended their range westward (Barnosky 1985a; Hicock et al. 1982; Worona and Whitlock 1995).

In contrast, the climatic influence of Cordilleran ice was not felt far south of its margins. A site located a few kilometers south of the Puget ice lobe features high pollen percentages of cold-tolerant alder and pine (Heusser 1977), suggesting that trees or shrubs grew near and perhaps on the glacier surface. The insensitivity of vegetation to local glacial conditions is also evident from the observation that vegetation change was not synchronous with ice retreat. Cordilleran ice receded rapidly from Washington after 14,500 years ago, but temperate taxa did not appear in the lowlands until about 2000 years later (Whitlock 1992). Apparently plant communities south of the ice responded more to continental-scale changes in climatic patterns and less to local glacier position.

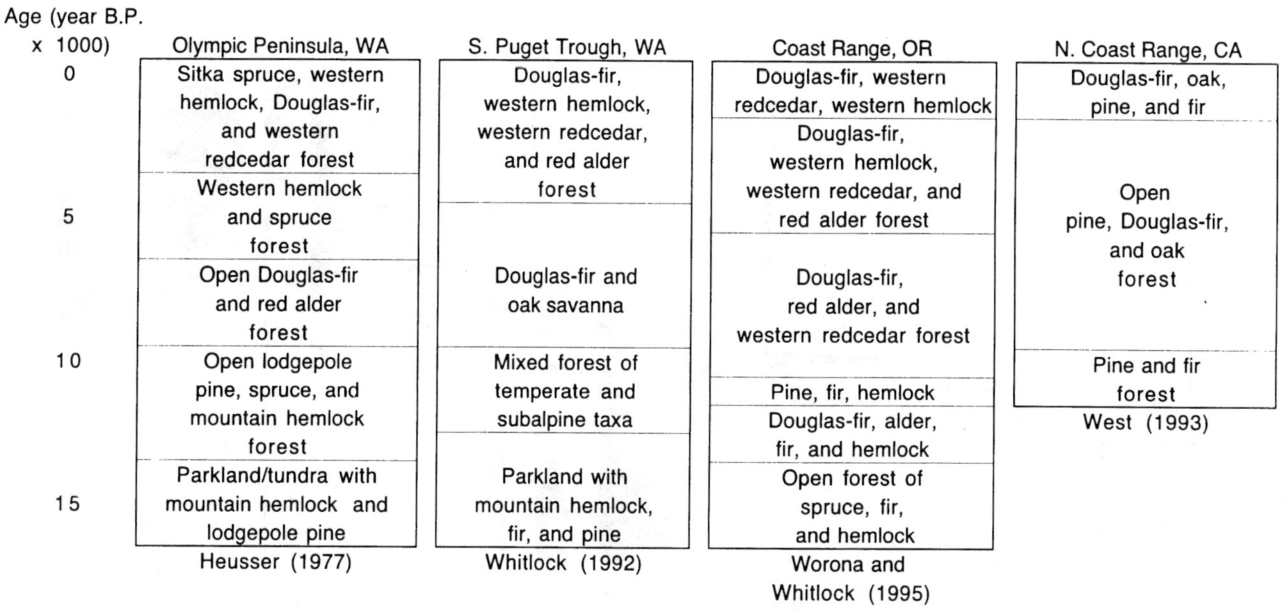

Age (year B.P. x 1000): 0, 5, 10, 15

Olympic Peninsula, WA	S. Puget Trough, WA	Coast Range, OR	N. Coast Range, CA
Sitka spruce, western hemlock, Douglas-fir, and western redcedar forest	Douglas-fir, western hemlock, western redcedar, and red alder forest	Douglas-fir, western redcedar, western hemlock	Douglas-fir, oak, pine, and fir
Western hemlock and spruce forest	Douglas-fir and oak savanna	Douglas-fir, western hemlock, western redcedar, and red alder forest	Open pine, Douglas-fir, and oak forest
Open Douglas-fir and red alder forest	Mixed forest of temperate and subalpine taxa	Douglas-fir, red alder, and western redcedar forest	Pine and fir forest
Open lodgepole pine, spruce, and mountain hemlock forest	Parkland with mountain hemlock, fir, and pine	Pine, fir, hemlock	
Parkland/tundra with mountain hemlock and lodgepole pine		Douglas-fir, alder, fir, and hemlock	
		Open forest of spruce, fir, and hemlock	
Heusser (1977)	Whitlock (1992)	Worona and Whitlock (1995)	West (1993)

Figure 9.6. Biogeochron sequences spanning the last 15,000 years for selected sites in Washington, Oregon, and California.

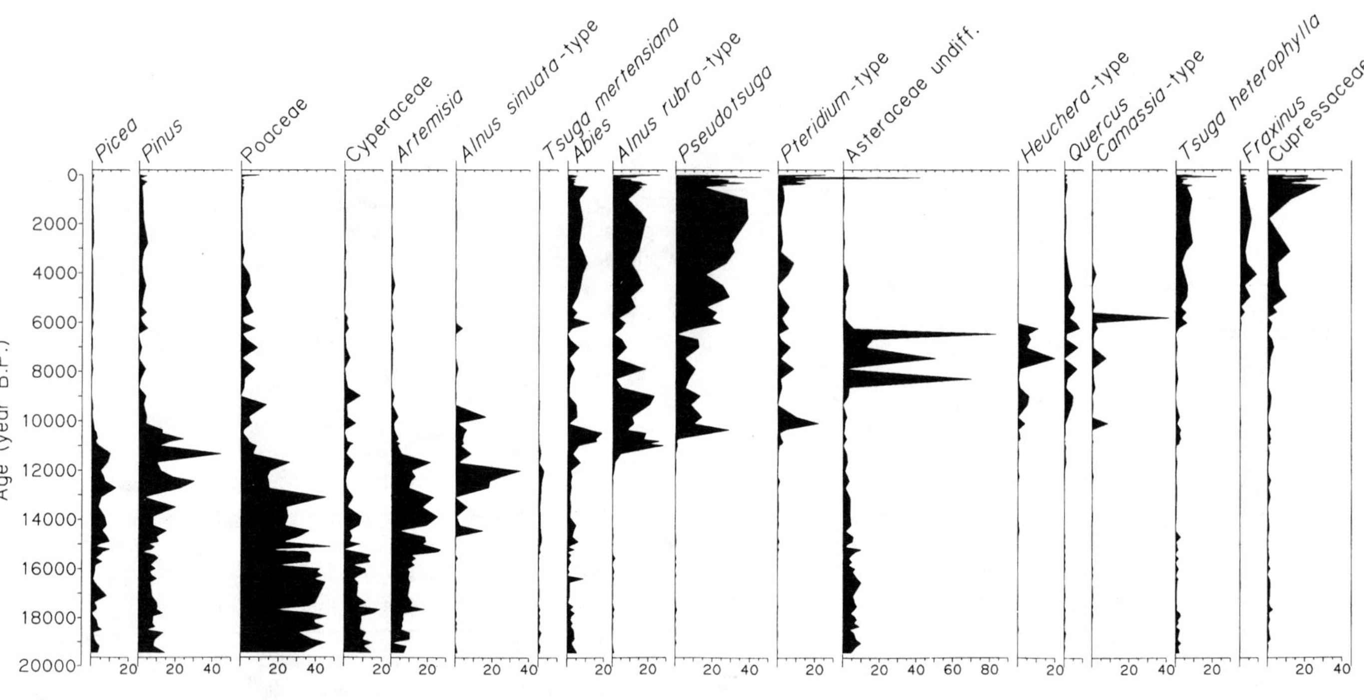

Figure 9.7. Simplified pollen diagram from Battle Ground Lake, southern Puget lowland, Washington. *Source:* Barnosky (1985a).

Reforestation After Ice

Soon after the Cordilleran ice sheet retreated, temperate trees colonized the emerging land (Hebda 1995; Whitlock 1992). From 12,000 to 6000 years ago, perihelion (when the earth is nearest the sun) occurred in summer rather than in winter as it does today, and this had a dramatic effect on the Northwest's climate. Summer radiation on the coastal rain forest was 8 to 10 percent higher 9000 years ago than today, so temperatures were higher and effective moisture lower. Increased summer radiation strengthened the East Pacific subtropical high-pressure system, exaggerating summer drought (Barnosky et al. 1987).

In parts of Haida Gwaii (Queen Charlotte Islands) and coastal Alaska, open tundralike communities developed following the ice retreat (Heusser 1985; Mathewes 1989). But over much of the deglaciated terrain, the earliest invader was lodgepole pine (Figures 9.5 and 9.8). This pine biogeochron persisted for nearly 2000 years in a region from southeastern Alaska to Washington State (Heusser 1985; Hebda 1995). Forests were initially open with willows (*Salix* spp.) and soapberries (*Shepherdia canadensis*), but as soils developed and climate warmed they closed (Mathewes 1973) and diversified. Lodgepole pine's success on the freshly deglaciated landscape may have reflected its superior ability to germinate and grow on poorly developed soils. The source area for pine's widespread expansion, so quickly after ice melted, has yet to be determined (Peteet 1991).

Several full-glacial coastal refugia have been suggested (Heusser 1989). One source may have been the regions south of the ice sheet, where lodgepole pine, mountain hemlock, and other species survived during glacial times. As the low sea levels during glacial periods exposed parts of the coastal shelf, conifers may have found refuge there. Though no direct evidence has yet been found for such continental shelf refugia, one can imagine that these areas may have provided the source for the pioneering conifer forests that moved into British Columbia and southeastern Alaska.

The glacial location of Douglas-fir has also been the subject of much speculation (Tsukada 1982; Worona and Whitlock 1995). In the central Coast Range of Oregon, Douglas-fir occurred in a mixed-conifer forest about 13,000 years ago (Worona and Whitlock 1995) (Figure 9.6), although we have no record of its whereabouts before that. In the Puget lowland, Douglas-fir expanded its range very rapidly between 11,000 and 10,000 years ago (Tsukada et al. 1981; Barnosky 1985a). The appearance of Douglas-fir at sites in the Puget lowland, however, shows no clear south-to-north pattern as might be expected by migration from a southern source. Douglas-fir may have spent the last glaciation in isolated pockets in the unglaciated hills of the Cascade Range or in communities along the now-submerged coastal shelf. Between 10,000 and 9000 years ago, Douglas-fir expanded onto eastern Vancouver Island and along the west

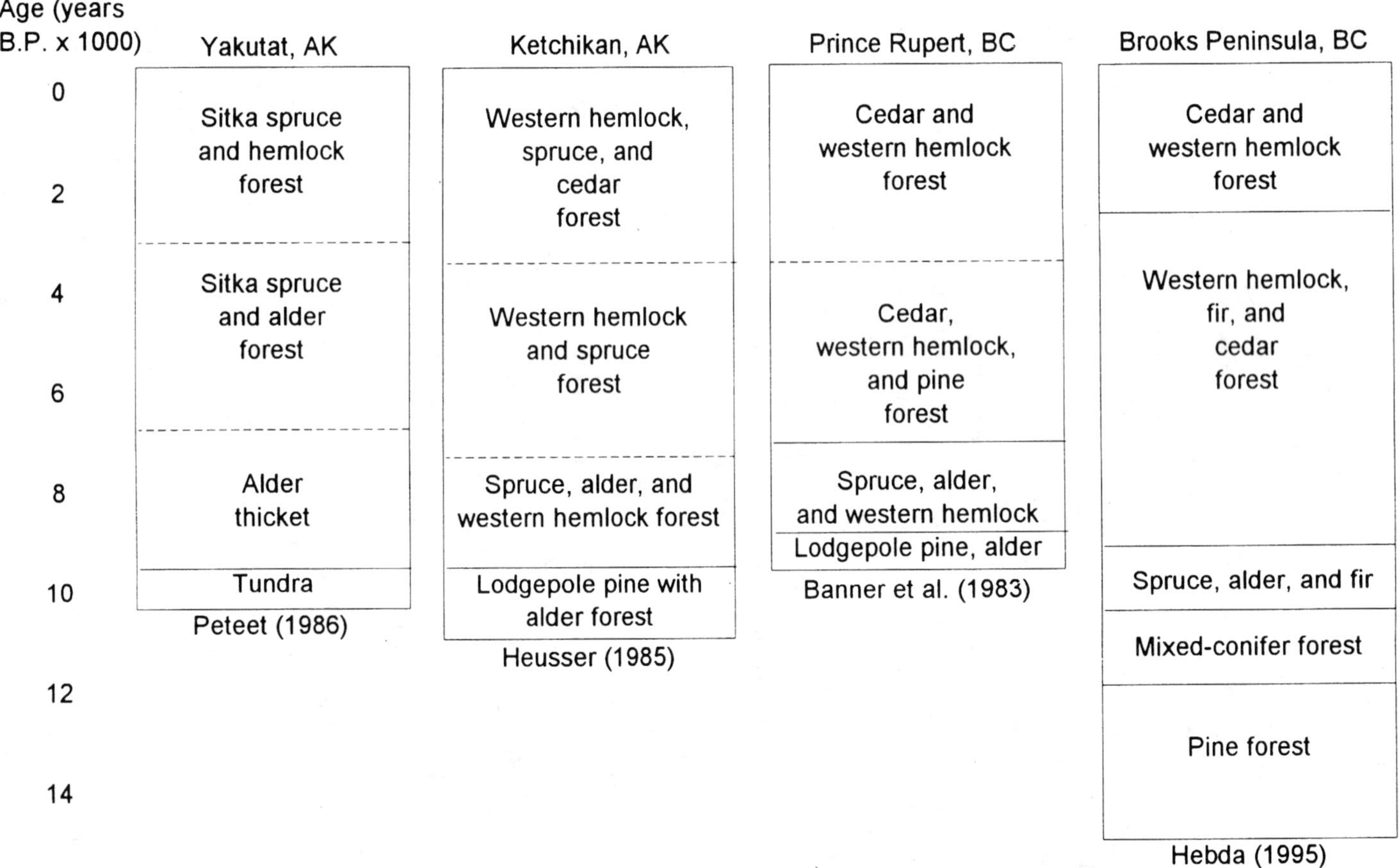

Figure 9.8. Biogeochron sequences following the last glaciation for selected sites in coastal British Columbia and Alaska.

side of the Olympic Peninsula (Heusser 1985). How far north this species spread in response to early Holocene warming will remain unknown until studies are carried out on British Columbia's central coast.

At various times between 12,000 and 10,000 years ago, a mixed forest of lodgepole pine, mountain hemlock, western hemlock, Sitka spruce (probably), and true fir appeared from Haida Gwaii to the central Oregon Coast Range (Heusser 1985; Whitlock 1992; Hebda 1995; Worona and Whitlock 1995). The admixture of high-elevation and northern coastal taxa, such as mountain hemlock, along with temperate species, such as western hemlock, has no extensive modern equivalents, making it difficult to imagine the nature of the vegetation or the climate. Mathewes (1993) argues that a brief and sudden cooling, called the Younger Dryas in Europe, was in part responsible for these combinations. Within the former glacial limits and somewhat south grew even more diverse conifer forests that included Douglas-fir, Sitka spruce, western hemlock, and grand fir (Whitlock 1992). These late-glacial forests are notable because they brought together many of the present-day species of the coastal temperate rain forest, but in unusual combinations (Figures 9.6 and 9.7). At the same time in the dry rain shadow of the northeastern Olympic Peninsula, communities dominated by shrubs and herbs developed on coarse-textured soils (Petersen et al. 1983).

The Warm and Dry Early Holocene

The vegetation types from 10,000 to about 7000 years ago are consistent with a period of severe summer drought as a result of greater-than-present summer solar radiation and lesser-than-present winter solar radiation. The southern coastal rain forest region supported Douglas-fir and alder, although western hemlock and spruce survived in moist settings (Heusser 1977). In relatively moist settings on Vancouver Island and on the west side of the Olympic Peninsula, for example, Douglas-fir combined with Sitka spruce, forming a forest that has no modern analog (Hebda 1983) (Figure 9.9). Alder was probably widespread in the early stages of vegetation recovery and as a riparian species. Bracken fern (*Pteridium aquilinum*) also occupied forest openings and regularly disturbed areas, especially in response to fires (Mathewes 1985). Indeed, the abundance of charcoal in the sediments of this period suggests that fires were more frequent than today (Cwynar 1987).

In the dry Willamette Valley and the southern Puget lowland, Garry oak was abundant and Douglas-fir and oak savanna was widespread (Barnosky 1985a; Sea and Whitlock 1995) (Figure 9.7). Sagebrush and grass communities, resembling today's interior steppe, dotted the warm and dry early Holocene landscape at the eastern limits of the coastal rain forest region in the Fraser Valley (Mathewes and Rouse 1975).

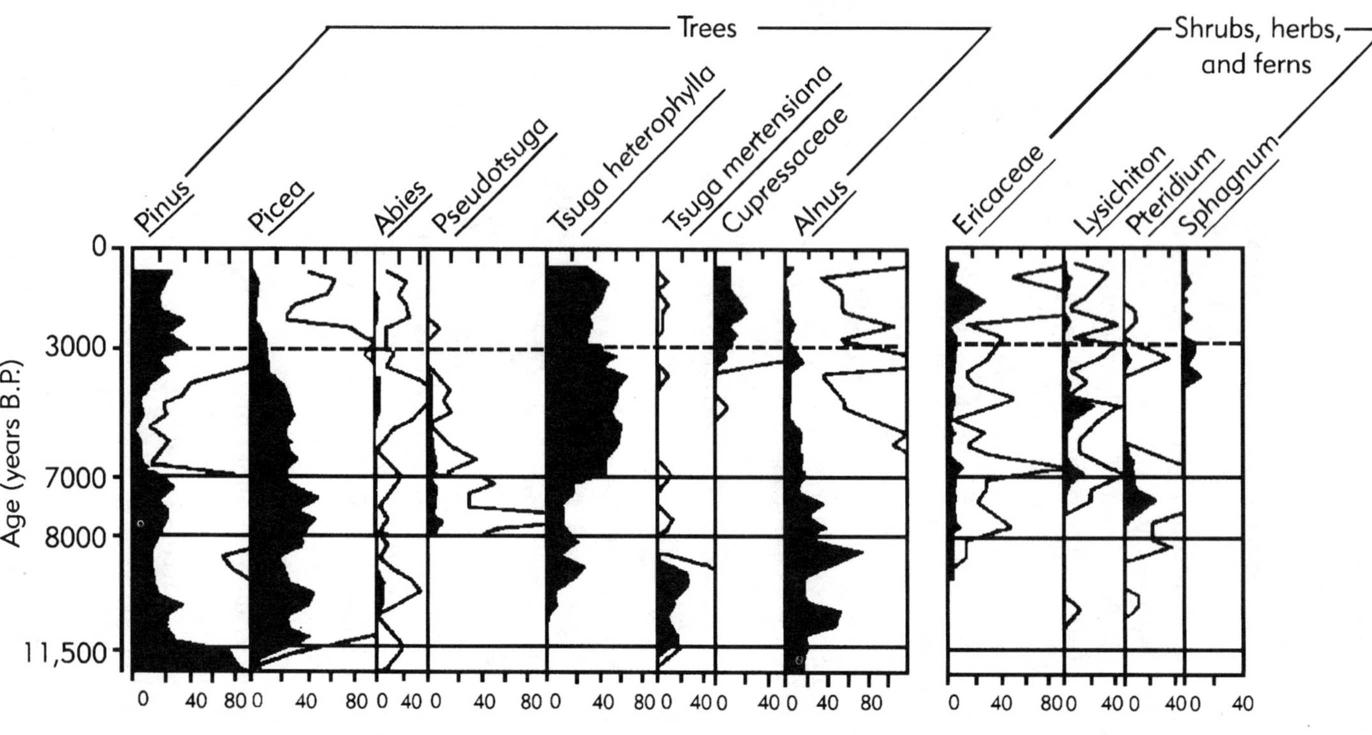

Figure 9.9. Simplified pollen diagram from Bear Cove Bog, north Vancouver Island. *Source:* Hebda (1983).

North of the present range of Douglas-fir (the mid-coast of British Columbia), climatic warming brought about the development of a distinctive assemblage dominated to varying degrees by Sitka spruce and alder and, in places, mixed also with western hemlock (Heusser 1985; Mathewes 1989) (Figure 9.8). This presumed spruce forest grew on Vancouver Island, on Haida Gwaii, on the adjacent British Columbia mainland, and in southeastern Alaska (Heusser 1985; Hebda 1995). Subalpine sites on Haida Gwaii, now surrounded by mountain hemlock, supported western hemlock (Pellatt and Mathewes 1994). Along the coast of Alaska, forest vegetation reached only as far as Icy Cape (Heusser 1985).

Other paleoenvironmental data provide independent evidence to suggest that the climate of the early Holocene was warmer and drier than present. In today's relatively dry regions of the coast and in areas east of the Cascade Range, lake levels were lower than today (Allen 1995; Barnosky 1985b), presumably because of intensified summer drought.

The Transition to Modern Vegetation and Climate

A major increase in moisture, driven by decreasing summer radiation, occurred throughout the range of today's coastal temperate rain forest, beginning before 7000 and continuing to 4000 years ago. This singular climatic change initiated an interval of major adjustment in coastal forests eventually leading to the evolution of the coastal rain forests we see today. The changes spanned some 3000 to 4000 years and can be interpreted as a time of transition from a relatively warm and dry to a moderate and moist climate state (Mathewes 1985; Whitlock 1992; Hebda 1995).

Moisture-favoring species—especially western hemlock but also spruce—proliferated from southeastern Alaska to north and west Vancouver Island and the Olympic Peninsula as drought-adapted species declined (Heusser 1985). From southeastern Alaska to north and west Vancouver Island and the Olympic Peninsula, Sitka spruce and western hemlock formed a distinct and widespread assemblage, as western hemlock displaced alder as a major pollen type and certainly expanded the role of hemlock in the forest (Heusser 1985; Peteet 1986; Hebda 1995). North of this forest assemblage in the Gulf of Alaska, Sitka spruce forests developed in previously unforested terrain as far as Icy Cape (Peteet 1986).

On the British Columbia mainland, spruce played only a minor role in the transitional forests. Instead western hemlock predominated—first with pine followed by western redcedar in the north (Banner et al. 1983) (Figure 9.8) and Pacific silver fir, then western redcedar on the south coast (Wainman and Mathewes 1987). In Washington and Oregon, western hemlock and western redcedar also increased, as hemlock became a major forest element along with Douglas-fir (Whitlock 1992; Worona and Whitlock 1995) (Figure 9.6).

The present-day differences between the dry and moist phases of the coastal rain forest were evident during the middle Holocene. Dry regions supported Douglas-fir and alder stands. On southeastern Vancouver Island, oak arrived between 7000 and 8000 years ago (Heusser 1985; Allen 1995) and spread quickly to become an important forest component over a greater area than it occupies today (Hebda 1995). These oak forests and savanna occurred where Douglas-fir predominates today, whereas Douglas-fir dominated stands in slightly moister settings. Signaling changes to come throughout the region, western redcedar became much more abundant in humid regions on the west side of Vancouver Island, such as the Brooks Peninsula (Hebda 1995), and slopes of the Fraser River valley (Wainman and Mathewes 1987).

Of special note during this transitional time are the coastal plant communities at Yakutat near the northern limits of the region (Peteet 1991). As at other southeast Alaska sites, Sitka spruce was a dominant conifer—at first glance apparently alone. Yet fossil needles reveal that western and mountain hemlock, despite very low pollen percentages, also grew here. The roles of both species increased significantly in the late Holocene, demonstrating that some of today's major forest taxa persisted as minor elements until environmental changes allowed their expansion. Species diversity may not have changed over the Holocene, but structural and ecological diversity certainly has, and only a few millennia ago.

Another response to the increased moisture in the mid-Holocene was the development and growth of bog ecosystems (Hebda 1995). Skunk cabbage wetlands were replaced by heath (Ericaceae) and sphagnum wetlands. These changes in wetland community types and in the nature of organic deposits were paralleled by changes on the floor of wet forests where more acid organic matter accumulated and bog forests arose (Banner et al. 1983). The soils of mesic (average moisture) stands may have been altered, too, ushering in the western hemlock–western redcedar biogeochron and the coming of age and spread of modern conifer rain forests.

The last major development in the evolution of the modern coastal rain forest took place between 5000 and 4000 years ago. In British Columbia, western redcedar and, at higher altitudes and northern latitudes, yellow-cedar expanded nearly to the limits of the region (Hebda and Mathewes 1984). Regional vegetation histories suggest that redcedar spread from the south and west. As Peteet (1991) has suggested, the species may simply have invaded the forests from local moist sites.

In the Olympic Peninsula, western hemlock and spruce became dominant (Heusser 1977). Forests of Douglas-fir, western redcedar, western hemlock, and grand fir replaced the oak woodland vegetation in southwestern Washington (Barnosky 1985a) (Figure 9.7). The west slopes of the Cascades experienced increases in Pacific silver fir, mountain hemlock, and western hemlock (Sea and Whitlock 1995). These changes reflect the adaptation of each region's forests to the onset of the moderate humid climate of the present day.

Although the 4000-years-ago horizon can be seen as a critical time of modernization (Hebda 1995), many sites in the coastal rain forest region continued to change between 3000 and 2000 years ago. This interval coincides in part with neoglaciation, a time when alpine glaciers extended beyond their early Holocene positions (Burke and Birkeland 1983). These adjustments do not seem to reflect similar regional climatic changes. Parts of Washington and northern California, for example, registered marked increases in late-successional species—a possible indication of further cooling that eventually led to present-day conditions (Whitlock 1992; West 1993). In contrast, Douglas-fir and true fir expanded in Oregon's Coast Range, implying drier climate. On south Vancouver Island, Garry oak declined, perhaps an indication of increased moisture. To understand the vegetational changes of the last few millennia—in what we sometimes mistakenly consider to be a rather uniform coastal temperate rain forest—we will need more research on the role of natural disturbance, short-term climate change, and human activity.

People and the Forest

The expansion of western redcedar not only dramatically altered the nature of coastal forests but doubtless played a critical role in the evolution of the remarkable and unique First Nations cultures of the coast (Hebda and Mathewes 1984). Humans have inhabited the coast for at least 10,000 years and depended on its natural resources (Chapter 10; Hebda and Frederick 1990). For the most part, these inhabitants subsisted on marine resources, leaving behind deposits known as middens, full of the detritus of sea creatures (Suttles 1990). Despite intensive exploitation, these marine resources were not depleted: apparently the people knew how to manage them on a sustainable basis (Hebda and Frederick 1990).

Our knowledge of the First People's use of the forest landscape is much less complete than for the marine environment. Land animals certainly were used from the earliest times (Davis 1990). Forest plant resources doubtless played an important role in day-to-day life, too, but the record of their use is poor. A sense of the importance of wood to the people of the coast comes from ethnobotanical studies (Chapter 11) and the archaeological record of tools possibly related to woodworking. As Hebda and Mathewes (1984) have shown, woodworking tools occur abundantly in archaeological deposits; hammerstones and possible antler and bone wedges are present almost throughout the known interval of human occupation. Specialized tools appeared only after 5000 years ago, and direct evidence of the first large wooden western redcedar structures appeared only 2000 to 3000 years ago. A fully developed woodworking technology, characterized by stone mauls and wood wedges, was in place by about 2500 years ago. There appears to be a clear connection between the rise of cedar pollen, presumably reflecting an abundance of trees, and the development of massive woodworking technology (Figure 9.10). Considering the enormous importance of western redcedar to the lifeways of coastal peoples, the nature of the coastal temperate rain forest and the cultures of its First Peoples may well have changed together.

Given the apparently sustainable management of marine resources, it is likely that forest resources were managed in a like manner. Although there is yet no study that documents the impact of the prehistoric Native population on the forest, Hebda (1995) notes the possibility that on Anthony Island in Haida Gwaii increased pine in the late Holocene may have resulted from local deforestation of western hemlock–spruce forests. Gottesfeld et al. (1991) suggest that increased pine in the Skeena River drainage may have resulted from landscape burning. Almost certainly natural fires and aboriginal burning in the southern coastal rain forest region maintained successional and open vegetation (C. Long 1996; C. Pearl, unpublished data, 1995). The impact of early First Peoples on the forest, especially at the local level, needs much more study.

In contrast to the impact of First Peoples, the European immigrant's use of the forest left a distinct mark in the lake and bog pollen record. When widespread logging occurred, early-successional red alder expanded dramatically. In the dry phase of the coastal rain forest, the resulting assemblage resembles the pollen signal of the Douglas-fir–alder biogeochron of the early Holocene. There is little similarity to late-Holocene rain forest assemblages and, by implication, the way the forests were utilized by First Peoples.

The Past as Key to the Future

Our account of the unique history of the coastal rain forest has important lessons about large-scale ecological processes and provides insights into its future. Our coastal rain forest is not the stable, long-lived, and widespread ecosystem that it might seem. Rather, it is only one of a series of biogeochrons that have existed on the Northwest Coast of North America since the last glaciation. Its history has been shaped by global climatic phenomena, such as changes in seasonal levels of solar radiation and the location of the jet stream, as well as by regional factors such as ice advances and soil characteristics.

Figure 9.10. Culturally modified western redcedar, showing cedar plank split off in the 1800s, west coast of Vancouver Island. (Photo courtesy of Royal British Columbia Museum staff.)

The fossil record reveals that changes may occur gradually or suddenly depending on the nature of the forcing phenomenon. No assumptions should be made concerning the response time of the forests following environmental change. Clearly, different parts of the region have experienced different histories. Some areas achieved their modern configuration in only the last two or three millennia; others have been relatively stable for four millennia or more. Time may hold different futures for many parts of the region, particularly with the profound disturbance by modern-day humans. How might future forest configurations arise? This too may vary. As we have seen, one or two species may expand their range with changing climate—as Douglas-fir did in the early Holocene. In today's fragmented forest, on the other hand, changes may come when local minor species gain competitive advantage—as western hemlock did in the mid-Holocene of southeastern Alaska.

The coastal rain forest region today faces environmental changes of intensity and scope perhaps greater than any experienced since the last glaciation. As a consequence, vegetation will transform into new states. Undoubtedly phenomena such as predicted global warming will force major adjustments in the extent and form of the rain forest. The possible impacts of global warming and the nature of present human activity on the landscape point to early-Holocene biogeochrons as useful analogs for insight into forest management strategies. Significant climatic change will most likely have serious impacts on moist forest ecosystems, especially bog forests and their associated biodiversity, as well as ecosystems in geologically sensitive areas (Brubaker 1991; Whitlock 1992; Hebda 1994).

The history of the region also suggests that certain areas, mainly those supporting drier forests, will be sensitive to anticipated warm and dry conditions. In the early Holocene, some of these regions supported vastly different vegetation, including nonforest plant communities, with differing disturbance regimes. Paleoecological studies have a critical role in determining the sensitivity of these communities to climate change and the mechanisms by which these changes may occur.

We must understand, too, that ecosystem adjustments will occur at the level of the species and population—not by some coherent response of the coastal temperate rain forest "superorganism" (Brubaker 1991). Though we might think we can anticipate potential winners and losers in the sweepstakes of change, our understanding of the ecological characteristics and adaptability of even major species is incomplete. Could wise scholars 6000 years ago have predicted the explosion of western redcedar? The forests of the future will best be served by maintaining the maximum possible natural biological and ecological diversity today. In this way we can be sure of providing an ecologically complex and species-rich reservoir for the development of future ecosystems.

Acknowledgments

The authors thank Rolf Mathewes, Dorothy Peteet, and Peter Schoonmaker for valuable reviews of this chapter. David Gillan of Victoria, British Columbia, drafted some of the figures. Some of the research was supported by National Sciences and Engineering Research Council of Canada grant 0090581 and grants from the Royal British Columbia Museum to Richard Hebda. Cathy Whitlock's research was supported by National Science Foundation grant ATM9307201.

References

Allen, G. B. 1995. "Vegetation and climate history of southeast Vancouver Island, British Columbia." M.Sc. thesis, School of Earth and Ocean Sciences, University of Victoria, B.C.

Alley, N. F. 1979. "Middle Wisconsin stratigraphy and climatic reconstruction, southern Vancouver Island, British Columbia." *Quat. Res.* 11:213–237.

Alley, N. F., and S. R. Hicock. 1986. "The stratigraphy, palynology, and climatic significance of pre-middle Wisconsin Pleistocene sediments, southern Vancouver Island, British Columbia." *Can. J. Earth Sci.* 23:369–382.

Armstrong, J. E., J. J. Clague, and R. J. Hebda. 1985. "Late Quaternary geology of the Fraser lowland, southwestern British Columbia." In D. Tempelman-Kluit, ed., *Field Guides to Geology and Mineral Deposits in the Southern Canadian Cordillera*. Vancouver: Geological Association of Canada.

Banner, A., J. Pojar, and G. E. Rouse. 1983. "Postglacial paleoecology and successional relationships of bog woodland near Prince Rupert, British Columbia." *Can. J. For. Res.* 13:938–947.

Barnosky, C. W. 1985a. "Late Quaternary vegetation near Battle Ground Lake, southern Puget Trough, Washington." *Geol. Soc. Am. Bull.* 96:263–271.

———. 1985b. "Late Quaternary vegetation in the southwestern Columbia Basin, Washington." *Quat. Res.* 23:109–122.

Barnosky, C. W., P. M. Anderson, and P. J. Bartlein. 1987. "The northwestern U.S. during deglaciation: Vegetational history and paleoclimatic implications." In W. F. Ruddiman and H. E. Wright, Jr., eds., *North America and Adjacent Oceans During the Last Deglaciation*. Boulder: Geological Society of America.

Berglund, B. E. 1986. *Handbook of Holocene Palaeoecology and Palaeohydrology.* New York: Wiley.

Brubaker, L. B. 1991. "Climate change and the origin of old-growth Douglas-fir forests in the Puget sound lowland." In G. Wall, ed., *Implications of Climate Change for Pacific Northwest Forest Management*. Department of Geography Publications Series, Occasional Paper 15. Waterloo, Ont.: University of Waterloo.

Burke, R. M., and P. W. Birkeland 1983. "Holocene glaciation in the mountain ranges of the western United States." In H. E. Wright, Jr., ed., *Late Quaternary Environments of the United States.* Vol. 2: *The Holocene*. Minneapolis: University of Minnesota Press.

Clague, J. J. 1989. "Introduction (Quaternary stratigraphy and history, Cordilleran ice sheet)." In R. J. Fulton, ed., *Quaternary Geology of Canada and Greenland*. Ottawa: Geological Survey of Canada.

Cwynar, L. C. 1987. "Fire and the forest history of the North Cascade Range." *Ecology* 68:791–802.

Davis, S. D. 1990. "Prehistory of southeastern Alaska." In W. Suttles, ed., *Handbook of North American Indians*. Vol. 7: *Northwest Coast*. Washington, D.C.: Smithsonian Institution.

Franklin, J. F., and C. T. Dyrness. 1988. *Natural Vegetation of Oregon and Washington*. Corvallis: Oregon State University Press.

Gottesfeld, A. S., R. W. Mathewes, and L.M.J. Gottesfeld. 1991. "Holocene debris flows and environmental history, Hazelton area, British Columbia." *Can. J. Earth Sci.* 28:1583–1593.

Graumlich, L. J., and L. B. Brubaker. 1986. "Reconstruction of annual temperature (1590–1979) for Longmire, Washington, derived from tree rings." *Quat. Res.* 25: 223–234.

Hebda, R. J. 1983. "Late-glacial and postglacial vegetation history at Bear Cove Bog, northeast Vancouver Island, British Columbia." *Can. J. Bot.* 61:3172–3192.

———. 1994. "The future of British Columbia's flora." In L. E. Harding and E. McCullum, eds., *Biodiversity in British Columbia: Our Changing Environment*. Vancouver: Canadian Wildlife Service.

———. 1995. "British Columbia vegetation and climate history with focus on 6 KA BP." *Geogr. Phys. et Quat.* 49:55–79.

Hebda, R. J., and S. G. Frederick. 1990. "History of marine resources of the Northeast Pacific since the last glaciation." *Trans. R. Soc. Canada,* series 1, 1:319–341.

Hebda, R. J., and R. W. Mathewes. 1984. "Holocene history of cedar and Native Indian cultures of the North American Pacific Coast." *Science* 225:711–713.

Hemstrom, M. A., and J. F. Franklin. 1982. "Fire and other disturbances of the forests in Mount Rainier National Park." *Quat. Res.* 18:32–51.

Heusser, C. J. 1977. "Quaternary palynology of the Pacific slope of Washington." *Quat. Res.* 8:282–306.

———. 1983. "Vegetational history of the northwestern United States, including Alaska." In S. C. Porter, ed., *Late Quaternary Environments of the United States*. Minneapolis: University of Minnesota Press.

———. 1985. "Quaternary pollen records from the interior Pacific Northwest coast: Aleutians to the Oregon–California boundary." In V. M. Bryant and R. G. Holloway, eds., *Pollen Records of Late-Quaternary North American Sediments*. Dallas: American Association of Stratigraphic Palynologists Foundation.

———. 1989. "North Pacific coastal refugia: The Queen Charlotte Islands in perspective." In G.G.E. Scudder and N. Gessler, eds., *The Outer Shores*. Queen Charlotte City: Queen Charlotte Islands Museum Press.

Heusser, C. J., and L. E. Heusser. 1981. "Palynology and paleotemperature analysis of the Whidbey Formation, Puget lowland, Washington." *Can. J. Earth Sci.* 18:136–149.

Hicock, S. R., R. J. Hebda, and R. J. Armstrong. 1982. "Lag of the Fraser glacial maximum in the Pacific Northwest: Pollen and macrofossil evidence from western Fraser lowland, British Columbia." *Can. J. Earth Sci.* 19:2288–2296.

Imbrie, J., E. A. Boyle, S. C. Clemens, A. Duffy, W. R. Howard, G. Kukla, J. E. Kutzbach, et al. 1992. "On the structure and origin of major glaciation cycles. 1: Linear response to Milankovitch forcing." *Paleoceanography* 7:701–738.

Krajina, V. J. 1965. "Biogeoclimatic zones and classification of British Columbia." *Ecol. West. N. Am.* 1:1–17.

Kutzbach, J. E., P. J. Guetter, P. J. Behling, and R. Selin. 1993. "Simulated climatic changes: Results of the COHMAP climate-model experiments." In H. E. Wright, Jr., J. E. Kutzbach, W. F. Ruddiman, F. A. Street-Perrott, T. Webb III, and P. J. Bartlein, eds., *Global Climates Since the Last Glacial Maximum*. Minneapolis: University of Minnesota Press.

Long, C. J. 1996. "Fire history of the central Coast Range, Oregon: A ca. 9000 year record from Little Lake." Unpublished.

Mathewes, R. W. 1973. "A palynological study of postglacial vegetation changes in the University Research Forest, southwestern British Columbia." *Can. J. Earth Sci.* 11:2085–2103.

———. 1979. "A paleoecological analysis of Quadra Sand at Point Grey, British Columbia, based on indicator pollen." *Can. J. Earth Sci.* 16:847–858.

———. 1985. "Paleobotanical evidence for climatic change in southern British Columbia during Late-glacial and Holocene time." In C. R. Harington, ed., *Climate Change in Canada 5: Critical Periods in the Quaternary Climatic History of Northwestern North America*. Ottawa: National Museums of Canada.

———. 1989. "Paleobotany of the Queen Charlotte Islands." In G.G.E. Scudder and N. Gessler, eds., *The Outer Shores*. Queen Charlotte City: Queen Charlotte Islands Museum Press.

———. 1993. "Evidence for younger Dryas-Age cooling on the North Pacific coast of America." *Quat. Sci. Rev.* 12:321–331.

Mathewes, R. W., and G. E. Rouse. 1975. "Palynology and paleoecology of early postglacial sediments from the lower Fraser River Canyon of British Columbia." *Can. J. Earth Sci.* 12:745–756.

Meidinger, D., and J. Pojar, eds. 1991. *Ecosystems of British Columbia*. Victoria: British Columbia Ministry of Forests.

Moore, P. D., J. A. Webb, and M. E. Collinson. 1991. *Pollen Analysis.* 2nd ed. Oxford: Blackwell.

Morrison, P., and F. J. Swanson. 1990. "Fire history and pattern in a Cascade Range landscape." Research Paper PNW-GTR-254-77. Portland: USDA Forest Service.

Pellatt, M. G., and R. W. Mathewes. 1994. "Paleoecology of postglacial tree line fluctuations on the Queen Charlotte Islands, Canada." *Ecoscience* 1:71–81.

Peteet, D. M. 1986. "Modern pollen rain and vegetational history of the Malaspina Glacier District, Alaska." *Quat. Res.* 25:100–120.

———. 1991. "Postglacial migration of lodgepole pine near Yakutat, Alaska." *Can. J. Bot.* 69:786–796.

Petersen, K. L., P. J. Mehringer, Jr., and C. E. Gustafson. 1983. "Late-glacial vegetation and climate at the Manis Mastodon site, Olympic Peninsula, Washington." *Quat. Res.* 20:215–231.

Sea, D. S., and C. Whitlock. 1995. "Postglacial vegetation and climate of the Cascade Range, central Oregon." *Quat. Res.* 43:370–381.

Stewart, W. N., and G. W. Rothwell. 1993. *Paleobotany and the Evolution of Plants.* 2nd ed. New York: Cambridge University Press.

Stuiver, M., and P. Reimer. 1993. "Extended 14C data base and revised CALIB 3.0 14C age calibration program." *Radiocarbon* 35:215–230.

Suttles, W., ed. 1990. *Handbook of North American Indians.* Vol. 7: *Northwest Coast.* Washington, D.C.: Smithsonian Institution.

Thompson, R. S., P. J. Bartlein, C. Whitlock, S. P. Harrison, and W. G. Spaulding. 1993. "Climatic changes in the western United States since 18,000 yr B.P." In H. E. Wright, Jr., J. E. Kutzbach, W. F. Ruddiman, F. A. Street-Perrott, T. Webb III, and P. J. Bartlein, eds., *Global Climates Since the Last Glacial Maximum.* Minneapolis: University of Minnesota Press.

Tsukada, M. 1982. "*Pseudotsuga menziesii* (Mirb.) Franco: Its pollen dispersal and late Quaternary history in the Pacific Northwest." *Jap. J. Ecol.* 32:159–187.

Tsukada, M., S. Sugita, and D. M. Hibbert. 1981. "Paleoecology of the Pacific Northwest I: Late Quaternary vegetation and climate." *Verhandlungen der Internationalen Vereinigung für Theoretische und Angewandte Limnologie* 21:730–737.

Viereck, L. A., C. T. Dyrness, A. R. Batten, and K. J. Wenzlick. 1992. *The Alaska Vegetation Classification.* PNW-GTR-286. Portland: USDA Forest Service Pacific Northwest Research Station.

Wainman, N., and R. W. Mathewes. 1987. "Forest history of the last 12,000 years based on plant macrofossil analysis of sediment from Marion Lake, southwestern British Columbia." *Can. J. Bot.* 65:2179–2187.

Waitt, R. B., and R. M. Thorson. 1983. "The Cordilleran ice sheet in Washington, Idaho, and Montana." In S. C. Porter, ed., *Late Quaternary Environments of the United States.* Minneapolis: University of Minnesota Press.

West, G. J. 1993. "The late Pleistocene-Holocene pollen record and prehistory of California's North Coast Ranges." In G. White, P. Mikkelsen, W. R. Hildebrandt, and M. E. Basgall, eds., *There Grows a Green Tree: Papers in Honor of David A. Fredrickson.* Publication 11. Davis: Center for Archaeological Research, University of California.

Whitlock, C. 1992. "Vegetational and climatic history of the Pacific Northwest during the last 20,000 years: Implications for understanding present-day biodiversity." *Northwest Env. J.* 8:5–28.

Worona, M. A., and C. Whitlock. 1995. "Late-Quaternary vegetation and climate history near Little Lake, central Oregon Coast Range, Oregon." *Geol. Soc. Am. Bull.* 107:867–876.

10. Pre-European History

WAYNE SUTTLES AND KENNETH AMES

□ □

When Europeans first explored the Northwest Coast of North America during the last quarter of the eighteenth century, they found themselves in a region that was very distinct in climate, flora, and fauna. They met peoples, moreover, who were equally distinct in culture. The Natives of the Northwest Coast were unlike any elsewhere in the New World and nearly unique among the world's peoples. They were foragers, living wholly on fish and game and wild plants, and yet they had relatively dense populations, permanent villages of great wooden houses, socially stratified societies, and highly developed art and ceremonies. Who were these people? How did their distinctive culture develop? Were their lives in any way shaped by this distinctive environment?

For answers we have three major sources of data: the accounts of early European observers and the materials they collected; the work done by scholars (mainly ethnographers and linguists) with the Native peoples themselves; and the work done by scholars (mainly archaeologists) on sites once used by the Native peoples and with materials recovered from these sites. The first two sources now give us a fairly clear picture of who these people were and what their lives were like at the time of first contact. From the third source, we now know that human history began in this region at least nine millennia ago, and we can see in outline the course of that history. In the first part of this chapter we will present some of the answers that we find in the first two sources of data. Then we will present some of the answers found in the third source.

Insights from History, Ethnography, and Linguistics

The natural area found by the eighteenth-century explorers was the coastal temperate rain forest; the cultural area was what anthropologists call the Northwest Coast. The two are not quite identical. In its most recent general treatment (Suttles 1990), the Northwest Coast is defined as extending from the mouth of the Copper River on the Gulf of Alaska to the mouth of the Chetco River near the Oregon–California border and inland to the Chugach and St. Elias ranges in Alaska, the Coast Mountains in British Columbia, and the Cascades in Washington and Oregon. Earlier works have taken the culture area as far south as Cape Mendocino in northern California. The cultural area is thus somewhat broader than the coastal temperate rain forest in its extent inland into somewhat drier environments. Still, as with the other cultural areas of the continent, the fit is suggestive (Kroeber 1939).

Who were the Native peoples of this region? They were, of course, the ancestors of thousands of people who are still there. When we use the past tense, it is only in reference to their pre-European history. What we mean is: who were they in an anthropological sense—biologically, linguistically, and culturally?

Biologically they did not seem to differ greatly from other Natives of the New World. If Northwest Coast men were more often seen with mustaches and beards than other Native Americans, it was probably more a matter of style than genetic difference. Within the region there was some variation—people tended to be taller and lighter-pigmented in the north. A survey of research on the human biology of the area (Cybulski 1990) suggests there were several relatively stable regional populations. But intermarriage must have occurred among these populations and with those in neighboring culture areas. Like all Natives of the New World, the Natives of the Northwest Coast were the descendants of people who came from Northeast Asia many thousand years ago. They probably received some genes from Siberia in the time after they first arrived, but popular notions that the Natives of the Northwest Coast have recent East Asian ancestry have no support.

Linguistically, the Northwest Coast is a paradox. Its languages seem diverse in one respect but similar in another. The region seems very heterogeneous if we count the languages spoken here and look at the evidence for their relatedness. There were over sixty languages spoken throughout a region corresponding to the coastal temperate rain forest; over 40 of those were spoken between the Oregon/California border and the Copper River delta in Alaska (Suttles 1985; Ecotrust et al. 1995; Plate 4). By "languages" we mean forms of speech not mutually intelligible—as opposed to "dialects," which are forms of a language intelligible to all speakers of the language. On the basis of evidence for common origin, most of these languages have been placed into language families or larger, less firmly established groups called phyla. But we are still left with seven different, apparently unrelated, lines of descent (Thompson and Kinkade 1990).

Starting in the north, we can identify an Eyak-Athapaskan Family—represented by Eyak, spoken at the mouth of the Copper River, and the Athapaskan (or Dene) subfamily, represented by a language near the mouth of the Columbia River and two more on the southern Oregon coast. (Other Athapaskan languages are spoken in the Yukon and Mackenzie basins, in northwestern California, and in the Southwest.) The Tlingit language, spoken in southeastern Alaska, shows evidence of common origin with Eyak-Athapaskan but at a more remote time, and it has therefore been combined with Eyak-Athapaskan to form a phylum called Na-Dene. It was once proposed that Haida, spoken mainly in Haida Gwaii (Queen Charlotte Islands), be included in Na-Dene, but linguists who have worked on Haida in recent years reject the proposal. We must now say that Haida is a language isolate—it has no known relatives. Moving southward, we find four family isolates: Tsimshian (consisting of three languages on the northern British Columbia coast and up the Nass and Skeena rivers); Wakashan (eight languages from Kitimaat to Cape Flattery); Salishan (twenty-three languages: one on Burke and Dean Channel on the central British Columbia coast, ten in the Strait of Georgia–Puget Sound Basin, four in southwestern Washington, and one on the northern Oregon coast, plus seven more east of the Cascades in the Fraser and upper Columbia drainages); and Chimakuan (two languages on the Olympic Peninsula). Finally, on the Columbia River and to the south we find four families (Chinookan, Takelman, Alsean, and Coosan) and a language isolate (Siuslaw)—all of which have been placed into a Penutian Phylum, which includes a number of other families spoken mainly in California. (It was once suggested that the Tsimshian Family belongs in the Penutian Phylum, but most linguists working on these languages now reject this proposal.) Linguistic taxonomy thus places the forty-odd languages of the Northwest Coast into two phyla (Na-Dene and Penutian), four families that we cannot place in any phylum (Tsimshian, Wakashan, Salishan, and Chimakuan), and one language that we cannot place in any family or phylum (Haida).

But when we ignore evidence for genetic (family-tree) relationship and simply look at phonology, grammar, and semantics, we find regional features that unite all or major parts of the Northwest Coast. All the languages of this region have complex phonological systems using similar or identical sounds that Europeans generally find impossible to pronounce. Languages that show no evidence of being related share grammatical principles (such as the ways that words and sentences are formed) and semantic principles (such as giving directions by reference to the shoreline and the flow of water and not having terms corresponding to north, south, east, and west).

The paradox that seven lines of descent seem to divide the area while other features more or less unite the area has implications for the history of the region. It was once supposed (as in Drucker 1963) that the large number of apparently unrelated linguistic groups on the Northwest Coast could only be the product of repeated migrations into the region, presumably from the interior of the continent. But this view was associated with the belief that the coast was first settled only 2000 or 3000 years ago. Now that we know human beings have been on the coast for at least 9000 years, we must acknowledge that all these linguistic groups have probably existed on the coast for millennia.

First, regional linguistic features imply that these seemingly unrelated languages must have been influencing one another for a very long time. Second, the distribution of the members of each family or phylum points to an early homeland on or near the coast. Within the Eyak-Athapaskan Family, for example, Eyak and Athapaskan are coordinate branches—that is to say, in the history of the family the earliest split divided a language that became modern Eyak from a language that went on to separate into all of the modern Athapaskan languages. Edward Sapir (1951) said that if we look for the most likely region of origin of a language family, we must give equal weight to its major divisions. Giving equal weight to Eyak and Athapaskan, we must identify the "center of gravity" and thus the probable original home of the family as on or near the northern Northwest Coast. Including Tlingit with Eyak-Athapaskan to form the Na-Dene Phylum strengthens the case for a north coast origin. To suppose an interior homeland for the "ancestral" languages that became the modern divisions spoken on the coast would require bringing them piecemeal to the coast by a series of hypothetical migrations. The principle of parsimony favors a coastal homeland. Four taxa—Haida and the Tsimshian, Wakashan, and Chimakuan families—are found only on the coast. The Salishan Family consists of several divisions, only one of which is east of the mountains, so its center of gravity is on the coast as well. Even the widespread Penutian Phylum seems to have its center of gravity in western Oregon.

And third, we must recognize that saying these seven language groups are unrelated does not mean they *cannot* be related. It means simply that the comparative method, developed for Old World languages, cannot show evidence of a common origin. This method may not be able to discern a relationship that dates back more than 5000 or 6000 years. Thus several or all of the seven lines may have a common origin so old that evidence has been obliterated by the passage of time. There are no known linguistic facts that would contradict a very early occupation of the Northwest Coast by speakers of three, two, or even one language followed by differentiation once they were there. We need no longer postulate waves of migrants coming out of the interior.

The peoples of the Northwest Coast also give us cultural reasons to believe they have been there a very long time. Their technologies, subsistence activities, and social systems suggest a long history of adaptation to this region. So does their intimate knowledge of their environment, as illustrated by Nancy J. Turner's work (Chapter 11). Here we will outline the culture of the region, indicating differences within it, and then address the question of the relationship between the culture and the environment.

The Northwest Coast peoples had permanent villages from which they moved seasonally to capture or gather various natural resources—salmon and other fish, sea mammals, shellfish, waterfowl, land mammals, and a variety of plant foods (roots, shoots, and berries). This sedentary or semisedentary life was made possible by a well-developed woodworking technology, capable of constructing large plank houses and dugout canoes, and by good storage methods (Figure 10.1). Dugout canoes provided transport to the most productive harvest sites. The storage methods made it possible to use the great abundances that occurred during short periods of time. The plank houses were dwellings where several nuclear families passed the winter, but they were much more than that. They were food processing and storage plants, where fish and other foods were dried or smoked and put away. The posts and often the facades of the houses of the elite were carved or painted to identify the superhuman protectors of their owners. They were built to hold guests and provide space for feasts, potlatches, and winter ceremonies. These houses, therefore, were also banquet halls, theaters, and temples.

The most striking feature of the Northwest Coast social systems is social inequality. Nearly everywhere there was a wealthy and privileged elite; then there were commoners, dependent on the elite; and then there were slaves—war captives and their descendants—who were the personal property of the wealthiest members of the elite. The Northwest Coast peoples were not organized by what people of European origin ordinarily think of as political principles. The basic social unit was the local group with its house site or sites, resource sites, and origin myth. Although each group was basically autonomous, in fact it was linked with other groups through marriage, kinship, and participation in ceremonial systems. Over much of the area, several local groups shared a winter settlement—a village or town—and formed the group often called the tribe.

From the west coast of Vancouver Island northward, the elite within the local group were individually ranked and the local groups within the tribe were ranked. This ranking primarily determined precedence in ceremonial activities. The head of the highest-ranking local group within such a tribe might in fact exercise great influence over other heads, but each of these could, with the support of his own people, act independently. There were no oaths of loyalty binding lower-ranking to higher-ranking "chiefs," and the highest-ranking

Figure 10.1. This cedar plank house on Haida Gwaii (Queen Charlotte Islands) was known as "House Where People Always Want to Go." In the foreground is Chief "Highest Peak in a Mountain Range" in ceremonial attire. (Photo courtesy of the Royal British Columbia Museum (Victoria): PN 701.)

"chief" had no formal authority over any other "chief" nor any force beyond his personal influence to bring to bear on any other. From the Strait of Georgia southward, there was little formal individual ranking or group ranking. As in the north, the "chiefs" were simply heads of local groups who were able, through personal ability and wealth, to influence others.

Despite similarities in social stratification and polity throughout the Northwest Coast, there were two markedly different principles of kinship organization—matrilineal in the north, bilateral in the center and south. On the northern coast, descent was reckoned through the female line, a man might marry his mother's brother's or father's sister's daughter, and he was succeeded by his sister's son. From the central coast southward, descent was reckoned through both parents, a man could not marry any known relative, and he was succeeded by his own son or, in some circumstances, a daughter.

Northwest Coast sculpture and painting are now generally recognized as one of the world's great art traditions (Figure 10.2). Northern and southern two-dimensional styles can be distinguished, but they appear to be created according to the same general principles (Holm 1990). The rediscovery of these principles has led to a great modern revival of Northwest Coast art. There are also northern and southern weaving styles that produced blankets that were both works of art and items of wealth.

Northwest Coast ceremonies too are famous. The potlatch, at which the hosts lavishly transfer wealth to their guests in exchange for recognition of status, has become the subject of a voluminous literature. The "secret society" performances of the central coast have been justly celebrated, as well, for their drama and stagecraft. But dramatic performances of a similar sort once took place from Alaska to the Columbia River; what distinguishes the central coast is its fantastic masks.

On the Northwest Coast there were cultural differences of the sort just indicated, but the area was not what linguistic diversity, as measured by the number of languages, might imply. Forty languages between the Winchuck River in Oregon and Copper River in Alaska might suggest forty different societies with forty different cultures, but this was certainly not the case. Speakers of adjacent languages, even apparently unrelated ones, were in frequent if not constant communication and often intermarried, maintaining far-reaching social and biological continua.

Within the cultural area as a whole we can discern several regional social systems. One included the Tlingit, Haida, Tsimshian, and Haisla of the northern coast. These peoples were organized by the principle of matrilineal descent into lineages and larger kin groups that identified everyone in the whole social system. A symbol of participation in this system was the labret (lip

Figure 10.2. Interior of the Whale House, Raven 3, Klukwan, about 1895. This richly carved interior displays the sophisticated symbolic representation that characterizes Northwest Coast art. (Photo courtesy of Alaska Historical Library (Juneau): PCA 87-10.)

plug) worn by women to show they were of respectable status and marriageable (Figure 10.3). South of this system, from the Heiltsuk (Bella Bella) to the central Oregon coast, people reckoned descent bilaterally, and the mark of respectability was some type of modification of the shape of the head performed on infants in the cradle. Within this bilateral area there seem to have been three or four regional networks: a central coast network consisting of a northern portion that included the Heiltsuk, Nuxalk (Bella Coola), and Kwakiutl (some of whom prefer to be identified as Kwakwa̲ka̲'wakw) and a southern portion that included the Nuu-chaa-nulth (the speakers of the Nootka and Nitinaht languages) and the Makah; a Georgia-Puget network embracing the Coast Salish of the Gulf of Georgia–Puget Sound region; and a lower Columbia network that included southwestern Washington and the northern Oregon coast. The peoples of southwestern Oregon probably participated in a regional network that extended into northwestern California. These regional social systems were not bounded and were probably not stable over any long period of time.

The Northwest Coast is also noted for its relatively large population for a foraging base. Estimating numbers in early times, before the great epidemics that began in the 1770s, is very difficult. Using accepted methods, very conservatively, Robert Boyd (1990; pers. comm. 1995) has estimated the pre-epidemic population of the whole Northwest Coast as around 182,000. His figures would give the regional social systems the following approximate populations at the time of contact: northern c. 45,000, central c. 36,000 (northern division c. 18,500, southern c. 17,300), Georgia-Puget c. 35,500, lower Columbia c. 35,800, southwestern Oregon c. 30,000. These figures are based on known historic numbers, epidemic history, and assumptions about mortality. For instance, it is thought that one-third of the population perished from the first smallpox epidemic. But epidemic mortality may have been much higher, and the precontact population may have been much greater.

The correspondence of cultural and natural areas implies a relationship between culture and environment. While no one today would assert that environment simply creates or shapes culture, we do recognize that environment offers opportunities and poses challenges that, for historic reasons, different cultures meet in different ways. The Northwest Coast offered a wealth of natural resources. In fact, anthropologists once supposed that the complexities of

Figure 10.3. The size of the labret (lip plug) worn by this Haida woman indicates her social rank, wealth, and age. (Photo courtesy of Royal British Columbia Museum (Victoria): PN-1053A.)

Northwest Coast culture existed simply because they were made possible by the rich environment in which they lived: as Native peoples had no problem making a living, they had the surplus and leisure to devise social hierarchies, patronize the arts, and give lavish feasts and potlatches.

But a closer look at the environment reveals the challenges. Indeed, while there were huge quantities of salmon and other resources, most of these were available only in certain places and for short periods of time. There were also periods when not much was available. Taking advantage of the abundances required being at the right place at the right time and with the right equipment to take and preserve each resource. Reliance on these restricted abundances required control of access, an organized labor force, and a system of exchange. The status of "chiefs" depended on their roles in just these things. They controlled the use of resources, they gathered followers and slaves, and they promoted exchange—through marriage ties, reciprocal feasting and potlatching, pure commerce, and warfare. Marriage ties especially promoted the exchange of food, wealth, and people. The wedding itself was based on exchanges between families. Exchanges continued through the marriage. And the children of the marriage provided ties of kinship and the basis for shifts in residence. Social systems, like technology, can be seen as an adaptation to the varying riches of the Northwest Coast environment (Suttles 1968, 1987).

The rain forest of the Northwest Coast was only one component in this environment; another was certainly the sea with its wealth of resources, and another was the great river systems, including those of the Fraser and Columbia, which flowed from the drier interior, through the rain forest zone, and supported some of the largest salmon runs. People of other regions harvest marine resources and salmon. But the most important gifts this environment offered its Native peoples were the challenge of coping with its great abundances and the means, especially western redcedar for houses and canoes, for doing so. The history of how the Native peoples achieved this must be reconstructed from the record of archaeology.

Insights from Archaeology

Archaeological research on the coast emphasizes reconstructing and explaining the developmental history of these cultures. When did people on the coast begin to rely heavily on stored salmon? When did they begin living in permanent villages? When did the coast's chiefly elite develop? Archaeologists have also been concerned with tracing the history of the Northwest Coast's famous art style. Other important research issues involve the coast's role in the early peopling of the Americas—was the migration route south of the Late Pleistocene glaciers?—and answering questions about the development of maritime economies along the west coast of North America and in the northern Pacific generally.

The coast's cultural history is divisible into two major periods: an early one spanning the time between about 11,000 years ago and 5500 years ago and a later period that ends in contact with Europeans. The following discussion is based on Ames (1994) and the sources cited there.

Early Period (11,000–5500 B.P.)

The earliest firmly dated archaeological sites on the coast tend to date about 10,000 years ago. These sites include Glacier Bay and Hidden Falls in southeastern Alaska and Namu on the central British Columbia coast (Carlson 1987). One possible site—the Manis Mastodon site in northwestern Washington—produced three dates spanning the period between 11,000 and 10,000 years ago but is not firmly established as an archaeological site (Fladmark 1982). It is likely that the coast was occupied at least by 11,000 to 11,500 years ago. This presumption is supported by the presence of Clovis materials on the south coast, as well as east of the Cascade Mountain range, and by recent geological research showing that extensive portions of the presently drowned continental shelf were exposed and ice free more than 12,000 years ago. Clovis is the oldest well-documented human culture in North America. It dates between 11,200 and 10,900 years ago. Clovis tools are found across much of North America south of the Late Pleistocene glaciers.

Fladmark (1979) has even suggested that the Northwest Coast was the primary route by which America's first inhabitants entered the continent. He has proposed that people moved south along the exposed portions of the continental shelf. At present this hypothesis is neither supported nor refuted by archaeological evidence—there are no data that bear directly on the question. There are a very few sites in southern Alaska and southern Oregon that show exploitation of coastal habitats before 8000 years ago. Presumably the intervening coastline was also inhabited and exploited. After that date, the number of excavated sites increase. Namu and Glenrose Cannery, located near Vancouver, British Columbia, are the major deeply stratified sites for the latter portions of the Early period (Matson 1976).

Early-period archaeological assemblages on the northern and central coast differ from those found on the southern coast. Northern and some central coastal sites (such as Namu) contain microblades and microblade cores similar to those found in Alaska and Northeast Asia, while southern coastal assemblages are dominated by an array of chipped stone tools, including laurel-leaf-shaped bifaces and cobble tools. These assemblages are similar to assemblages found east of the Cascade Range on the Columbia Plateau at this time. Subsistence data, including faunal remains, artifact assemblages, and site locations, indicate that from at least 8000 years ago a broad array of resources were taken from an equally broad array of environments, including land, rivers, beaches, and the sea (Matson 1992). These resources (taking all these sites together) include wapiti, deer, salmon, smelt, flatfish, freshwater and saltwater shellfish, seals, and perhaps dolphin. Features such as hearths or postholes that might indicate residential patterns are rare.

The climate and environments of the coast during the Early period differed from those of later times. The Early period is marked by deglaciation and major climatic changes. Between about 12,000 years ago and 8000 years ago, mean annual temperatures increased by as much as 8° to 10°C, reaching levels as much as 2°C above present mean annual temperatures. In some portions of the coast at least, this period was also marked by less rainfall than at present. Deglaciation returned enormous quantities of water to the world's oceans, which had been as much as 150 meters below their current levels. Sea level rose rapidly, flooding large portions of the continental shelf between 12,000 and 8000 years ago. Sea level rose more gradually after 6800 years ago, achieving its approximate modern position 5000 years ago.

In short, the coastal environment occupied by the region's early inhabitants was markedly different from the present one. It was probably much less stable and predictable, necessitating very flexible economies and social arrangements. Portions of the coast that are presently heavily forested may have been more open—and thus more useful to hunter-gatherers than is deep forest. It should be noted here that tools for working wood, particularly wedges of wapiti antler, are present in tool assemblages as early as 10,000 years ago.

In sum, then, Early-period peoples were probably very mobile hunter-gatherers, with a very flexible adaptation to the coast's changing environment, exploiting a wide range of habitats and resources, but not emphasizing any single one. Their population densities were probably quite low and their technology geared to ease of movement and maximum utility.

Pacific Period (5500 B.P. to Contact)

The Pacific period follows the Early period and is marked by the evolution of Northwest Coast culture as encountered in the eighteenth century. The period is known by a variety of names—Developmental, Late, Pacific—depending on the author. We prefer Pacific. This period, by whatever name, is always divided into three subperiods.

Early Pacific Subperiod (5500–3500 B.P.)

The beginning of the Pacific period is marked by the disappearance of microblade technology in the north and the appearance of large shell middens all along the coast between about 5500 and 4500 years ago (Figure 10.4). Shell middens are accumulations of mollusk shells mixed with organic and nonorganic debris produced by human activity. They are excellent environments for the preservation of bones, so Northwest Coast archaeologists have much better data on subsistence for the Pacific period than for the Early period.

Figure 10.4. Shell middens like this one—excavated at Toquot, British Columbia—offer a primary source of archaeological remains from prehistoric settlements of the coastal temperate rain forest. (Photo courtesy of Allen MacMillan.)

The appearance of shell middens heralds significant changes in settlement, residential, and subsistence patterns, including increased exploitation of mollusks. These changes may have resulted from the stabilization of sea levels, population growth, coastwide changes in salmon productivity, or from other factors still unknown. It is clear that during this subperiod the basic structure of the ethnographically documented Northwest Coast subsistence economy and technology developed. There is debate about whether salmon storage began playing a significant economic role during this or during the next subperiod. There is debate, too, over whether permanent elites were present on the coast before the end of this subperiod. There is evidence for relatively elaborate funeral rituals between 3500 and 2000 years ago, and some individuals during that time received special treatment when they were buried: ethnographic status markers, including lip plugs, are present by 3500 years ago. The earliest examples of Northwest Coast art motifs date between 4000 and 3500 years ago. There is very little evidence about residences or other aspects of life.

Although coastal economies depended on a wide range of resources, they focused on salmon (or other fish where salmon were not available), sea mammals, other fish such as smelt and flatfish, large terrestrial mammals such as wapiti and deer, and plant foods. Berries and ferns were widely exploited. Roots and nuts were consumed, as well, on southern portions of the coast. The coast's environment is not uniform. It is a complex patchwork of small habitats and local ecosystems tied together by having some organisms in common. As a result, local subsistence economies on the Northwest Coast might vary considerably from bay to bay. This diversity is fundamental to understanding the Northwest Coast as a human environment.

Middle Pacific Subperiod (3500–1500 B.P.)

This is a period of profound economic, social, and cultural development. The weight of available evidence indicates that widespread reliance on stored salmon developed between 3500 and 3000 years ago. The earliest villages and rectangular houses date between 3200 and 2600 years ago. The Paul Mason site in northern British Columbia, the coast's best-preserved early village, yields evidence for twelve houses dating between about 3200 and 2800 years ago (Coupland 1985). Large samples of excavated burials from Prince Rupert Harbor in northern British Columbia, and the Gulf of Georgia in southern British Columbia, clearly indicate the coastwide presence of an elite by about 2500 years ago. Grave goods include such wealth and prestige items as copper ornaments, shell (including dentalium) and stone beads, whalebone clubs, and stone labrets. The acquisition and making of some of these items required long-distance trade and specialized artisans. The burial and artifactual data also indicate widespread warfare—there is evidence, for example, of injuries caused by warding off blows (Cybulski 1993). Sites are much more numerous and generally much larger than previously, suggesting more intensive occupation and larger populations. The art style was well developed by 2200 years ago. The regional interaction spheres that were evident during the nineteenth century appear to have formed or crystallized during this period. Subsistence evidence, beyond indicating salmon storage, suggests that people continued to expand food production throughout the period and in some areas developed heavy reliance on sea mammals. In most places on the coast, sea mammals played a significant economic role second only to fishing. From the southern Haida Gwaii (Queen Charlotte Islands) to northwestern Washington, whaling was important. In sheltered waters, the sea otter was the primary prey; in river mouths and harbors, seals and sea lions were hunted.

The Middle Pacific subperiod reveals excellent evidence for use of the coast's evolving rain forest habitats. Hebda and Mathewes (1984) have shown that the coast's redcedars did not reach their modern distribution until this period (Chapter 9). Moreover, cedars of a size suitable for canoes and large houses would not have been present across much of the region. The presence of rectangular structures indicates that the large houses of cedar and spruce timbers and planks documented by observers during the nineteenth century were being constructed. There is also evidence for wooden boxes. A variety of woodworking tools, including stone adzes, mauls, and stone and antler wedges, become common in archaeological sites. There is indirect evidence on the northern British Columbia coast for large freight canoes that required an entire cedar log to construct. Wet sites (where normally perishable artifacts are preserved in waterlogged deposits) along the coast have produced a wealth of basketry as well as fragments of carvings, planks, and other items—all indicating that the Northwest Coast's famous woodworking skills were flourishing in the Middle Pacific subperiod (Croes 1976).

Late Pacific Subperiod (1500 B.P.)

It is generally thought that the ethnographically documented societies and cultures of the coast were present by the beginning of the Late Pacific subperiod. The archaeological evidence, in fact, overwhelmingly indicates cultural continuity during the entire Pacific period for the coast as a whole. This is not to say that significant social and cultural changes did not occur, however. The beginning of the period is marked by a profound change in funerary ritual (Cybulski 1993). During the previous two subperiods, a portion of the population was buried in shell middens in both residential and nonresidential sites. In some instances at least, these interments appear to have been immediately behind the dead's household. These graves are our major source of evidence about the social changes described above as well as other important data about health and diet. This practice ceases almost completely along the coast about 1300 years ago, though younger midden burials are known. This shift probably indicates the beginnings of the burial practices of the late eighteenth and early nineteenth centuries. The reasons for the shift are unknown. On the southern coast, the Late Pacific subperiod is also marked by shifts in technology—probably related to fishing equipment and to the development of new fishing techniques. There are indications all along the coast of increased warfare. In the north, there may have been parallel changes in social and household organization. The Northwest Coast art style appears to have taken on its historic form about 900 years ago in the north. There are tantalizing suggestions that the roles of decoration and art motifs in the south, relative to their roles in the north, may have changed.

There is a proliferation of large, heavy-duty, woodworking tools during the

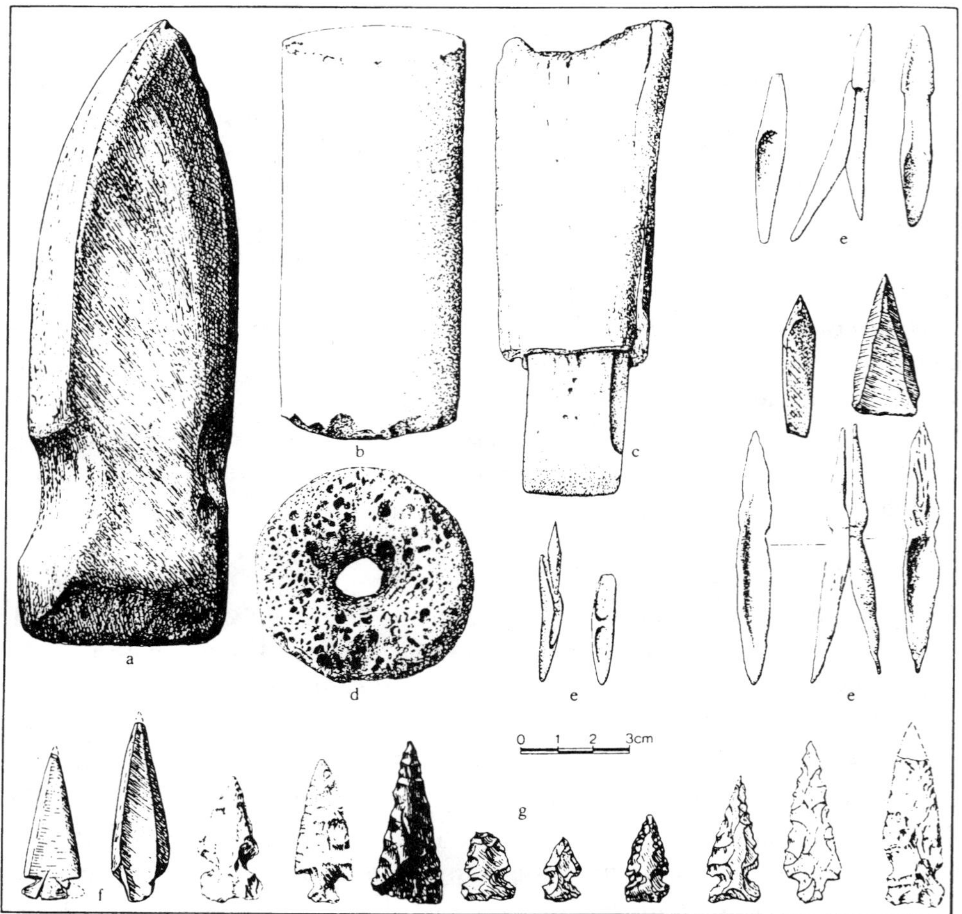

Figure 10.5. Artifacts typical of the Late Pacific subperiod. *Source:* Carlson (1987).

Late Pacific subperiod—heavy adzes and mauls, pile drivers, and other such tools—indicating an expanded use of the region's forests (Figure 10.5). The world-famous Ozette wet site on the northwest Washington coast preserves the wealth of wooden tools used by a Northwest Coast household: literally thousands of fish hooks, harpoon shafts, arrow shafts, handles, clubs, pegs, boards, boxes, weaving frames, textiles, baskets, and so on all indicate high levels of skill (Samuels 1991). The Olympic rain forest and other woodlands provided the raw materials for this wealth.

European explorers and traders entering Northwest Coast waters in the eighteenth century may have been preceded by Old World epidemic diseases, particularly smallpox, and by trade goods. When Cook's expedition sailed along the Oregon coast in 1789, for example, they encountered people with smallpox scars. Despite the profound impacts of the last two centuries, Northwest Coast

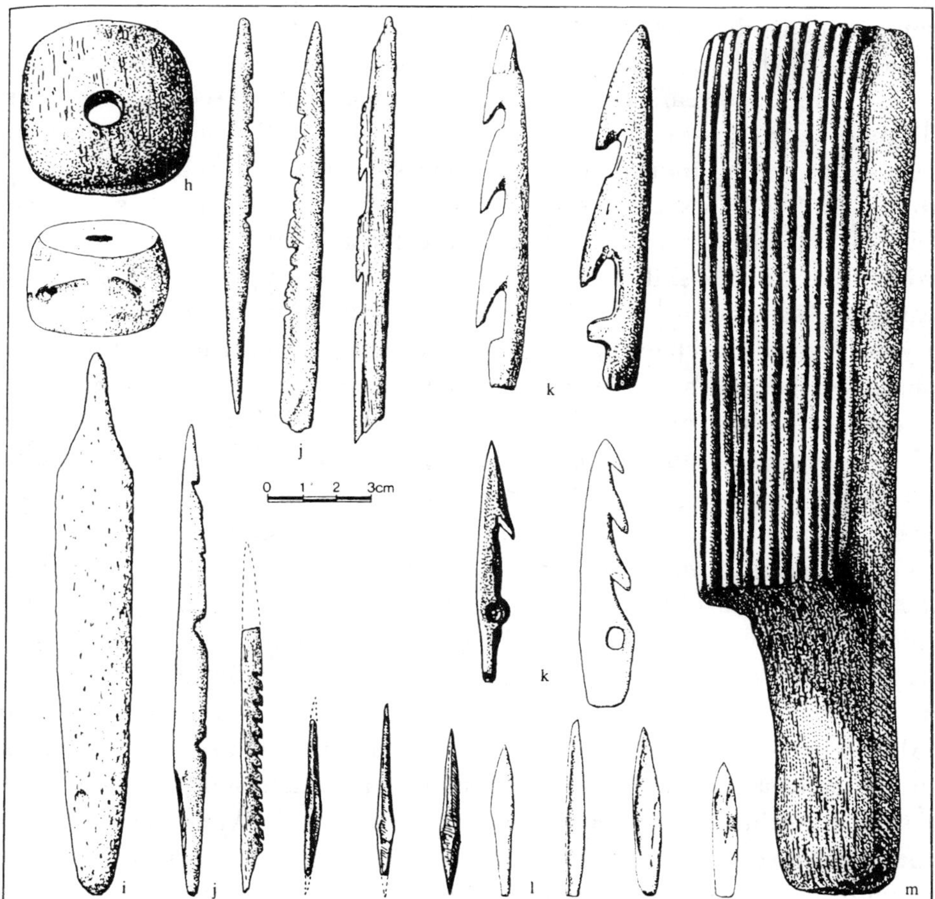

Figure 10.5. *Continued*

Native history is preserved not only in archaeology but in the oral traditions of the coast's people.

Looking for Answers

At the beginning of this chapter we asked who the people of the coast were, how their distinctive cultures evolved, and how their lives were shaped by the distinctive environment of the coast. Actually, the first of these questions should be reworded because these people are very much still here: Who *are* these people? The Native peoples of the coast have clearly been here for a very long time and, despite their linguistic diversity, have shared a great number of common social and cultural traits.

How did their distinctive culture develop? The archaeological record shows that the coast's distinctive cultures were the products of long, dynamic histories. The coast's earliest occupants exploited its rivers, bays, and waters, but their distinguishing subsistence economy—heavy reliance on stored salmon and a wealth of other resources—does not appear to have taken its final form until perhaps 3500 years ago. At about that same time, we see the coast's first villages of rectangular houses. Burial data clearly show that status distinctions were made among individuals—particularly in who did and did not wear labrets. By 2500 years ago, social stratification had developed. These changes were accompanied by evidence of warfare, extensive trade and interaction, and pronounced cultural shifts, such as changes in funerary practices around 500 years ago. Despite dramatic local events (some sites were abandoned for a century or more) the archaeological evidence overwhelmingly indicates cultural continuity: the coast's people appear to have been there for a very long time—certainly before the rain forest achieved its modern form and distribution.

We come now to our third question: Were their lives in any way shaped by this distinctive environment? The rain forest was a source of raw materials—the coast's people are justly famous for their carving and their skill as carpenters and artisans of wood. Though wood seldom survives, Middle and Late Pacific wet sites clearly indicate that these skills were well developed by the Middle Pacific subperiod. Woodworking tools recovered from midden sites indicate that use of forest resources continued to develop through the Late Pacific subperiod, as heavy-duty tools such as large adzes and pile drivers appear in the archaeological record. Western redcedar was also a source of fiber for fabrics and basketry. Studies at Ozette show that certain woods were carefully selected according to their qualities for particular uses. While the rain forest provided a diversity of resources, it was not an especially productive ecosystem from the human standpoint, and its productivity declines from south to north, reinforcing the need of humans to exploit riverine and marine resources to the hilt.

But perhaps we should really be asking: What was the impact of the coast's people on the evolving rain forest? As Nancy J. Turner (Chapter 11) and others have shown, some of the coast's people manipulated local environments by controlled burning. Their requirements for wood, moreover, may not have been insignificant. A single large house may have required as much as 70,000 board feet of lumber. One such structure near Portland, Oregon, was used continuously for 400 years and would have required between 500,000 and 1 million board feet of lumber during that period for maintenance and repair. And that is just one house, 55 feet wide and 120 feet long, home to forty-five to sixty people in a valley with a population of between 3000 and 10,000 people who would have required 50 to 180 or more such structures. In addition to houses, demand for wood included firewood, canoes, bark, tools, and other uses. The coast's people, therefore, are likely to have had as significant an effect on the development of the rain forest as the rain forest had on them.

The relationship between humans and their environment is subtle and complex. The ecosystem of Northwest Coast peoples stretched from the abyssal depths of the North Pacific to high alpine environments. Although it is difficult to pinpoint any part of that vastness as the most important, the rain forest may have been the most distinctive. The rain forest provided the means to make the central items in Northwest Coast technology: their houses and their canoes. Yet the rain forest also pinned people to the coast: it was difficult to penetrate, and it did not contain a lot of food. It reinforced, particularly in the north, the need to go to sea, while providing the means to do so.

References

Ames, K. M. 1994. "The Northwest coast: Complex hunter-gatherers, ecology and social evolution." *Ann. Rev. Anthropol.* 24:209–229.

Ames, K. M., and H. Maschner. In press. *The Northwest Coast*. London: Thames & Hudson.

Boyd, R. T. 1990. "Demographic history, 1774–1874." In W. Suttles, ed., *Northwest Coast.* Washington, D.C.: Smithsonian Institution.

Carlson, R. L., ed. 1982. *Indian Art Traditions of the Northwest Coast.* Burnaby, B.C.: Archaeology Press, Simon Fraser University.

Carlson, R. L. 1987. "Cultural and ethnic continuity on the Pacific Coast of British Columbia." Paper presented at the annual congress of the Pacific Science Association, Seoul, Korea, 1987.

Coupland, G. 1985. "Household variability and status differentiation at Kitselas Canyon." *Can. J. Archaeol.* 9:39–56.

Croes, D. R. 1976. "The excavation of water-saturated archeological sites (wet sites) on the Northwest Coast of North America." *Mercury Series Paper #50.* Archaeological Survey of Canada, Ottawa, Ontario.

Cybulski, J. S. 1990. "Human biology." In W. Suttles, ed., *Northwest Coast.* Washington, D.C.: Smithsonian Institution.

———. 1993. "A Greenville burial ground: Human remains in British Columbia coast prehistory." In *Archeological Survey of Canada*. Ottawa: Canadian Museum of Civilization.

Drucker, P. 1963. *Indians of the Northwest Coast*. New York: Natural History Press.

Easton, N. A. 1985. "The underwater archaeology of Straits Salish reef-netting." M.A. thesis, University of Victoria, B.C.

Ecotrust, Pacific GIS, and Conservation International. 1995. *The Rain Forests of Home: An Atlas of People and Place.* Part 1: *Natural Forests and Native Languages of the Coastal Temperate Rain Forest.* Portland, Ore.

Fladmark, K. R. 1979. "Routes: Alternative migration corridors for early man in North America." *Am. Antiq*. 44:55–69.

———. 1982. "An introduction to the prehistory of British Columbia." *Can. J. Archaeol.* 3:131–144.

Hebda, R. J., and R. W. Mathewes. 1984. "Holocene history of cedar and Native Indian cultures of the North American Pacific Coast." *Science* 225:711–713.

Holm, B. 1990. "Art." In W. Suttles, ed., *Northwest Coast.* Washington, D.C.: Smithsonian Institution.

Kroeber, A. L. 1939. *Cultural and Natural Areas of Native North America*. Publications in American Archaeology and Ethnology 38. Berkeley: University of California.

Matson, R. G. 1976. "The Glenrose Cannery site." *Mercury Series Paper #52.* Archaeological Survey of Canada, Ottawa, Ontario.

———. 1992. "The evolution of Northwest Coast subsistence." *Res. Econ. Anthropol. Suppl.* 6:367–430.

Matson, R. G., and G. Coupland. 1995. *The Prehistory of the Northwest Coast*. San Diego: Academic Press.

Samuels, S. R. 1991. *House Structure and Floor Midden: Ozette Archeological Project Research Reports.* Vol. 1. Pullman: Washington State University.

Sapir, E. 1951. "Time perspective in aboriginal American culture: A study in method." In D. G. Mandelbaum, ed., *Selected Writings of Edward Sapir*. Berkeley: University of California Press. Originally published in 1916.

Suttles, W. 1968. "Coping with abundance: Subsistence on the Northwest Coast." In R. B. Lee and I. DeVore, eds., *Man the Hunter*. New York: Aldine. Reprinted in Suttles (1987).

———. 1985. *Native Languages of the Northwest Coast.* Color map. Portland: Press of the Oregon Historical Society.

———. 1987. *Coast Salish Essays*. Vancouver: Talonbooks.

———, ed. 1990. *Handbook of North American Indians.* Vol. 7: *Northwest Coast*. Washington, D.C.: Smithsonian Institution.

Thompson, L. C., and M. D. Kinkade. 1990. "Languages." In W. Suttles, ed., *Northwest Coast.* Washington, D.C.: Smithsonian Institution.

11. Traditional Ecological Knowledge

NANCY J. TURNER

Within the past few decades there has been a growing movement worldwide toward recognition of the importance and validity of traditional, or local, knowledge. This movement is both ethical and practical. Indigenous peoples have been oppressed in many ways, removed from their lands, or restricted within small areas of their traditional territories, or sometimes outside of them, forbidden to speak their language or practice their culture, including the use of traditional resources and traditional land and resource management (Berger 1991; Jennings 1993). In practical terms, the loss of indigenous knowledge is a loss to all humanity. (See Schultes 1988; Linden 1991; Gadgil et al. 1993; Durning 1993.) The wisdom and knowledge of indigenous peoples is needed now, more than ever, to help us develop more sustainable ways of living. Western science alone has not been successful in accomplishing this.

The movement to recognize indigenous cultures was given major impetus by the Brundtland Report, *Our Common Future* (World Commission on Environment and Development 1987). Many other recent publications address the need to recognize and validate traditional cultures and traditional knowledge. (See, for example, Berkes 1993; Clarkson et al. 1992; Durning 1993; Doubleday 1993; Freeman and Carbyn 1988; Inglis 1993; Johnson 1992; Knudtson and Suzuki 1992; Tyler 1993; Wavey 1993; Williams and Baines 1993.)

But there are still many misunderstandings about traditional ecological knowledge, and there is substantial resistance to its recognition. Nakashima (1993:100) attributes this to "an elitism and ethnocentrism that runs deep in much of the western scientific community." Nevertheless, traditional knowledge has received major recognition as being complementary to and equivalent to scientific knowledge. (See Ford 1979; Posey 1990; Colorado and Collins 1987; Colorado 1988; Williams and Baines 1993.) Certainly this is evident in British Columbia, where Nuu-Chah-Nulth representatives were included on the Scientific Panel for Sustainable Forest Practices in Clayoquot Sound, appointed by the provincial government. The Nuu-Chah-Nulth were formerly known as Nootka, or Westcoast Peoples, and their traditional territory extends along the west coast of Vancouver Island. This panel of eighteen scientists and Nuu-Chah-Nulth representatives was appointed in response to recommendations by the Commission on Resources and Environment (CORE). Its task was to review current forest practices in Clayoquot Sound and make recommendations on how these practices can be improved to become "the best in the world." Its work incorporated traditional knowledge in significant ways (Clayoquot Scientific Panel 1994, 1995).

Indigenous peoples are cultural groups residing in a particular locality for a long period of time, usually many generations, and generally depending on the resources of that area for their sustenance. They typically have a distinct language, culture, and religion and, moreover, have a custodial concept of land and resources. Their social relations often involve collective management of resources and group decision making. Indigenous peoples cannot be assumed to be a single entity—in contrast to industrial or postindustrial society—and not all individuals or segments within an indigenous society share identical worldviews or follow the same philosophies. Each culture, each group, is different. Each has its own traditions, its own environment, its own institutions and strategies for sustenance, and its own history.

Traditional ecological knowledge is holistic and not easily subject to fragmentation. As a result, all of the themes discussed here are interrelated. In the words of the Nuu-Chah-Nulth elders on the Clayoquot Scientific Panel, *"hishuk ish tsawalk"* ("everything is one"). Furthermore, although many of the references cited here are historical, the concepts continue: traditional knowledge is still known, still applied by many people, and still being passed on.

This chapter discusses traditional ecological knowledge systems of indigenous peoples of the Pacific coastal temperate rain forest—especially the traditional knowledge of indigenous peoples of the Northwest Coast Culture Area. As a major component and reflection of traditional lore, knowledge about plants and their cultural importance in this region is emphasized. Characteristics of ecological knowledge in these traditional cultures fall into three broad categories (Table 11.1):

- Philosophy or worldview
- Practices and strategies for resource use and sustainability
- Communication and exchange of knowledge and information

Table 11.1. Traditional Ecological Knowledge of the Northwest Coast

Worldview	Strategies for Sustainable Living	Exchange of Knowledge
• belief in the spirituality and innate power of all things • respect for other life forms and entities • ideological systems that enforce sustainable use of resources (social sanctions, sharing) • concepts of interactive relationships with other life forms • close identification with ancestral lands	• knowledge and application of sustainable practices: inventory, monitoring, use of ecological indicators; environmental modification; harvesting strategies • understanding of major principles of ecology: relationships among all life forms and the environment; ecological succession • adaptation to change in resource availability and living conditions	• exchange of knowledge and resources within and among groups • language: classification and naming of culturally important features • development of social structures and institutions to promote sustainable living • development of culturally appropriate ways of teaching and learning about the environment and traditional knowledge and attitudes

Worldview

As a fundamental attribute, traditional attitudes reflect a belief in the innate power and spirituality of all things in the environment, both biological and nonbiological. This belief is expressed in almost every aspect of traditional life. Examples can be drawn for every cultural group and for many facets of life—from the narratives that are recounted (Barbeau 1961; Boas 1921, 1966; McIlwraith 1948; Swanton 1905; Kennedy and Bouchard 1983) to the winter dances and many other ceremonies and rituals that are such an integral part of Northwest Coast cultures.

Because of this central concept, respect for other life forms is an essential component of the worldview. It is expressed, for example, in words of praise, acknowledgment, and thanks offered to plants and animals as they were harvested and during their preparation and use:

> Kwagiulh berry-picker: "I have come, you Supernatural Ones, you, Long-Life Makers, that I may take you, for that is the reason why you have come, brought by your creator, that you may come and satisfy me. . . . Do not blame me for what I do to you when I set fire to you the way it is done by my ancestor who set fire to you . . . when you get old . . . that you may bear much fruit. . . . Look! I come now dressed with my large basket and my small basket that you may go into it,

> Healing-Women. . . . I mean this, that you may not be evilly disposed towards me, friends. That you may only treat me well." [Boas 1930:203]

In these words the concept of interactive relationships is also expressed—the recognition that the plant (or animal or object) has the power to influence the life of the person using it. If appropriate respect is not shown, the person might be harmed; if praise and gratitude are expressed, the person can expect to receive help and good fortune. Boas (1930) provides many Kwagiulh examples of similar addresses to all types of life forms used by people. An example of similar regard for a physical entity is the Saanich belief that pointing at the Malahat Mountains, on southern Vancouver Island, said to be a mark of disrespect, might cause thunder and rain (Elsie Claxton, pers. comm. 1989).

Ideological systems have developed that enforce sustainable use of resources through education and through social pressures against noncompliance; everywhere there are sanctions against waste, wanton killing, and destruction. There are also imperatives to share resources—and for the other life forms, too, to share themselves with humans. Ceremonial preparations for Nuu-Chah-Nulth whaling and other forms of hunting, for example, were intended to supplicate the animal being hunted to recognize and acknowledge the needs of humans and yield itself willingly to the hunters (Richard Atleo, pers. comm. 1994). Charles Hill-Tout has described these concepts in relation to the First Salmon ceremony of the Pemberton Lillooet (Lil'wat):

> No one would dream of taking a lauwa [sockeye] salmon before this ceremony had been performed. . . . The significance of these ceremonies is easy to perceive when we remember the attitude of the Indians towards nature generally, and recall their myths relating to the salmon, and their coming to their rivers and streams. Nothing that the Indian of this region eats is regarded by him as mere food and nothing more. Not a single plant, animal, or fish, or other object upon which he feeds, is looked upon in this light, or as something he has secured for himself by his own wit and skill. He regards it rather as something which has been voluntarily and compassionately placed in his hands by the goodwill and consent of the spirit of the object itself, or by the intercession and magic of his culture-heroes; to be retained and used by him only upon the fulfillment of certain conditions. These conditions include respect and reverent care in the killing or plucking of the animal or plant and proper treatment of the parts he has no use for, such as the bones, blood and offal; and the de-

> positing of the same in some stream or lake, so that the object may by that means renew its life and physical form. [Hill-Tout 1905:140–41].

Similar concepts and ceremonies were widely observed by Northwest Coast groups. The Pemberton Lillooet are classed as Interior Salish, but their territory is transitional between the coast and the interior, and they have many coastal cultural traits.

Indigenous peoples closely identify with their ancestral lands because of their deep associations with their resources and, undoubtedly, because of their long-term occupation of particular areas—probably thousands of years in the case of Northwest Coast peoples. Narratives recounting origins of peoples and local associations abound (as in Nuxalk origin stories, McIlwraith 1948:295–351). Geographical landmarks and ancient historical traditions binding people to their lands are widely recognized. Examples for southern Vancouver Island include the Saanich stories of the Star Husbands (Brown 1873), the Origin of Salmon (Jenness n.d.), and the Flood (Philip Paul, pers. comm. 1991). In these stories from ancient times, several specific places are mentioned (Elk Lake, Knockan Hill, Portage Inlet, Discovery Island, and Lau wel new, or Mount Newton). The Nuu-Chah-Nulth traditions about transformed rocks in the Clayoquot area (documented by Bouchard and Kennedy 1990) are another example.

Gwaganad, in an eloquent statement before the British Columbia Supreme Court, expressed her views as a Haida that serve as the essence of these basic concepts of identity with land and place:

> So I want to stress that it's the land that helps us maintain our culture. It is an important, important part of our culture. Without that land, I fear very much for the future of the Haida nation. Like I said before, I don't want my children to inherit stumps. I want my children and my grandchildren to grow up with pride and dignity as a member of the Haida nation. I fear that if we take that land, we may lose the dignity and the pride of being a Haida. [Gwaganad 1990]

The intimate association of First Nations with their homelands was never adequately recognized by the colonial government and subsequent agencies who wrested control of the lands and resources from aboriginal peoples across North America. Possibly this is because most of these people willingly left their own homelands—England, Scotland, France, and other distant places—in their own search for self-determination, power, spiritual fulfillment, or wealth. As a result, they may have had little understanding or sympathy for peoples who were so intimately and permanently attached to their homeplaces. It is no wonder that aboriginal peoples have fought so persistently to regain control of their traditional lands. The words of the Gitksan Chief, Delgam Uukw, explain the importance of the land:

> For us, the ownership of territory is a marriage of the Chief and the land. Each Chief has an ancestor who encountered and acknowledged the life of the land. From such encounters come power. The land, the plants, the animals and the people all have spirit—they must all be shown respect. That is the basis of our law. [Gisday Wa and Delgam Uukw 1989]

Strategies for Sustainable Living

Long-term residence in one locality—coupled with total dependence on local resources for survival and close observation and monitoring—has led to detailed knowledge and understanding of ecological relationships and environmental indicators that help to optimize the use of resources. There are numerous examples. Take, for instance, the technology for procuring salmon. Salmon weirs set across spawning streams were not just efficient structures for capturing fish. They allowed close observation and selection of fish and enabled those who were managing this resource to ascertain that the runs were strong enough for harvesting. If not, fewer fish were taken. Often the largest fish were released to continue upstream to spawn. Salmon weirs and traps were used in hundreds of major spawning streams up and down the coast. The removal of these weirs was forced on aboriginal peoples without any understanding of their complete function.

Traditional management practices were also applied in vegetation burning practices. Meadows and certain areas of forest were burned over periodically to maintain early seral species and to enhance the productivity of forage for deer as well as certain plant foods, particularly berry and root vegetable species (Turner 1991; Lewis and Ferguson 1988). Elders maintain that since this type of burning has been suppressed by the Forest Service, the production and quality of wild berries and roots have declined noticeably. Soapberries in the upper Bella Coola Valley, for instance, are said to be less abundant than formerly (Margaret Siwallace, pers. comm. 1984). Other strategies for sustainable resource use include selective harvesting of root vegetables such as camas and riceroot bulbs, springbank clover rhizomes, and Pacific silverweed roots. Only the largest roots are taken; the smaller ones, or fragments, are returned to grow for the next season. Beds of root vegetables were also frequently nurtured; large rocks and weedy invaders were removed, and the tilling of the soil accompanying root digging apparently also enhanced production. (See Turner and Bell 1971; Turner and Kuhnlein 1982, 1983; Alice Paul, pers. comm. 1975.) The pointed wooden root-digger was an ideal implement for this type of harvesting (Figure 11.1). Greens, too, were harvested selectively. Cow-parsnip shoot growth, for example, was said to be enhanced by continuous, selective harvesting (Kuhnlein and Turner 1987).

Figure 11.1. Barbara Wilson (Haida) displays a yew-wood root-digging stick carved by Ron Wilson (Gitsga). (Photo by N. Turner.)

Gathering of materials and medicines was also done carefully and systematically. To obtain cedar bark, it was customary to remove bark from only one-third of the circumference of the tree. In this way, the tree would survive and continue to grow. The thousands of "culturally modified trees" in the coastal temperate rain forest are a testimony to the success of this practice. Similarly, where possible, boards were split from standing cedar trees with the use of adzes, mauls, and wedges of yew wood, leaving the cedar trees alive. Spruce roots and cedar roots for baskets also were dug selectively, since it was recognized that taking too many roots from a tree would be harmful. The bark of many trees was used for medicine. And almost always, the bark pieces were cut in a narrow vertical strip from the sunrise side or river side of the tree (Turner and Hebda 1990). This was to allow the tree to continue to grow, and the sunrise side was said to heal most quickly (Tom Sampson, pers. comm. 1987; Daisy Sewid-Smith, pers. comm. 1994).

Close observation of life cycles and seasonality of plants and animals allowed the use of ecological indicators to determine harvesting times and anticipate levels of resource abundance. For the Sliammon (Coast Salish), the arrival of the sandhill crane in March indicated the onset of herring spawning. The full blooming of oceanspray, usually in late June, announced the peak of plumpness and flavor of butter clams (Kennedy and Bouchard 1983; Elizabeth Harry, pers. comm. 1993). Saanich (Straits Salish) shellfish harvesters watched the crows on the beach closely; if the crows stopped coming down to the beach to forage for food, the Saanich people stopped their own foraging, assuming that the shellfish would not be safe (Elsie Claxton, pers. comm. 1990). For Kwagiulh hunters, twin fawns indicated a plentiful year for deer; if only single fawns were born, hunters would restrict the numbers of deer they took (Daisy Sewid-Smith, pers. comm. 1994). Many other ecological relationships were known, including the feeding patterns of many animals: deer browse on old man's beard lichens (*Usnea* and *Alectoria* spp.); bears enjoy young skunk-cabbage shoots and rhizomes in the early spring; geese and ducks often forage on the roots of silverweed and wild clover (Turner 1995; Turner and Kuhnlein 1982; Turner et al. 1983; Compton 1993).

For Northwest Coast peoples, as for other hunter-fisher-gatherer societies, diversity was a key to sustainable resource use. Diets were based on a wide variety of plants and animals, and many different species were used as materials and medicines (Table 11.2). Plants and animals played other important cultural roles (as amulets, for example), but their use for food, materials, and medicine was the most intensive.

Even a conservative estimate, allowing for overlap in the use of some species in more than one category, indicates that at least 400, and probably closer to 500, different plant species and about 100 animal species were used

Table 11.2. Estimated Numbers of Plant and Animal Species Used Traditionally by Northwest Coast Peoples

Foods (Including Beverages and Condiments)

Estimated Total: Over 200 Species

Category	Number of Species	Examples
Fruits	~50	blueberries and huckleberries, salal, salmonberries, high-bush cranberries
Root vegetables	~25	Pacific silverweed, springbank clover, riceroot, wapato, spiny wood fern
Green vegetables	~20	cow parsnip, fireweed, giant horsetail, thimbleberry shoots, seaweed
Other plant products	~10	Labrador tea, licorice fern, western hemlock (inner bark)
Mammals	~20	deer, elk, seals, whales, bear, mountain goats
Birds	~20	ducks, geese, grouse, seabirds, gull eggs
Fish	~35	salmon (five species), herring (and spawn), eulachon (and "grease"), halibut, cod
Invertebrates	~35	clams, mussels, sea urchins, octopus, crabs, barnacles, sea cucumber

Medicines

Estimated Total: About 150 Species

Category	Number of Species	Examples
General tonics	~20	grand fir, red alder, devil's club, Labrador tea, yellow pond lily, black twinberry
Purgatives, laxatives, emetics	~10	cascara, goatsbeard, red elderberry
Salves, poultices, and washes for skin ailments	~30	spruce pitch, cottonwood bud resin, salmonberry bark, wild lily-of-the-valley, broad-leaved plantain, salal, skunk cabbage
Medicines for colds, coughs, tuberculosis, other respiratory ailments	~20	yarrow, licorice fern, red alder bark, Indian consumption plant, Pacific crabapple bark, common juniper, western hemlock
Aids for internal ailments (digestive tract, internal injuries)	~25	devil's club, grand fir, red alder, western hemlock
Gynecological medicine	~15	salal, water parsley (Ditidaht), red elderberry (Saanich)
Rheumatism and arthritis	~20	stinging nettle, devil's club, crowberry, lodgepole pine bark, Pacific yew
Miscellaneous other medicines	~20	black twinberry, conehead liverwort, yellow pond lily
Medicinal animal products	~10	eulachon grease (tonic), banana slugs (poultice), deer fat (salve), white-winged scoter broth (laxative)

(*continues*)

Table 11.2. *Continued*

Materials in Technology

Estimated Total: About 200 Species

Category	Number of Species	Examples
Wood for construction manufacture	~25	western redcedar, Pacific yew, red alder, Sitka spruce, oceanspray, crabapple, willow, devil's club
Wood and other materials for specialized fuel	~10	Douglas-fir bark, redcedar wood and shredded bark, dried bracken rhizome fibers, western hemlock
Fibrous plant materials	~15	inner bark of redcedar and yellow-cedar; redcedar withes; leaves of cattail, basket sedge, and bear grass; tule stems; gooseberry roots; spruce roots; bull kelp stipes
Other plant products used in technology	~35	pitch (adhesive, waterproofing), grand fir boughs (bedding), skunk cabbage leaves (drying berries), salal (cooking pits), horsetail (abrasive)
Skins	~20	ermine, deer, bear, elk, mountain goat (skins and wool)
Antler/bone/shell	~30	elk, deer bone and antler (harpoon points, awls); mussel shell (chisels); horse clam shell (ladle); dentalium and abalone (decoration, currency)
Feathers	~20	hummingbird, flicker, bald eagle, heron, Canada goose (down for bedding)
Miscellaneous animal materials	~15	sea lion stomach (floats), ling cod stomach (pouch), sea lion whiskers (mask decoration), sinew of deer and whale for cordage, octopus for bait

in fundamental ways in the region. The use of diverse resources is itself an adaptive strategy—it allows recourse to other foods if one type turns out to be in short supply for whatever reason. Berries, for example, undergo cycles with fruit production varying from one year to the next. If salal berries were poor in one season, people could almost always acquire more stink currants, or huckleberries, or thimbleberries to make up their dietary requirements. Similarly, if deer were scarce in one year, people might rely more heavily on fish. In extreme cases, species were exploited that would not normally be used. But even these strategies represent important knowledge that permitted survival (Turner and Davis 1993).

Living patterns of Northwest Coast peoples also reflect adaptive strategies. The location of most winter village sites just above the shoreline—often in quiet bays and near river estuaries—provided protection from the elements and convenient access to transportation routes and many needed resources. During the growing season an established round was followed by families and small groups who would move to a variety of locations in order to undertake hunting, fishing, and gathering as the different resources became available. From the first salmonberries in the spring to the last "winter huckleberries" (evergreen huckleberries) at the end of the season, the timing of resources was understood and followed. Division of labor—women gathering and preparing food and fibrous materials, men hunting, fishing, and working with wood—also helped to optimize resource use, since these activities were generally done concurrently. The wide array of gear, implements, vessels, and containers applied in these various tasks represents ingenious adaptive strategies (Stewart 1977; Turner 1979; Turner et al. 1983) (Figures 11.1–11.3). Processing and preservation techniques for resources—pit cooking, smoking, dehydrating, and rendering and underwater preserving for tart foods such as crabapples—also allowed optimal resource use and good nutrition.

Figure 11.2. Baskets of numerous styles and materials are an important reflection of traditional ecological knowledge. Back center: redcedar bark (Kwagiulh); back left: yellow-cedar bark (Haida); front left and back right: split, coiled redcedar root decorated with bitter cherry bark and reed canary grass stalks (Pemberton Lilooet); and front right: wrapped twined baskets of tall basket sedge, American bullrush, and raffia (Nuu-Chah-Nulth). (Photo by R. Turner.)

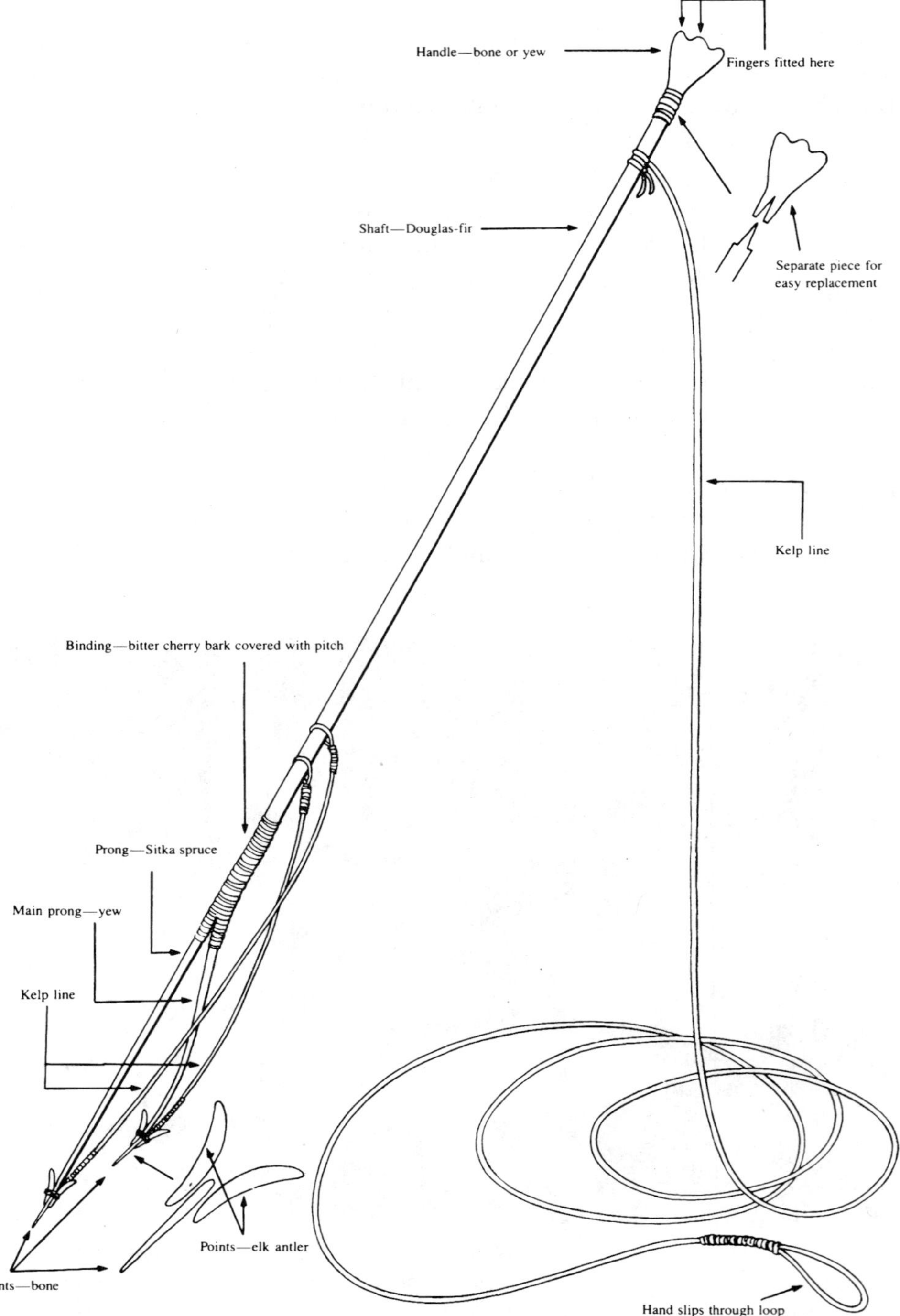

Figure 11.3. Ditidaht two-pronged salmon harpoon, showing diverse materials and intricate construction. Drawn by Elizabeth J. Stephen; reproduced with permission of the Royal British Columbia Museum from Turner et al. (1983).

Another aspect of sustainable resource use is the ability to adapt to changing circumstances. Indeed, the ability to adopt and assume new knowledge, new resources, and new technologies is an important feature of traditional knowledge systems. Certainly the aboriginal peoples of the Northwest Coast have faced fundamental changes to their lives in recent history, most of them highly disruptive. Yet many individuals and communities have retained the essential elements of their worldview and lifestyles, a testament to the strength of these philosophies. Adaptation has sometimes been positive. The adoption of metal tools, for example, allowed the woodworking artistry of Northwest Coast peoples to flourish. The introduction of the potato as a valued root vegetable, integrated readily into food production systems, has also been generally beneficial (Suttles 1987). New media for art, and new shapes and styles for basketry, have provided different opportunities for aboriginal people while allowing them to maintain their cultural identity. Many aboriginal groups have adapted their original intergroup trading economies to participate in broader commercial activities in modern times: the Haida and others became major producers of potatoes for trade; the Nuu-Chah-Nulth and Straits Salish harvested red laver seaweed for the Asian people of Victoria; the Heiltsuk, who traditionally traded herring eggs, seaweed, and other products to the Nuxalk, have harvested these products and others in modern commerce. Many First Nations people have used their knowledge and enjoyment of the outdoor lifestyle to earn a living in the cash economy by picking forest greenery and pine mushrooms, harvesting geoduck clams and other shellfish, and participating in the commercial fishery.

Social institutions of Northwest Coast peoples also reflect adaptive strategies. Family and kinship groups generally allow people to better share goods and services, and to barter and trade resources, both in times of scarcity and times of abundance (Suttles 1987: nos. 2, 4). One social institution constructed around sustainable living is h̲ah̲uulhi of the Nuu-Chah-Nulth. Roy Haiyupis characterizes h̲ah̲uulhi as follows:

> Hahoolthe [h̲ah̲uulhi] . . . indicates . . . that the hereditary chiefs have the responsibility to take care of the forests, the land and the sea within their hahoolthe and a responsibility to take care of their mus chum or tribal members. [Nuu-Chah-Nulth elder; Clayoquot Scientific Panel 1995; Haiyupis 1992:1]

With this system, virtually all areas of the land, water, and shoreline fall under the specific "ownership" of one individual, usually a hereditary chief. This person receives the benefits of using the area and harvesting its resources. But, in turn, this person has the primary responsibility of maintaining and sustaining it and sharing the resources equitably among other members of the community. Thus each salmon stream, each shellfish beach, each silverweed and clover patch, each prime berry-picking area, was (and still is to some extent) recognized as belonging to an individual and his (or her) family and was therefore closely observed and maintained by them. The person who is to inherit each area is carefully schooled from an early age to take on the responsibility of caring for it. And when ownership is actually assumed, this person continues to consult with elders and other knowledgeable people in making management decisions about the area and its resources. There is both individual attention and long-term continuity in this system.

Similar systems of ownership and stewardship of prime areas and major resources existed, to a greater or lesser extent, throughout the Northwest Coast region. The ownership and rights to use these resources were validated through public recognition at feasts and potlatches and other ceremonial occasions. Ceremonies, such as the first salmon rites and the sacred cedar-bark ceremony (Sewid-Smith et al. in press), were also a means of recognizing the importance of resource sustainability and formally showing respect for the other life forms upon which people depend.

Communication and Exchange of Knowledge

The success of sustainable resource use depends on the effectiveness of communication, exchange, and perpetuation of knowledge from one generation to the next both within and among cultural groups. Such communication is reflected in numerous ways. Systems of classification and nomenclature of culturally important resources and phenomena have been developed in every language and culture. These are discussed in general terms by Berlin (1992), who with his colleagues developed a series of "universal principles" for folk biological classification and nomenclature. Work by Hunn (1982) and Turner (1974) on traditional systems of naming and classifying plants in the Pacific Northwest has determined that for temperate hunter-fisher-gatherer groups, at least, there is a definite tendency to name and classify those species that are culturally important; species of less importance often are not named except in a very general way (Hunn et al. in press).

On the Northwest Coast, there is almost always, in the case of culturally significant plants and animals, a one-to-one correspondence with scientific species (or groups of closely related species in a genus). Sometimes, with species of high significance or differentiation, more than one name is applied—either to different growth stages or to different forms of the same species. Thus the different color forms of salmonberry (golden, ruby, dark red) are generally distinguished with specific names. For salal, different names might be applied to the green, leafy bushes—to be used as lining in pit cooking—and to berry-bearing bushes. Cow parsnip and giant horsetail have variable names applied to the vegetative and fertile components, each of which is differentiated in its use (Turner and Efrat 1982; Turner et al. 1983; Compton 1993). Similarly, there may be separate names for young Douglas-fir and mature, old-growth individuals. Western redcedar, used in many different ways, often has a series of associated names—for young trees, older trees, wood, outer bark, inner bark, withes, and roots. For "prototypical" species of high cultural importance, the names may sometimes undergo expansion of reference. In Haida, the name for Sitka spruce also applies generally to evergreen trees. The Ditidaht name for Pacific silverweed also refers to a general class of edible roots. In the Nuxalk language, the name for black huckleberry is equivalent to the word for "all berries."

The names of plants and animals often reflect their uses and cultural aspects. Pacific yew is often called "bow" or "bow plant," and big-leaf maple is called "paddle plant" in some Salishan languages. The Ditidaht name for Sitka spruce means "scaring," because the boughs are used to "threaten" people with their sharp needles during winter dancing. Yellow pond lily is called "west wind plant" in Hesquiaht, because it is associated with the west wind; the floating leaves align themselves on the water in the same direction when the west wind blows, and the plant is used as a charm to invoke the west wind. Ecological relationships are often reflected in names, as well, as in the Ditidaht name for coltsfoot: "elk's food."

Knowledge about obviously edible species, such as salmon, deer, blueberries, and strawberries, might be acquired independently by diverse groups of people. The knowledge of other food sources, however, such as the inner bark of trees, as well as specific applications of various medicines, probably originated through sharing and exchange. For example, devil's club has many common uses throughout the region (Turner 1982; Compton 1993). Similarly, specific preparation methods and basketry techniques are likely to have been shared widely while at the same time being adapted for the specific requirements of individuals and groups.

Tracing the names for plants and animals is another means of determining how resources and knowledge about them have been shared. The Straits Salish name for licorice fern, for example, is apparently borrowed from Ditidaht; the Carrier name for high-bush cranberry is borrowed from the Nuxalk language (and possibly originally from Wakashan languages); the Skidegate and Massett Haida names for soapberry are borrowed from Tsimshian and Tlingit, respectively (Turner 1981). There are many records of traditional trading and exchanges. For example, black ducks (scoters) and halibut were exchanged between the Ditidaht peoples of Whyac and Clo-oose (Turner et al. 1983). Such exchanges still occur today. Recent Nuxalk potlatch gifts, for example, included jarred soapberries, wild berry jam, and dried seaweed (Margaret Siwallace Memorial Potlatch, October 1993).

Finally, the development of culturally appropriate ways of teaching and learning is a major element in the perpetuation of traditional ecological knowledge in both its philosophical and practical aspects. Stories with encapsulated messages were, and are, an effective means of transferring information and teachings. Since the Northwest Coast peoples did not use written language, all information had to be transferred orally or through participatory learning and observation. Children learned listening skills from an early age. Indeed, some were specifically trained as historians and orators, so that they could publicly recount the information about cultural traditions at feasts and potlatches and other ceremonial occasions. Many stories teach about traditional ecological knowledge. One is the Hesquiaht story of the Yellow-Cedar Sisters:

> Three young women were down on the beach drying salmon. Raven came along and wanted their salmon, so he kept asking them if they were afraid to be there by themselves—if they were afraid of bears, or wolves or other such animals. They kept saying "no" to everything he asked them about until he said "owls." At this they said, "Oh, don't even talk about owls to us; we are afraid of owls." Raven went away, but hid in the bushes nearby and began to imitate owl sounds. The women were so frightened they ran away into the woods. They kept running until they came partway up the side of a hill. They were so tired, they decided to stop and rest. They said to themselves, "We'd better stand here now on the side of the mountain; they will call us ʔalhmapt." And they turned into yellow-cedar trees. Raven snuck out and ate all their dried salmon. That is why yellow-cedars are always found on the mountainsides, and why they are such nice looking trees, with smooth trunks and few branches, because they used to be attractive young women with long shining hair. [Told by Alice Paul of Hesquiaht; see Turner and Efrat (1982:33) and Figure 11.4]

Children were also taught from an early age not to waste food, not to play with parts of animals, and not to kill wantonly. Often people were warned of the specific consequences of inappropriate behavior. There is a Nisga'a story about a boy who was amusing himself by cutting the backs of salmon, pouring hot pitch onto them, and setting the pitch on fire so that he could watch the burning fish struggling up the creek at night. The consequence of this disrespectful action was a volcanic eruption that destroyed the village and covered a major part of the Nass Valley with lava (Joint Nisga'a/B.C. Parks 1992:8; Chester Moore, pers. comm. 1993). Another example of a story that teaches—

Figure 11.4. The Yellow-Cedar Sisters. (Drawn by Molly Turner.)

one of literally hundreds from the Northwest Coast area—is the Nuxalk (Bella Coola) story called "The Woman Who Befriended a Wolf" (McIlwraith 1948, I:691):

> Not many hundred years ago . . . a woman [named Ksninsnimdimut] was picking blueberries. As she started to climb a steep bank to a shelf where she saw that the fruit was plentiful, she heard one of the berries speaking to its companions.
>
> "Let's hide," it said, "that foul-mouthed woman is coming." The berry was speaking of her . . . [as] "foul-mouthed" because of

> her habit of munching berries as she picked instead of putting all in her basket for future consumption.
>
> She hurried up the bank so swiftly that many of the berries were unable to hide, and she saw them in their human forms, a host of goggle-eyed little boys sitting on the berry shoots (Figure 11.5). On that occasion, she obtained more than any of her companions. Thanks to her sight of the berries in human form, she was thenceforth able to see them in their hiding places and was accordingly always fortunate. She respected the wishes of the fruit, never eating as she picked, but chewing dried salmon instead. [This same woman later encountered a wolf giving birth and assisted her in delivering the cubs. In gratitude, the wolf gave this midwife the power of a shaman.]

These lessons reflect the traditional teachings of the power and sacredness of all things, as well as the need for respect and gratitude for the resources used and the favors sought.

In the Future

Traditional ecological knowledge of indigenous peoples in the coastal temperate rain forest region can be related to worldviews, to practices and strategies for sustainable living, and to methods of communication and exchange of knowledge. These concepts are inextricably linked. The indigenous peoples of this forest region had a deep and broad understanding of natural ecological systems, an understanding that is underlain by their traditional spiritual beliefs. Moreover, they used this knowledge to practice sustainable management and harvesting as well as optimization of resources for food, material, and medicine.

One strategy for sustainable living was diversification of resource use. In total, some 400–500 plant and animal species were used as food, technological material, and medicine. The division of labor between men and women, the resource processing techniques, the development of social institutions promoting stewardship and resource exchange—all are practices leading to sustainable resource use.

Traditional ecological knowledge was, and is, learned and transmitted through firsthand instruction from one generation to the next by means of stories, ceremonial traditions, and participatory learning. It was also incorporated into, and reflected by, indigenous systems of biological and ecological classification and naming. A strength of traditional ecological knowledge is

Figure 11.5. "A host of goggle-eyed little boys sitting on the berry shoots." (Drawn by Molly Turner.)

that it can adapt to changing conditions and situations so long as they are compatible with the underlying worldview.

Although the traditional ecological knowledge of indigenous peoples is becoming more widely recognized in Canada and around the world, there is need for its further recognition and application in mainstream society. The work of the Clayoquot Scientific Panel (1995) is one example of how traditional ecological knowledge can be recognized and used. Other examples of its incorporation can be seen in the development of Gwaii Haanas reserve on Haida Gwaii (Queen Charlotte Islands) by the Haida Nation in collaboration with the Canadian Parks Service and the development of the Kitlope watershed on the central coast of British Columbia, where the knowledge and philosophies of the Haisla and Henaaksiala peoples will direct the comanagement of this spectacular new reserve.

Acknowledgments

The knowledge and wisdom of many people and many coastal First Nations are incorporated here. I would like to acknowledge the specific contributions of Elsie Claxton, Tom Sampson, and the late Philip Paul (Straits Salish); the late Elizabeth Harry (Sliammon); Dr. Richard Atleo, Roy Haiyupis, Larry Paul, Stanley Sam, and the late Alice Paul (Nuu-Chah-Nulth); Daisy Sewid-Smith,

Chief Adam Dick, and Kim Recalma-Clutesi (Kwagiulh); Cyril Carpenter (Heiltsuk); the late Margaret Siwallace (Nuxalk); Chester Moore (Nisga'a); Barbara Wilson and Ron Wilson (Gitsga); Gwaganad (Diane Brown), Guujaaw, and the late Florence Davidson (Haida).

I am grateful to Dr. Marianne Boelscher Ignace, Ron Ignace, Laurie Montour, Judith Mitchell, Karen Chester, Dr. Brian Compton, Randy Bouchard, Dorothy Kennedy, and Eugene Hunn for valuable information and/or discussions that helped me develop this chapter. Peter K. Schoonmaker, Fikret Berkes, and my husband, Robert Turner, offered invaluable editorial suggestions. The salmon harpoon (Figure 11.3) was drawn by Elizabeth J. Stephen, and is reproduced with permission of the Royal British Columbia Museum from Turner et al. (1983). My daughter, Molly Turner, drew the illustrations in Figures 11.4 and 11.5. Scientific names for the plants mentioned in this chapter may be found in Pojar and MacKinnon (1994).

References

Barbeau, M. 1961. *Tsimsyan Myths.* Bulletin 174. Ottawa: National Museum of Canada.

Berger, T. R. 1991. *A Long and Terrible Shadow: White Values, Native Rights in the Americas, 1492–1992.* Vancouver: Douglas & McIntyre; Seattle: University of Washington Press.

Berkes, F. 1993. "Traditional ecological knowledge in perspective." In Julian T. Inglis, ed., *Traditional Ecological Knowledge: Concepts and Cases*. Ottawa: International Program on Traditional Ecological Knowledge, International Development Research Centre, and Canadian Museum of Nature.

Berlin, B. 1992. *Ethnobiological Classification: Principles of Categorization of Plants and Animals in Traditional Societies*. Princeton, N.J.: Princeton University Press.

Boas, F. 1921. *Ethnology of the Kwakiutl.* 35th Annual Report, Parts 1 and 2. Washington, D.C.: Bureau of American Ethnology.

———. 1930. "Religion of the Kwakiutl Indians." In *Columbia University Contributions to Anthropology*. Vol. 10, Parts 1 and 2. New York: Columbia University Press.

———. 1966. *Kwakiutl Ethnography.* Edited by H. Codere. Chicago: University of Chicago Press.

Bouchard, R., and D.I.D. Kennedy. 1990. *Clayoquot Sound Indian Land Use*. Report prepared for MacMillan Bloedel Ltd., Fletcher Challenge Canada, and British Columbia Ministry of Forests. Victoria: British Columbia Indian Language Project.

Brown, R. 1873. "The Indian story of Jack and the Bean Stalk." In *The Races of Mankind, Being a Popular Description of the Characteristics, Manners and Customs of the Principal Varieties of the Human Family*. Vol. 1. London: Cassel, Petter & Galpin.

Calvert, S. G. 1980. "A cultural analysis of faunal remains from three archaeological sites in Hesquiat Harbour, B.C." Ph.D. diss., Department of Anthropology and Sociology, University of British Columbia, Vancouver.

Clarkson, L., V. Morrissette, and G. Regallet. 1992. *Our Responsibility to the Seventh Generation: Indigenous Peoples and Sustainable Development*. Winnipeg: International Institute for Sustainable Development.

Clayoquot Scientific Panel. 1994. *Report of the Scientific Panel for Sustainable Forest Practices in Clayoquot Sound*. Victoria, B.C.

———. 1995. *First Nations Perspectives: Sustainable Forest Practices in Clayoquot Sound*. Victoria, B.C.

Colorado, P. 1988. "Bridging native and western science." *Convergence* 21(2–3):49–59.

Colorado, P., and D. Collins. 1987. "Western scientific colonialism and the re-emergence of Native science." *Practice: The Journal of Politics, Economics, Psychology, Sociology and Culture*, Winter issue.

Compton, Brian D. 1993. "Upper North Wakashan and Southern Tsimshian ethnobotany: The knowledge and usage of plants and fungi among the Oweekeno, Hanaksiala (Kitlope and Kemano), Haisla (Kitamaat) and Kitasoo peoples of the central and north coasts of British Columbia." Ph.D. diss., Department of Botany, University of British Columbia, Vancouver.

Doubleday, N. C. 1993. "Finding common ground: Natural law and collective wisdom." In J. T. Inglis, ed., *Traditional Ecological Knowledge. Concepts and Cases*. Ottawa: International Development Research Centre.

Durning, A. T. 1993. "Supporting indigenous peoples." In L. R. Brown, ed., *State of the World 1993*. New York: Norton.

Ellis, D. W., and L. Swan. 1981. *Teachings of the Tides: Uses of Marine Invertebrates by the Manhousat People*. Nanaimo, B.C.: Theytus Books.

Ellis, D. W., and S. Wilson. 1982. *The Knowledge and Usage of Marine Invertebrates by the Skidegate Haida People of the Queen Charlotte Islands*. Skidegate, B.C.: Queen Charlotte Islands Museum Society.

Ford, R. I. 1979. "The science of Native Americans." In *Yearbook of Science and the Future 1979*. Chicago: Encyclopedia Britannica.

Freeman, M. R., and L. N. Carbyn. 1988. *Traditional Knowledge and Renewable Resource Management in Northern Regions*. Edmonton: International Union for the Conservation of Nature, Commission on Ecology, and Boreal Institute for Northern Studies.

Gadgil, M., F. Berkes, and C. Folke. 1993. "Indigenous knowledge for biodiversity conservation." *Ambio* 22(2–3):151–156.

Gisday Wa and Delgam Uukw. 1989. *The Spirit in the Land: The Opening Statement of the Gitksan and Wet'suwet'en Hereditary Chiefs in the Supreme Court of British Columbia, May 11, 1987*. Gabriola Island, B.C.: Reflections.

Gunther, E. 1973. *Ethnobotany of Western Washington*. Seattle: University of Washington Press. Originally published in 1945.

Gwaganad. 1990. "Speaking in the Haida way." Statement made before the Honorable Mr. Justice Harry McKay in British Columbia's Supreme Court on November 6, 1985, on the application by Frank Beban Logging and Western Forest Products Ltd. for an injunction to prohibit Haida picketing of logging roads on Lyell Island,

South Moresby. In V. Andrus, C. Plant, J. Plant, and E. Wright, eds., *Home! A Bioregional Reader*. Gabriola Island, B.C.: New Society Publishers.

Haiyupis, R. 1992. "The social impacts of governmental control of our chief's Ha Hoolthe and natural resources." Unpublished manuscript cited with permission.

Hill-Tout, C. 1905. "Report on the ethnology of the StlathumH [Lillooet] of British Columbia." *J. Anthropol. Inst. G.B. and Ire.* 35:126–218.

Hunn, E. N. 1982. "The utilitarian factor in folk biological classification." *Am. Anthropol.* 84(4):830–847.

Hunn, E. N., N. J. Turner, and D. French. In press. "Ethnobiology and subsistence." In D. Walker, ed., *Handbook of North American Indians*. Vol. 12: *Plateau*. Washington, D.C.: Smithsonian Institution.

Inglis, J. T., ed. 1993. *Traditional Ecological Knowledge: Concepts and Cases*. Ottawa: International Development Research Centre.

Jenness, D. n.d. "Origin of salmon." *Coast Salish Field Notes, M.S. 1103.6*. Ottawa: Ethnology Archives, National Museum of Civilization.

Jennings, F. 1993. *The Founders of America: How Indians Discovered the Land, Pioneering in It, and Created Great Classical Civilizations; How They Were Plunged into a Dark Age by Invasion and Conquest; and How They Are Reviving*. New York: Norton.

Johnson, M., ed. 1992. *LORE: Capturing Traditional Environmental Knowledge*. Ottawa: International Development Research Centre.

Joint Nisga'a/B.C. Parks Committee. Minga Nisga'a/Mismaaksgum Gat. 1992. *Anhluut'ukwsim LaXmihl Angwinga'asanskwhl Nisga'a*. Nisga'a Memorial Lava Bed Park. Summary. Vancouver: B.C. Parks.

Kennedy, D., and R. Bouchard. 1983. *Sliammon Life, Sliammon Lands*. Vancouver: Talonbooks.

Knudtson, P., and D. Suzuki. 1992. *Wisdom of the Elders*. Toronto: Stoddard.

Kuhnlein, H. V., and N. J. Turner. 1987. "Cow-parsnip (*Heracleum lanatum* Michx.): An indigenous vegetable of Native People of northwestern North America." *J. Ethnobiol.* 6(2):309–324.

Lewis, H. T., and T. A. Ferguson. 1988. "Yards, corridors, and mosaics: How to burn a boreal forest." *Human Ecol.* 16(1):57–77.

Linden, E.. 1991. "Lost tribes, lost knowledge." *Time*, Sept. 23, pp. 44–56.

McIlwraith, T. 1948. *The Bella Coola Indians*. Vols. 1 and 2. Toronto: University of Toronto Press.

Nakashima, D. J. 1993. "Astute observers of the sea ice edge: Inuit knowledge as a basis for Arctic co-management." In J. T. Inglis, ed., *Traditional Ecological Knowledge: Concepts and Cases*. Ottawa: International Development Research Centre.

Pojar, J., and A. MacKinnon, eds. 1994. *Plants of Coastal British Columbia*. Vancouver: Lone Pine Publishing.

Posey, D. A. 1990. "The science of the Mebêngôkre." *Orion* 9(3):16–21.

Schultes, R. E. 1988. "Primitive plant lore and modern conservation." *Orion* 7(3):8–15.

Sewid-Smith, D. (Mayanilh), Chief Adam Dick (Kwaxsistala), and N. J. Turner. In press. "The sacred cedar tree of the Kwakwaka'wakw people." In *Essays Accompanying the Alcoa Foundation Hall of Native Americans*. Pittsburgh: Carnegie Museum of

Natural History.

Stewart, H. 1977. *Indian Fishing: Early Methods on the Northwest Coast*. Seattle: University of Washington Press.

Suttles, W. 1951. "The economic life of the Coast Salish of Haro and Rosario Straits." Ph.D. diss., University of Washington, Seattle.

———. 1987. *Coast Salish Essays*. Vancouver, B.C.: Talonbooks.

———, ed. 1990. *Handbook of North American Indians*. Vol. 7: *Northwest Coast*. Washington, D.C.: Smithsonian Institution.

Swanton, J. R. 1905. *Haida Texts and Myths*. Bulletin 29. Washington, D.C.: Smithsonian Institution, Bureau of American Ethnology.

Turner, N. J. 1974. "Plant taxonomic systems and ethnobotany of three contemporary Indian groups of the Pacific Northwest (Haida, Bella Coola, and Lillooet)." *Syesis* 7, supp. 1.

———. 1979. *Plants in British Columbia Indian Technology*. Handbook 38. Victoria: British Columbia Provincial Museum.

———. 1981. "Indian use of *Shepherdia canadensis* (soapberry) in western North America." *Davidsonia* 12(1):1–14.

———. 1982. "Traditional use of devil's-club (*Oplopanax horridus;* Araliaceae) by Native Peoples in western North America." *J. Ethnobiol.* 2(1):17–38.

———. 1991. "Burning mountain sides for better crops: Aboriginal landscape burning in British Columbia." In K. P. Cannon, ed., *Archaeology in Montana* (Montana Archaeological Society, Bozeman), special issue, vol. 32 (2).

———. 1995. "Plants of Haida Gwaii." Unpublished manuscript.

———. 1996. *Food Plants of Coastal First Peoples.* Victoria: Royal British Columbia Museum; and Vancouver: University of British Columbia Press.

Turner, N. J., and M.A.M. Bell. 1971. "The ethnobotany of the Coast Salish Indians of Vancouver Island." *Econ. Bot.* 25(1):63–104 and 25(3):335–339.

———. 1973. "The ethnobotany of the southern Kwakiutl Indians of British Columbia." *Econ. Bot.* 27(3):257–310.

Turner, N. J., and A. Davis. 1993. "When everything was scarce: The role of plants as famine foods in northwestern North America." *J. Ethnobiol.* 13(2):1–28.

Turner, N. J., and B. S. Efrat. 1982. *Ethnobotany of the Hesquiat Indians of Vancouver Island*. Cultural Recovery Paper 2. Victoria: British Columbia Provincial Museum.

Turner, N. J., and R. J. Hebda. 1990. "Contemporary use of bark for medicine by two Salishan Native elders of southeast Vancouver Island." *J. Ethnopharmacol.* 229:59–72.

Turner, N. J., and H. V. Kuhnlein. 1982. "Two important root foods of the Northwest Coast Indians: Springbank clover (*Trifolium wormskioldii*) and Pacific silverweed (*Potentilla anserina* ssp. *pacifica*)." *Econ. Bot.* 36(4):411–432.

———. 1983. "Camas (*Camassia* spp.) and riceroot (*Fritillaria* spp.): Two liliaceous root foods of the Northwest Coast Indians." *Ecol. Food and Nutr.* 13:199–219.

Turner, N. J., J. Thomas, B. F. Carlson, and R. T. Ogilvie. 1983. *Ethnobotany of the Nitinaht Indians of Vancouver Island*. Occasional Paper 24. Victoria: British Columbia Provincial Museum.

Turner, N. J., L. C. Thompson, M. T. Thompson, and A. Z. York. 1990. *Thompson Eth-*

nobotany: Knowledge and Usage of Plants by the Thompson Indians of British Columbia. Memoir 3. Victoria: Royal British Columbia Museum.

Tyler, M. E. 1993. "Spiritual stewardship in aboriginal resource management systems." *Environments* 22(1):1–7.

Wavey, R. 1993. "International workshop on indigenous knowledge and community-based resource management: Keynote address." In J. T. Inglis, ed., *Traditional Ecological Knowledge: Concepts and Cases*. Ottawa: International Development Research Centre.

Williams, N. M., and G. Baines, eds. 1993. *Traditional Ecological Knowledge: Wisdom for Sustainable Development*. Canberra: Centre for Resource and Environmental Studies, Australian National University.

World Commission on Environment and Development. 1987. *Our Common Future*. Oxford: Oxford University Press.

Validating Vernacular Knowledge: The Ahousaht GIS Project

EDWARD BACKUS

Governments and corporations make resource management decisions guided by western science. Information gathered in surveys or formal research is deemed valid because trained observers with accepted university credentials have collected it and submitted it for review by professional peers. In fish, wildlife, and timber management, data collected in this way are used to make harvest decisions or compiled to reveal trends. Over the years, western societies attempt to accumulate reliable management knowledge by a variant of this process.

Public agencies, unfortunately, rarely return the information they collect to communities. Even when they do, communities often find that agency information is at odds with what they know and believe to be true. Knowledge of a different sort, often multigenerational and multicultural, resides in rural, resource-dependent communities. From local scientists (amateur or professional) to resource harvesters (fishers, oyster culture specialists, loggers, mushroom gatherers) and generalist observers, communities possess a wealth of insight into local conditions. This "vernacular" knowledge contrasts with agency information, but the two can be complementary.

The public agencies mandated to inventory resources can afford neither the people nor the time to collect detailed information everywhere throughout all seasons of the year. Instead, they rely on statistical sampling and index surveys. The knowledge in local communities is considerable—and residents who can collect information far outnumber agency employees—but neither knowledge nor people are easy to organize.

Agency data and local knowledge are both place-specific. Although public agencies use computerized geographic information systems (GIS) to organize, analyze, and display information from the field, until recently GIS technology was too costly and cumbersome for most local organizations. How can local knowledge be incorporated into agency information systems? How can communities gain access to public agency information? Are there ways to bridge the cultural gap that often separates agencies and communities? These are important questions for communities of the coastal temperate rain forest.

Interrain Pacific (founded in 1993 as Pacific GIS) helps nongovernment organizations in coastal rain forest communities develop the capability to use GIS and expand their access to public GIS information. Recently we have sought ways to incorporate local knowledge into locally based GIS systems.

In northern Clayoquot Sound on Vancouver Island, Interrain Pacific and a Canadian partner, the B.C. Conservation Mapping Consortium, are collaborating with the Ahousaht Band Council to develop a village-based, council-controlled geographic information system. The goal is to establish capacity in the Native village of Ahousaht to integrate information from provincial ministries with the knowledge of village elders. The Ahousaht traditional territory encompasses a large portion of Clayoquot Sound—in recent years the scene of intense controversy over how coastal forests and fisheries should be managed. The Ahousaht wish to regain control over resource management in their territory, perhaps through comanagement arrangements with public agencies. Their goals include the revitalization and protection of intergenerational cultural knowledge—the intellectual property of the Ahousaht people—and integration of their traditional knowledge about forests and fisheries with the inventory data collected by public agencies.

Fisheries programs in Ahousaht are characterized by a mix of Native and western techniques. The Ahousaht wish to strengthen the part played by village-based knowledge in the management of fisheries. The Ahousaht fisheries program staff already has a capacity to compile stock and habitat information, and relate it to historical knowledge, that far outstrips government knowledge of the fisheries resources of the Ahousaht territory.

GIS can help bridge the gap that has divided western and Native cultures. The ability to store geographic information in a multimedia format on CD-ROM disks means that the oral history of places, resources, and people can be recorded and linked to maps and images—thus preserving the knowledge of elders in a form that future generations can use. This capacity is especially important in communities where the traditional oral transfer of knowledge has dwindled under pressure from the dominant non-Native culture.

In 1994, the Ahousaht Band Council hired two GIS trainees. Interrain Pacific and the Mapping Consortium provide support to the band for GIS training, public data access and compilation, analysis, integration with a global positioning system (GPS) used to collect field data, and the capability to print large maps. The Ahousaht GIS staff has established contacts with government ministries and timber corporations to gain access to GIS data resources. Interrain Pacific has also arranged a donation of GIS educational software to the Ahousaht (Maaqtusiis) community school.

A special focus of the project is to teach GIS skills to the Ahousaht Fisheries

Enforcement and Salmon Enhancement staffs. The GIS staff spends time in the field with the fisheries managers to understand their jobs and learn how GIS can support them. The objective is to gather field data that can be compiled in the GIS, which then generates analyses and maps that both the field staffs and the Ahousaht Council can use to support decisions of many types, enhancing the council's knowledge base.

The ability to incorporate elders' knowledge in GIS is a key to community acceptance. The Ahousaht GIS staff is developing a program promoting geographic and resource literacy in the community school using parts of the GIS database created for the whole territory. To Ahousaht youth, real data about the Ahousaht territory offer a much more compelling introduction to basic geography than studies of the provinces of Canada. When a computerized resource includes videos of elders relating the history of the physical, cultural, and resource geography as one body of knowledge, students have a powerful opportunity to explore their connection with place.

These resources have comparable potential in non-Native communities, where long-term residents whose livelihoods are tied to the land and water possess special knowledge of how things work. This knowledge can be tapped and organized by local nongovernment organizations that seek to expand community access to local knowledge. Advocates for the inclusion of local knowledge in resource management are challenging public agencies to craft approaches that invite local involvement in management decisions. Resources can be neither sustained nor conserved over the long term without knowledge of ecosystem function and resilience at the local scale. That knowledge is a mix of scientific, empirical, and traditional Native understanding. Today the means exist to help communities tap their own expertise, organize it, and apply it in their own self-interest.

Edward Backus is the president of Interrain Pacific, based in Portland, Oregon.

□ □ □ □ □ □ □ □ □ □ □ □ □ □

Local Science in Willapa Bay, Washington

KATHLEEN SAYCE

Many people who live in the Willapa Bay watershed in Washington State have a direct and personal stake in science. As a lifelong naturalist here, I count myself among them. My neighbors are local fishermen, farmers, and oyster growers whose work depends on science, on paying attention to changes in the environment, and on accurate interpretation of available information. To them, geology, meteorology, oceanography, and ecology are not academic disciplines pursued in remote institutions, but vital concerns of their day-to-day lives. Their success or failure, and at times even their survival, depend on an eclectic blend of science and personal knowledge of local conditions.

Some residents work the tidal cycles of the bay in aquaculture and fishing. Local knowledge of the estuary's tidal peculiarities has accumulated for over a century. Knowledge of currents and tides informs understanding of when oyster larvae "set," where burrowing shrimp settle, and how flotsam and jetsam drift about the bay. Tidal circulation guides sediment deposits, shapes scouring patterns, and determines how beaches erode and low terraces form around the bay. Changes in currents, in the positions of channels and entrances, in the depths of subtidal and intertidal flats, occur much faster than U.S. Coast and Geodetic Survey charts for the bay can be updated—and faster, too, than the U.S. Coast Guard's channel markers can be repositioned. Nevertheless, safe navigation of local waters and safe access to the intertidal zone are impossible without accurate information.

Amateur and professional naturalists have lived in Willapa's communities for generations. Their firsthand observations of native species and newcomers represent an irreplaceable local resource. Records for birds are probably most complete, followed by plants, fish, marine invertebrates, and phytoplankton. Volunteers and amateurs can make lasting contributions to this kind of knowledge. Habitats in Willapa's wetland and upland areas are under constant pressure from nonnative plant and animal species. Documenting the condition of the bay at a point in time and then tracking changes—as in the spread of spartina in the tidelands or the arrival of purple loosestrife, hydrilla, and yellow nutsedge, to name a few floral invaders—is a task that both professionals and volunteers have taken on.

A local group called Willapa Watershed Volunteers arose in response to the absence of information about certain species and general watershed conditions. The program is modeled after Master Gardener programs run by state Cooperative Extension Services and volunteer programs organized by the Environmental Protection Agency. Participants study stewardship concepts and local natural history, then donate time to community projects. Volunteers monitor beach drift, evaluate streams for salmon restoration programs, develop educational videos and booklets, label stormwater drains, help with household hazardous waste collection, and maintain trails and remove weeds in public parks, among many other projects.

The community benefits are numerous. First, watershed volunteers learn more about the natural and human history of the region and pass that information on to their families and friends. Second, they learn that science is not a mysterious or arcane pursuit but a straightforward method of examination and analysis; they participate directly in science when they design and carry out monitoring projects. The demystification of science and emphasis on logic are especially important in a society in which the deliberate misstatement of facts for personal or political gain is all too common. Third, reconnecting people to the natural world is vital for rural communities that seek a sustainable path. The majority of Americans live in urban areas. Most will never appreciate the importance of a vital rural populace and healthy ecosystem if the rural minority lose their connections with the land.

Rural residents are often ambivalent when outsiders arrive to study "their" home. Willapa is no exception. But some research performed in Willapa has yielded results with significance far beyond the bay. U.S. Geological Survey researcher Brian Atwater, among others, found evidence for past earthquakes and tsunamis in the salt marshes of Willapa Bay and in nearby estuaries and rivers. His findings changed coastal land-use goals throughout the Northwest; they also changed local perceptions of the risks from these natural phenomena from a skeptical "if" to a concerned—and better prepared—"when."

Some scientists have encountered, and in a few cases have provoked, local resentment. Part of the clash may be cultural. University researchers do not always understand the discrepancies between fashionable—and fundable—research topics and local needs for better information. Monitoring is underfunded and unfashionable in universities and public agencies alike. Yet the dearth of solid baseline information on all but a few conditions in Willapa Bay means that monitoring projects remain among the highest local priorities. Willapa's need in this respect is not unique; it points to a widening gap between national goals and local needs in science research.

In Willapa Bay, programs that monitor plankton and the conditions of oysters have trouble securing funding. These programs, begun years ago, help to determine seasonal productivity in the bay and show long-term trends in ecosystem response to changes in climate and land use. Most academic scientists dismiss them as nonscience. (This dismissal does not keep researchers from requesting the data sets for inclusion in a variety of other studies.) Local researchers find themselves in a curious position: their results, repeatedly derided as monitoring and not "true" science, are nonetheless in great demand from scientists.

People who work and live in an ecosystem for the long term know that you cannot walk away from ecological problems. Ecologists around the world are learning that serious ecological disruptions to once-healthy systems cannot be dismissed as something "we will learn to live with." Willapa residents know this. No agency speaks for the long view; few large timberland owners in Willapa are concerned with problems off their own land; and newcomers too often seem concerned with getting their five acres, or fifty, and then closing the door on the rest of the world. Who speaks for Willapa and for the long term—for the time that will come after *our* time in this place?

Our survival in this place, and on this planet, comes from understanding interdependence and sustaining natural biological diversity. It comes from education and from supporting the livelihoods of people with a strong attachment to place and to nature. Finally, survival comes by avoiding choices that destroy the capacity of nature to renew itself. Local science is the first step.

Kathleen Sayce, a botanist and marine biologist, directs the science program of the Willapa Alliance, based in South Bend, Washington.

□ □ □ □ □ □ □ □ □ □ □ □ □ □

Oregon's Watershed Health Program

Mary Lou Soscia

In 1993, the state of Oregon created the Watershed Health Program as part of a new natural resource strategy acknowledging the importance of watersheds to Oregon's livability and economic health. The program grew from a realization that many Oregon watersheds can no longer satisfy the demands of a growing population and economy. New listings of endangered species, growing disputes over water rights, and degraded water quality were signs that better watershed management was seriously needed.

The 1993 legislature allocated $10 million from state lottery receipts to two areas: the Grande Ronde watershed in northeastern Oregon and the south coast and Rogue basins in southwestern Oregon. Snake River chinook were already listed under the Endangered Species Act in the Grande Ronde basin; in the south coast and Rogue basins, coastal coho and winter and summer steelhead stocks were on the verge of being listed. The Watershed Health Program was intended to promote cooperation between state agencies and voluntary action on the local level to improve watersheds. The Strategic Water Management Group, thirteen state and four federal agencies and the governor's office, was given oversight authority for local watershed councils.

Since the pilot initiative got under way in January 1994, over 200 miles of streambanks have been planted with more than 570,000 trees and with native grasses to shade stream channels and hold runoff. Woody debris has been placed in streams to replicate natural habitat. More than 200 miles of fences have been erected along streams to protect their banks from livestock. Thirty-six fish screens have been installed in the south coast region to prevent fish from being stranded in ditches and fields. The pilot initiative also set up a process to engage citizens in voluntary efforts to improve watersheds, and completed sixty watershed education projects in participating communities.

The cornerstone of this effort was the creation of more local watershed councils to work with local, state, tribal, and federal agencies to help solve local watershed problems. As a result, about fifty local councils have been set up throughout Oregon to manage watersheds and to fund and carry out protection and restoration projects. As in all pilot projects, however, there were lessons to be learned from the early stages of the Watershed Health Program. The boldness of the program made a certain amount of confusion inevitable.

Too much money was spent too soon. The program had to spend close to $7 million on projects in its two target areas in less than two years. This caused vastly unrealistic expectations about the support available to watershed groups in other areas. Residents in other watersheds in the state had organized watershed councils and had projects ready to be funded; they resented the delay. We learned that funding should be committed in a measured way, with early support for watershed council start-up and watershed assessment. And government funding should be used as a catalyst to encourage local funding, not as a sole source of support.

State funding for the program was tied to a political calendar. Many communities were hesitant to invest a lot of time and effort in something that might prove to be a temporary whim of government. Somehow, political commitments to watershed health must be good for the long term and extend beyond the usual election cycle.

Existing efforts were not well integrated into the pilot initiative. Oregon had a successful seven-year-old program, the Governor's Watershed Enhancement Board, which funded individual demonstration and education projects. Policy discussions held early on with this and related government initiatives could have led to decisions characterized by synergy and efficiency.

Like many initiatives, the Watershed Health Program suffered from too many decision-making layers. The central oversight group was seen as unnecessary by local groups whose plans had already gone through a lengthy process of local review. Watershed councils by definition represent a bottom-up approach to decision making. The participation of government in local councils assumes a partnership at the local level. A top-down decision-making structure works against the goals of watershed councils.

Local watershed councils were expected to develop assessments and design good projects in an unrealistically short time. We found that when people start working together, trust cannot be rushed. Government efforts intended to support local watershed councils must give them time to develop the trust people need to work together effectively.

Too much emphasis was placed too soon on projects. Funding was not allocated for watershed assessment, monitoring, or public education and involvement. Funding should be designated at the outset for essential efforts to understand watersheds, to share information with the residents of the community and political leaders, and to monitor the results of the work undertaken.

The state staff for the program was top-heavy. Twenty-three staff were hired—few with any experience in Oregon state government or watershed work. Many were located in the field, creating great confusion with existing field staff at the local level. Funds should have been deployed at the local level to staff watershed councils. Established structures, such as soil and water conservation districts, could have provided the needed support.

Governor John Kitzhaber and the Oregon legislature have made a strong commitment to provide long-term support to voluntary local watershed councils. The Governor's Watershed Enhancement Board has assumed responsibility for many activities of the Watershed Health Program; one component is a grants program that makes funds available for local councils, watershed assessment and monitoring activities, and education projects. Many of the watershed councils founded during the Watershed Health Program are moving forward with a strong base of local support, demonstrating the role that locally based partnerships can play in community sustainability.

Mary Lou Soscia was the program manager for Oregon's Watershed Health Program and is currently the watershed restoration coordinator with the Columbia River Inter-Tribal Fish Commission in Portland, Oregon.

□ □ □ □ □ □ □ □ □ □ □ □ □ □

The Tillamook Bay National Estuary Project

MARILYN SIGMAN

Tillamook Bay is one of twenty-one federally sponsored environmental planning projects. The National Estuary Program, created by amendments to the

Clean Water Act in 1987 and administered by the Environmental Protection Agency, seeks to address the water quality and living resource issues in estuaries of national significance through bottom-up coordinated planning by federal, state, and local agencies and stakeholders. The goal is a comprehensive conservation and management plan for Tillamook Bay, based on a consensus on the highest-priority problems and the most feasible actions that can be taken to solve them. Although the focus is on the estuary, the program necessarily includes its associated watershed.

The relatively small Tillamook estuary (18 square miles) and watershed (550 square miles) provide a model for planning in coastal temperate rain forest watersheds because the environmental problems there are typical of many in the ecoregion—high sedimentation rates, degradation of salmon habitat, and fecal coliform pollution from agricultural runoff and residential areas. Also typical is the degree to which the local economy is resource-based—a factor that sets the context for restoration planning.

The Tillamook estuary averages only 6 feet in depth today, but historically it was deeper: because of sedimentation, the bay holds roughly one-third the water at high tide that it held in 1867. It still supports shellfish, salmon and trout, groundfish, and numerous migrant bird species, including a number of threatened and endangered species.

Five rivers—the Miami, Kilchis, Wilson, Trask, and Tillamook—flow into the estuary. Four of them form an alluvial plain at its southeastern end. All five rivers support anadromous fish runs. Chinook salmon stocks are healthy, but coho salmon stocks are declining faster than in other Oregon north coast watersheds and are considered critically depressed. Although some of the few wild chum salmon populations left in Washington and Oregon use the estuary, these chum stocks are considered depressed.

Major resource-based industries in the Tillamook watershed include forest products, dairy farming, commercial fish and shellfish harvest, and tourism. Intensive dairy farm development in the lower watershed makes Tillamook County's agricultural sector larger than that of any other county in Oregon and gives it the highest agricultural income per acre. Dairy products are processed at a cooperatively owned factory in the watershed. Extensive pasturelands in the lower watershed were created by clearing floodplain forests, by filling wetlands, and by diking, channel diversions, and fills.

Tourism is a major industry: recreational sportfishing and clamming attract many visitors. Commercial harvest of the bay's nonnative oysters and native clams has been an important economic activity, although what could be a year-round harvest is restricted for part of each year due to bacterial contamination.

The National Estuary Program is funding a four-year planning project from 1994 through 1998. This federal funding is tied to requirements for a structured process to move from defining problems to making plans for action. The program process provides a strong supportive (but not controlling) role for the federal government through the provision of financial incentives, access to information about other communities grappling with similar environmental problems, and formal deadlines for strategic planning and decision making.

The project faces challenges different from those faced by grassroots planning efforts. Many people, particularly in rural areas, distrust government agencies in general and have trouble believing they will permit local people to participate meaningfully in decisions about public resource management. The National Estuary Program's emphasis on science-based planning also encounters barriers to acceptance; the pool of professional scientists and scientifically literate people in coastal communities is small.

Two initial goals for the project are to restore Tillamook Bay, by reversing the impacts of sedimentation, and to protect and enhance anadromous fish habitat. A third goal concerns the need to balance the local economy with environmental protection.

Draft and final plans are due to be presented to the Environmental Protection Agency in 1997 and 1998, respectively. For the future of coho salmon, the Tillamook Bay project will be crucial. Runs of as many as 150,000 coho to spawning grounds in the Tillamook basin in the 1930s have declined to less than 300 in recent years. A 1995 report on the status of coho salmon stocks by the Oregon Department of Fish and Wildlife said the condition of freshwater habitat in the Tillamook basin was "poor to mediocre . . . an unprecedented low point for stream recovery, and salmonid productivity."

Some participants in the project have begun to express hope that the goals of economic development and restoration can be linked. Dick Russell, chairman of the project management committee and a local banker, has worked to develop a regional economic strategy that recognizes forest products, tourism, and environmental technologies as key industries for the Tillamook watershed and the region.

"The NEP can do a lot for us," he told JoAnne Booth in *Ruralite* magazine. "We can't restore the bay. We can't go back to what it was like in, say, 1916, and decide we want it to be like that, with the channels thick with salmon. The community has to have a vision of what water quality we're willing to live with."

The challenge for the Tillamook project will be to define that shared vision about the future of the bay and the watershed. The fate and quality of life for the human and nonhuman communities of Tillamook Bay are not only interwoven: they depend on this vision.

Marilyn Sigman, former project director of the Tillamook Bay National Estuary Project, is executive director of the Montana Natural History Center in Missoula, Montana.

□ □ □ □ □ □ □ □ □ □ □ □ □ □

The Coastal Studies and Technology Center

MIKE BROWN AND NEAL MAINE

Citizens have few opportunities to participate in the decisions that affect ecosystems and watersheds. Most adults lack the skills and experience needed to participate knowledgeably. When opportunities for participation do arise, decisions often relate to crisis situations that have been developing over long periods of time. And as experience with endangered species listings in the U.S. Pacific Northwest has shown, seldom is a crisis resolved by a quick fix.

"Ecology centers" established in schools in the coastal temperate rain forest region offer communities a practical use of public resources to help young people gain real-world experience as citizens and members of a larger community. An interdisciplinary center in the community school offers students opportunities to partner with more experienced citizens, public agencies, local governments, and private business to learn how people and nature interact in local watersheds. The skills and knowledge these students gain are products of experience, not lessons to be saved until they grow up.

The Coastal Studies and Technology Center (CSTC) at Seaside High School in Seaside, Oregon, is one example of a center created to demonstrate this approach. It is a nonprofit organization established under the school district umbrella. The center seeks to present opportunities for young citizens to contribute to their community, region, and world.

In light of this mission, the center's educational programs focus on brokering opportunities for young citizens to develop skills and knowledge needed to participate directly in projects already under way in their community. Formerly academic subjects come to life. Language arts skills are put to work in the final report on a community wetlands project. Basic statistics make sense out of data collected on shoreline vegetation at the request of a local or national agency.

Everyone agrees that the skills, knowledge, and experience needed to make music differ from those needed to listen to music; playing a sport is nothing like watching it; viewing art and creating art are worlds apart. Yet almost all educational programs emphasize passive studies of natural resources and the environment. Ecology can be taught as an art—a useful problem-solving approach relevant to a host of individual and community questions. Students should know from the start, from their own experience, that ecology cannot always produce an answer, but it can be very useful in making decisions that contribute to community sustainability.

The center's approach provides a clearinghouse through which young citizens can participate in their own community in a way appropriate to their level of development: doing work that counts, helping to improve access for other youth, gaining actual experience through citizen participation that earns them respect.

Centers create partnerships with various entities in the community. The CSTC maintains such partnerships with the local city government, local offices of state and federal agencies, other schools that have developed centers, and Portland State University. This approach also allows school centers to accumulate tools and technologies that become resources for young citizens. These resources are made available not only to students, but to the community as a whole through the efforts of a special projects coordinator.

When the city of Cannon Beach, Oregon, began a wetlands inventory, the CSTC helped a fourth-grade teacher at the local elementary school to write and win a grant in support of a study on one wetland site. The goal was to produce an interpretive proposal for the city's parks and recreation committee.

The fourth-graders spent an intense year working with the city's public works department, with community members who had a special interest in this wetland, and with a local wetlands ecologist. With the grant their teacher had won, the students hired a graphic specialist to help them design public information about the site. Designs that the students themselves created became the basis for this specialist's work.

The students and their graphic "consultant" looked at interpretive signs already in the community to help plan their options. Students approved each step in the design process. When the mock-up was ready, the class made a presentation to the parks and recreation committee. Each of the twenty-four students took part in the presentation. The committee responded enthusiastically and the sign was approved. Elementary school students had made a tangible, and visible, contribution to their community.

Teachers and secondary school students from the CSTC, working in partnership with National Marine Fisheries Service (NMFS), provided technical support for a major salmon habitat restoration project. A portion of a jetty in Trestle Bay was slated for removal to improve water circulation in the bay and provide access for juvenile salmon.

During the summer before the start of the school year, the CSTC director invited teachers from other high schools in the region to join the Trestle Bay project as a cooperative effort. Teachers teamed up with the NMFS staff to inventory the conditions in the bay prior to jetty removal and to collect samples.

Teachers helped sieve the samples to remove invertebrates and then learned basic identification from the NMFS staff. In turn, they passed this skill to students at each of the schools. Twenty to thirty students at each of four regional high schools began "picking the samples" to identify and count invertebrates. The data were compiled and submitted to NMFS. This project won EPA's Region 10 Environmental Youth Services Award, and two students from participating high schools traveled to Washington, D.C., to receive the award from President Clinton.

Linking education to meaningful work is critical. Authentic experience-based learning can help students begin to build their own sense of place, gain insight into the long-term sustainability of their "home," and make tangible contributions to their communities. School-based organizations that offer these experiences use community as curriculum and provide the experience of citizenship. Both have an important place in efforts to build sustainable communities. Students are not observers but members of both human and natural communities. The best way for them to learn this is to let them participate as full citizens.

Mike Brown is the director, and Neal Maine is special projects coordinator, for the Coastal Studies and Technology Center in Seaside, Oregon.

12. "The Great Raincoast": The Legacy of European Settlement

William G. Robbins

□ □

To understand the relationship between culture and the natural world along America's North Pacific slope one must consider a broader geography of human activity, especially the social, economic, and political enterprises beyond the region. For the last 200 years larger global forces, especially those associated with an expanding world capitalist system, have increasingly influenced the course and direction of change along the "Great Raincoast of North America." This area of lush vegetation and magnificent forests extends from northern California through coastal British Columbia and the Alaska Panhandle; it embraces much of the region that has been called "Ecotopia" (Garreau 1981, after Callenbach 1975). The phrase "Great Raincoast of North America" was coined by Richard Maxwell Brown, who addressed the effects of climate on the development of a Pacific Northwest identity in a paper presented at the University of Oregon in April 1979.

In contrast to the Native American inhabitants, the European newcomers, who gradually extended their control over the Pacific slope as the nineteenth century advanced, brought with them a strikingly different cultural vision. It embraced a social imagination and core of beliefs—an "economic culture," Donald Worster called it—that viewed the natural world as capital, obliged humankind to use that capital for self-advancement, and insisted that the social order should promote the accumulation of personal wealth (Worster 1979). My point in this chapter is not to argue for or against the Native American land ethic and environmental relations, but to underscore the influence of marketplace and then industrial capitalism in reshaping the ecology and landscape of the North Pacific coast for the last 200 years.

All economic systems, Worster reminds us, are first of all systems of ideas. The cultural values of what we have come to know as the modern industrial marketplace first took root among the rising European bourgeoisie of the 1600s and 1700s and expanded until industrial capitalism emerged in modern times as the most influential voice in global affairs. In the process it created a new cultural consciousness, habits of thought and perception, that make its functioning seem entirely rational and reasonable. Its widespread acceptance over much of the globe has led to the view that the continued manipulation of nature, increases in productivity, and maximizing economic output are acceptable and appropriate routes to the good society. The consequences of these cultural beliefs, Worster concludes, are "clearly written in the historical record of England, the United States, and every nation that has been brought under the industrial system" (Worster 1993:178).

For the North Pacific slope that story begins with the early Spaniards who sailed the western seas in the seventeenth and eighteenth centuries and eventually leads to Captain James Cook and the publication of his *Voyages* in 1784. Cook's account, in particular, hinted to a larger audience about the bright prospects of turning the natural abundance of the region to advantage in distant markets. John Ledyard of the Royal Marines, who was on board one of Cook's vessels in 1778, reported that the *Resolution* and the *Discovery* were outfitted with new spars and other timbers from the vast forests along the shores of Nootka Sound. "This country will appear most to advantage," he observed, for the "variety of its animals, and the richness of their furs" (Ledyard 1783:69–70). Passing northward from the mouth of the Columbia River a few years later, British sea captain George Vancouver (Figure 12.1) praised what he termed the "luxuriant landscape":

> The country now before us . . . had the appearance of a continued forest extending as far north as the eye could reach, which made me very solicitous to find a port in the vicinity of a country presenting so delightful a prospect of fertility. [Vancouver 1960:210]

Finally, ashore at Tulalip on June 4, 1792, Vancouver took possession "of all the countries we had lately been employed in exploring, in the name of and for, His Britannic Majesty, his heirs and successors." By virtue of that act, according to the geographer Barry Gough, the Northwest Coast became "a dominion, a future sphere of empire." Those who followed, he concludes, "brought with them their morals, ideologies, knowledge, technology, plants, and animals. They also brought diseases, rum, and guns. They brought with

Figure 12.1. Captain George Vancouver's HMS *Discovery*. Photo of a painting by F. P. Thursby. (Photo courtesy of Oregon Historical Society, #OrHi 26774.)

them powers to build and powers to destroy" (Gough 1980:126, 147: Vancouver is quoted by Gough on p. 126).

While the land-based newcomers to the region shared similar perceptions about the *potential* meaning of the region's natural abundance, these first narrative descriptions are grudging and ambivalent. A careful reading of the accounts left by the early explorers suggests both frustration about the tangled density of rain forest undergrowth and humility toward the immensity of the elemental forces of nature. Two American explorers and empire makers, Meriwether Lewis and William Clark, expressed the difficulties and discomfort of their winter season at the mouth of the Columbia River:

> *November 12, 1805*. "Send out men to hunt they found the woods So thick with Pine & [decayed] timber and under groth that they could not get through."
>
> *November 13*. "I walked to the top of the first part of the mountain with much fatigue as the distance was about 3 miles thro' intolerable thickets of Small Pine, arrow wood a groth much resembling arrow wood with briers, growing to 10 & 15 feet high interlocking with each other & Furn, aded to this difficulty."
>
> *November 19*. "Up the Chinnook river . . . I observed in maney places pine of 3 or 4 feet through growing on the bodies of large trees which had fallen down, and covered with moss and yet part Sound."
>
> *December 1*. "Sent out men to hunt and examin the country, they soon returned . . . and informed me that the wood was so thick it was almost impenetrable." [Moulton 1990].

A few years later the trader Alexander Ross, in the employment of John Jacob Astor's American Fur Company, reported the party's difficulties in falling trees near the mouth of the Columbia. After many days work with the ax, a tree would finally give way:

> But it seldom came to the ground. So thick was the forest, and so close the trees together, that in its fall it would often rest its ponderous top on some other friendly tree; sometimes a number of them would hang together, keeping us in awful suspense, and giving us double labor to extricate the one from the other, and when we had so far succeeded, the removal of the monster stump was the work of days.

In two months of labor, Ross reported that the group had cleared less than an acre of land: "In the mean time three of our men were killed by the natives, two more were wounded by the falling trees, and one had his hand blown off by gunpowder" (Ross 1986:91–92).

Further up the Columbia River, a Northwest Fur company official, Ross Cox, wrote enthusiastically and "with pleasure" about the potential "productions of the country, amongst the most wonderful of which are fir trees." The northward-coursing Willamette River valley, Cox confided to his journal, enjoyed a "remarkably mild" climate and bottomlands that possessed "a rich and luxuriant soil, which yields an abundance of fruits and roots." For the resident Kalapuya people, he called for Christian missionary societies to extend "their exertions to the northwest coast of America" to assist the native populations who were "still buried in deepest ignorance" (Cox 1957). In their reaffirmation of the inferior status of Northwest native people, Cox and others, especially the Americans, perpetuated familiar Eurocentric stereotypes and established for their purposes a moral basis for seizing native lands.

The first commercial effort to exploit and profit from the great stands of Douglas-fir, cedar, spruce, and hemlock—as in virtually every other entrepreneurial venture along the coastal waters of the Northwest—must be attributed to the acquisitive abilities and instincts of the Hudson's Bay Company. With a water-powered sawmill upriver from Fort Vancouver, the company established a modest but profitable lumber export trade with Honolulu and with Spanish missions in California by the 1830s (Ficken 1987). While far-flung operations like the Hudson's Bay Company presented themselves as the benevolent bearers of civilization, historical geographer Donald Meinig reminds us that these trading outposts also represented visible declarations "of the power of the Europeans to intrude upon the territories of others. They were political as well as economic outposts" (Meinig 1991:6).

Indeed, the writings of both chief trader John McLoughlin and field governor George Simpson praised the efficacy and virtues of imperial dominion over the Northwest. The peripatetic and imperial-minded Simpson believed that it was the responsibility of the company to bring native people within the embrace of the market system. On a trip through the lower Columbia River country in 1824, Simpson took note of the region's abundance of salmon, its fertile soils, and its "salubrious" climate. At the site that eventually became Fort Vancouver, he observed that "a trading Establishment properly managed" ought to prosper:

> The place we have selected is beautiful as may be inferred from its Name and the Country so open that from the Establishment there is good travelling on Horseback to any part of the Interior; a Farm to any extent may be made there, the pasture is good and innumerable herds of swine can fatten. [Merk 1931:40, 74, 87, and 92]

The pace of cultural and physical change in the region remained modest during those first few decades of the nineteenth century—that is, until the formal agreement between England and the United States in 1846 to extend their common boundary westward along the forty-ninth parallel to the waters of the Pacific. A swelling tide of EuroAmericans—most of them trekking overland, others coming by sea to the Willamette and Puget lowlands—overran and pushed aside an already decimated native population from the most valuable agricultural lands during the 1840s and 1850s. The interlopers then imposed upon their freshly established homelands new bounds for reckoning the landscape, new definitions of natural phenomenon, and new perceptions about a common environment. Imbued with a commercial ethos that viewed the natural world in terms of its commodity potential, the incoming settler population set about—where cleared land and existing technology allowed—imposing familiar flora and fauna in their newly adopted physical settings. The narrative accounts reflect deeply held convictions, aspirations, and economic ambitions toward the landscape of the coastal region. In the process, American and British subjects alike transplanted to the Northwest the cultural prescriptions and the commercial practices and technologies from the industrializing world of the eastern United States and England.

Robert Cantwell's acclaimed book, *The Hidden Northwest*, praises the peaceful disposition of the boundary issue which, in his view, "would depend on the wishes of the people who lived there." The earlier Treaty of Joint Occupation (1818), he observes, lasted twenty-eight years without resort to violence: "A territory the size of Europe (without Russia) was governed almost without dispute; indeed, no part of America was settled so peacefully" (Cantwell 1972:68). Native Americans in the United States and Canada's First Nations, who represented a majority of the region's population in 1846, were invisible humans in Cantwell's calculus. But his interpretation of the course of events is at one with most of the conventional literature on the subject: it ignores the unmitigated exercise of power, the massive appropriation of the native land base both north and south of the international boundary, and the incidents of local genocide, especially in Oregon's Rogue River valley and in eastern Washington Territory (Robbins 1986a). Scholars on the Canadian side of the border have been more successful than their American counterparts in penetrating that veil of peaceful dispossession and recognizing that the violence against the white intruders should be viewed in the context of a people defending their homeland.

For the American side of the border, the emigrant guidebooks that began to appear in the 1840s further embellished the already rich descriptive literature about the Pacific Northwest. Lansford W. Hastings' celebrated *Emigrants' Guide to Oregon and California* extolled the "grand and majestic stream" of the valley of the Columbia, with its upper tributaries situated in a "fertile and delightful" country that produces "an abundance of grass and timber." The Willamette River, too, according to Hastings' pamphlet, watered "one of the most fertile and delightful regions." The mild climate of the valley was without "excess of cold . . . [or] heat," making it unnecessary to house or feed livestock during the winter months. And then came the ultimate lure to would-be emigrants: "Very few portions of the world . . . afford a greater variety, and quantity, or a better quality of timber" (Hastings 1932). Writers like Hastings left little to the imagination, especially for a population already imbued with the notion of profiting from nature's abundance.

The California gold rush was a momentous event for the fledgling settler communities at the far northwestern edge of the continent. San Francisco quickly emerged as the imperial center of a vast trading network that extended the length of the Pacific coast and beyond. As such, the city developed a voracious appetite for a great variety of agricultural foodstuffs and especially large quantities of timber. San Francisco's entrepreneurs also were active in investing in the immense forests that shrouded the waterways along the Northwest Coast from Coos Bay and the lower Columbia River to the great stands of timber that surrounded the waters of Puget Sound. The great forested wealth of the coastal region attracted the likes of lumbermen Frederick Pope and William Talbot to Puget Sound in the early 1850s and, following in the wake of the Fraser River gold rush of 1858, Edward Stamp to Alberni Inlet and Sewell Prescott Moody to New Westminster in the early 1860s (Holbrook 1971; MacKay 1982; Barman 1991; Clark 1970). For lumber entrepreneurs, the coastal rain forest was an environment lush with promise; it required only capital, technical expertise, and labor to set up operations along one of the region's extensive timbered waterways to mill the lumber for sale in distant markets.

Driven by human and animal power, these first commercial inroads on the coastal forests probably did not appreciably diminish the individual's sense of powerlessness in the forest environment. Logging with ax and crosscut saw, with human energy and teams of oxen, was slow and laborious—nearly superhuman in terms of the labor required—and few in that time or place questioned the essential rightness of it (Figure 12.2). Human skill and animal power were juxtaposed against the inert force of huge, felled logs, the latter usually resting a short distance upslope from a convenient waterway. The labor was strenuous, dangerous, and required a sizable work force to float the huge logs to strategically located mill sites. Although the most productive of these early sawmills were steam-powered, the volume of manufactured lumber was relatively small. Hence the limited application of industrial technology at the manufacturing end did not have an appreciable impact on the exploitation of the region's magnificent forest stands.

From stump to mill, coastal logging operations involved falling the trees and then yarding the felled timber to a landing; from that point the logs were moved to the site of the sawmill. The work in the woods was elemental: pairs of fallers would use springboards, double-bitted axes, crosscut saws, and wedges to fall the trees. Buckers, who usually worked alone, would then saw the timber into appropriate lengths. At this point, jacks or oxen were used to move the logs to water's edge where workers skillfully used riverine or tidal currents to transport the timber to mills (Figure 12.3; Rajala 1989).

Figure 12.2. Logging with ax and crosscut saw was slow and laborious. (Photo courtesy of Oregon Historical Society.)

Figure 12.3. Rivers were used to transport lumber to markets. (Photo courtesy of Oregon Historical Society.)

Annie Dillard has captured the white settler's reckoning with the huge trees in her novel *The Living*, a story set in early Bellingham:

> Falling the enormous firs was the easy part, if they missed the cabin; it took two men four or five hours. The hard part was chopping through the same trunk again and again, to make pieces small enough to move with oxen or burn, and then fighting the stumps. Rooney kept the fires going year round. He and other settlers did not have land; they had smoking rubble heaped higher than their heads. Surely, Rooney thought, soil that grew such oversized trees could grow boss wheat and corn, or anything, if a man could just claw his way down to it. [Dillard 1992:16].

With their limited technology, those who worked in the forest environment remained in awe of the majestic trees, taking only the best timber and leaving numerous culls that inadvertently provided abundant cover for animal life and seed for regeneration. According to USDA Forest Service studies, ox-team logging did not seriously disrupt forest ecosystems and the degree of regeneration was very high. Richard White has argued that the forest recovered quickly from these early logging efforts, not because of human sensitivity to the natural world, but because the technology itself was relatively unobtrusive (Munger 1927).

But then a revolutionary force—steam power—entered the forest environment. The distant forces of an industrializing world—an expanding market demand for lumber along with the increased mechanization of production processes everywhere—spurred the adaptation of the machine to woods operations. Although steam-powered sawmills had operated along the North Pacific slope since the 1850s, it was not until the 1880s that operators began adapting steam-driven machines to the task of hauling felled timber. The increasing use of the steam engine or "donkey" to yard and load logs vastly stepped up both the pace of activity in the woods and disturbance to the ecology of coastal forests (Figure 12.4). Early in the twentieth century loggers added an aerial twist that enabled them to haul logs to landings with one end suspended in the air. Speed and productivity were part of this new industrial environment, and the adaptation of steam power to the logging end of the production system cut the technological gap between logging and milling operations.

George Emerson, an early lumberman in Washington's Grays Harbor area and the first to use the steam donkey north of the Columbia River, was enthusiastic about its performance:

> When one considers [that] they . . . require no stable and no feed, that all expense stops when the whistle blows, no oxen killed and no

> teams to winter, no ground too wet, no hill too steep, it is easy to see they are a revolution in logging. [Emerson 1907:20]

With "no ground too wet, no hill too steep," the donkey engines began what Richard White has called "the creation of a new forest" (White 1980). From the perspective of a century later, we can say that the steam donkey offered mixed blessings for the future: it greatly increased log production *and* it dramatically expanded cultural influences in forest ecosystems. As such, the steam donkey would appear to be an appropriate fit for one of Henry David Thoreau's 1854 aphorisms: "Our inventions are improved means to unimproved ends."

The most visible symbol of the industrial world—the railroad—had an even greater and more dramatic influence in altering the landscape of the coastal temperate rain forest. The mechanical harbinger of the new age first appeared in the woods around Puget Sound in the early 1880s and then in southern British Columbia the following decade. Railroads liberated the transport of logs from the restrictions of natural geography. Now operators could build lines into hitherto inaccessible areas. In this sense logging railroads replaced waterways and brought sweeping changes to the coastal landscape, vastly stepping up and expanding industrial intrusion into the forest environment. "The railroad," historian William Cronon argues, "left almost nothing un-

Figure 12.4. The "steam donkey," like this one photographed in northern California, vastly increased the impact of logging on the coastal landscape. (Photo courtesy of Oregon Historical Society, #OrHi 85651.)

changed: that was its magic" (Cronon 1991:73). As an instrument of industrial production, the railroad freed loggers from topographical and seasonal constraints; the consequences were dramatic increases in the volume of timber delivered to the mills and equally spectacular changes to the forest landscape.

The technological prowess of the new machines gave further conviction and assurance to capitalists who had been nurtured with booster propaganda about the region's abundance. Newspapers and magazines had always extolled the virtues and the larger meaning of the region's abundance. David Newsom, writing in the monthly journal *West Shore* in 1876, was euphoric about the "endless source of wealth" represented in the region's "vast coal fields, timber, fisheries," and mineral resources. Although those natural riches were little appreciated as yet, he observed that "capital and brains [were] needed to utilize them, and erect factories, machine shops, foundries, fisheries, ship yards, . . . and to push out commerce to foreign lands." Four years later another *West Shore* writer observed that the "forests of Washington and Oregon are practically unlimited and inexhaustible." And in subsequent issues, *West Shore* even suggested that the physical character of the North Pacific slope was no detriment to progress: "No country, considering its rugged aspect, is more easily subjugated. It takes but a short time, and little expense, except labor, for a settler to make a home" (*West Shore* 1880).

Widely circulated publications like *West Shore* unquestionably helped to fuel the period of rampant speculation in coastal timberlands that took place on both sides of the international boundary around the turn of the century. With the end of the great pine forests of the Great Lakes states in sight, *West Shore* (1881) noted that "the Pacific coast [had] an inexhaustible fund upon which to draw." Finally, in the midst of the speculative frenzy in timberlands in the early twentieth century, a companion publication, *Pacific Monthly* (1904), issued a rousing call to action:

> The whole Pacific Coast today is a field laden with innumerable opportunities. . . . Here we have the garden spot of the world—everything that Nature can give or man's trained imagination can conceive. Beautiful and fertile valleys, glorious and majestic mountain and river scenery, wonderful forests, mines of gold, silver, copper, nickel, and what not, rivers teeming with delicious fish. . . .We have it; we are in possession of it—this garden spot, this land pregnant with hidden resources, possibilities that almost stagger the imagination, opportunities!! It is for YOU—YOU—for US.

The *Pacific Monthly*'s call was heeded. At the outset of the twentieth century the Northwest Coast had taken on the aura of an investor's frontier as lumber capitalists from eastern North America, the Great Lakes states, and England turned their attention to the region's forest bounty. The timber buyers who flocked to the coastal rain forests of British Columbia were of a kind with those south of the border. The race to gain access to timberland peaked in 1903 when the provincial government, in an effort to increase revenue, made Crown lands available to Canadian and foreign interests through a generous timber leasing system. The result of the government's decision was an inrush of rascals and land agents and rampant speculation until authorities discontinued the practice. When the worst of the abuses ended at about the time of World War I, more than 11 million acres—80 percent of all Crown forestland in British Columbia—was leased, most of it in the hands of large syndicates (MacKay 1986; Barman 1991; Robbins 1986b).

Similar conditions prevailed in Washington and Oregon. There an epidemic of speculation preceded President Theodore Roosevelt and forester Gifford Pinchot's huge March 1907 additions to the federal forests in the United States. By the time the extensive national forest system was in place, however, thousands of acres of valuable coastal forestland already had been transferred to the private sector, much of it concentrated in the hands of a few large ownerships. The Weyerhaeuser interests led the way in 1900 with the purchase of 900,000 acres of magnificent timber from the Northern Pacific Railroad, itself one of the largest owners of forestland. The Weyerhaeuser purchase established a precedent: ownership and control of timberland, not the operation of sawmills, would be the wave of the future (Ficken 1979).

These changes in the structure of the logging industry—toward the corporate domination of timberland—paralleled the arrival of transcontinental railroads to the region: the Northern Pacific to Portland and Tacoma in 1883; the Canadian Pacific to Vancouver in 1886; the Great Northern to Everett and Seattle in 1893. The great rain forests of the coastal region were now linked both by water and rail to an expanding and dynamic North American and, in the case of British Columbia, Atlantic economy. For American lumbermen, who were witnessing the rapid depletion of the great pine forests of the Great Lakes states, the new transportation infrastructure established the necessary conditions for large-scale production along the North Pacific slope (Rajala 1989).

The introduction of capital-intensive operations to harvest and move timber had social as well as ecological dimensions. At the start of the twentieth century, fluctuating markets and, especially in the case of British Columbia, the revenue needs of the provincial government were the sole restraints on timber harvests. British Columbia has led all Canadian provinces in lumber production since 1917; the state of Washington led the United States in lumber production between 1910 and 1940; Oregon has held first rank since then (Figure 12.5). The fast and relatively efficient steam engines sped the rate of timber harvesting, contributed to the rapid growth of small logging towns, and brought a

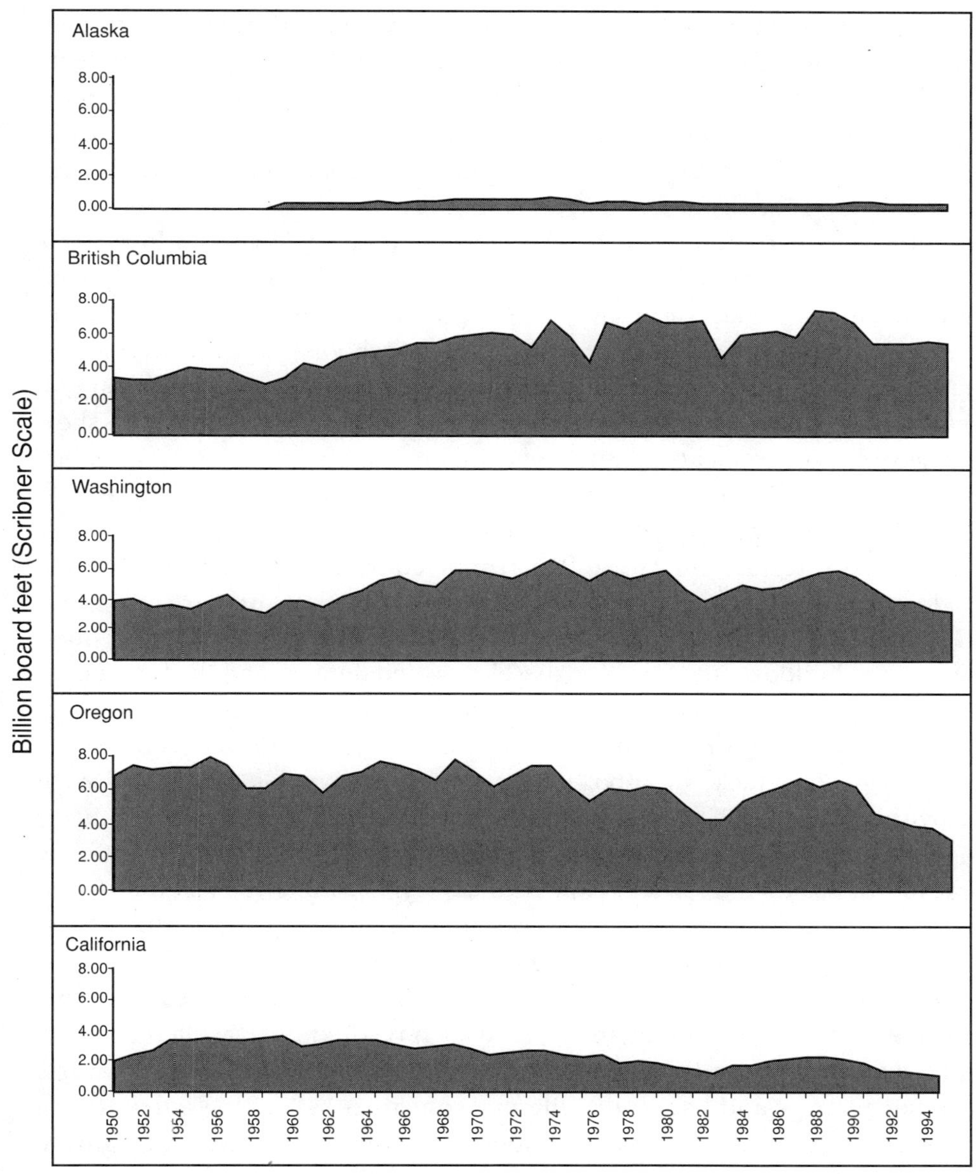

Figure 12.5. Aggregate timber harvest from the North Pacific slope, 1950–1994, by state and province. *Source:* Compiled by Interrain Pacific from various public agencies.

boom phenomenon to western Washington, to western Oregon, and to Vancouver Island and southern British Columbia. The new, highly productive technology also contained the seeds of social and environmental destruction in the areas of heaviest harvesting: when the timber was gone—as happened in Washington's Grays Harbor County by the 1930s—the consequences were treeless landscapes and equally devastated communities.

As the twentieth century advanced, loggers and foresters developed new practices to accommodate the ever more efficient and productive equipment they used in the woods. Clearcut harvesting, a method once considered avant-garde, eventually became the norm. Forestry schools at the University of British Columbia, the University of Washington, and Oregon Agricultural College (now Oregon State University) developed "scientific" arguments to show how clearcutting replicates natural phenomena such as forest fires. In effect, as cutting practices were adapted to technology, the machine became the chief determinant in reshaping the forest landscape.

Labor shortages during World War II spurred the development of the most productive technological innovation of them all—the revolutionary chain saw. The simultaneous adaptation of internal combustion engines to logging operations—the bulldozer or "Cat," the diesel-powered donkey, and the log truck—further increased productivity. And then the incredible postwar building boom provided marketing opportunities unlike any the industry had ever experienced. The "ticky-tacky little boxes" that Malvina Reynolds sang about went up by the square-mile section in southern California. Meanwhile, far to the north of that southern suburban sprawl, "daylight was being let into the swamp," to use an old logger's expression, at an alarming rate. In the United States private timberlands were cut over first. Then, increasingly, the harvest rate was stepped up on public lands as well.

And that brings us to the present state of affairs: cutover forestland, disrupted ecosystems, declining anadromous fish runs, and devastated communities that once depended on these resources for their livelihood. What has been happening—and I see little difference on either side of the international boundary—is a steady process of deindustrialization as the acreage of private and publicly held timber dwindles. As the twentieth century draws to a close, NASA satellite photographs clearly show that the forest fabric of the Pacific Northwest is thoroughly fragmented and includes "more holes than forest" (Canada's Future Forest Alliance 1993).

In this sense, the coastal rain forest region along the North Pacific coast is at one with other areas of the globe within the embrace of the global capitalist marketplace. And everywhere market economies tend to perpetuate certain historic features—especially the tendency toward a commodity-driven, land-use calculus that appears to recognize no limits. Since the coming of Europeans to the Pacific Northwest, the dominant culture has played out a production-driven endgame based on the assumption that more fish could be pumped from the rivers, more kilowatt-hours could be generated from the dams, and more board-feet of timber could be grown in the forests—if only rational engineering and scientific approaches were brought to the task. It has not worked. We would do well to heed the wisdom of forester Edward I. Kotok, who cautioned nearly fifty years ago that too much emphasis was being placed on the end product rather than "the maintenance or creation of a healthy, well-balanced biological complex that by its nature is conducive to favorable vegetative growth, water relations, and the support of animal life" (Kotok 1950:471).

References

Barman, J. 1986. *The West Beyond the West.* Seattle: University of Washington Press.

———. *The West Beyond the West: A History of British Columbia.* Toronto: University of Toronto Press.

Callenbach, E. 1975. *Ecotopia*. Berkeley: Banyan Tree Books.

Canada's Future Forest Alliance. 1993. *Brazil of the North.* New Denver, B.C.

Cantwell, R. 1972. *The Hidden Northwest.* New York: Lippincott.

Clark, N. 1970. *Mill Town: A Social History of Everett, Washington.* Seattle: University of Washington Press.

Cox, R. 1957. *The Columbia River.* Edited by E. I. Stewart and J. R. Stewart. Norman: University of Oklahoma Press.

Cronon, W. 1991. *Nature's Metropolis: Chicago and the Great West*. New York: Norton.

Dillard, A. 1992. *The Living*. New York: HarperCollins.

Emerson, G. H. 1907. "Logging on Grays Harbor." *Timberman* 8:20.

Ficken, R. E. 1979. "Weyerhaeuser and the Pacific Northwest timber industry, 1899–1903." *Pacific Northwest Q.* 70:146, 153.

———. *The Forested Land: A History of Lumbering in Western Washington.* Seattle: University of Washington Press.

Garreau, J. 1981. *The Nine Nations of North America.* Boston: Houghton Mifflin.

Gough, B. M. 1980. *Distant Dominion: Britain and the Northwest Coast of North America, 1579–1809.* Vancouver: University of British Columbia Press.

Hastings, L. W. *The Emigrants' Guide to Oregon and California.* Princeton: Princeton University Press. Originally published in 1845.

Holbrook, S. 1971. *Holy Old Mackinaw: A Natural History of the American Lumberjack.* New York: Macmillan.

Kotok, E. I. 1950. "The ecological approach to conservation programs." In *Renewable Natural Resources,* sec. IV. (Copy in library of Blue Mountains Natural Resources Institute, La Grande, Ore.)

Ledyard, J. 1783. *A Journal of Captain Cook's Last Voyage to the Pacific Ocean.* (1963 photocopy, Oregon State University library, Corvallis.)

MacKay, D. 1982. *Empire of Wood: The MacMillan Bloedel Story.* Vancouver: University of British Columbia Press.

———. *Empire of Wood.* Seattle: University of Washington Press.

Meinig, D. W. 1991. "Spokane and the inland empire: Historical geographic systems and a sense of place." In D. H. Stratton, ed., *Spokane and the Inland Empire: An Interior Pacific Northwest Anthology.* Pullman: Washington State University Press.

Merk, F., ed. 1931. *Fur Trade and Empire: George Simpson's Journals.* Cambridge: Harvard University Press.

Moulton, G. M., ed. 1990. *The Journals of the Lewis and Clark Expedition.* Vol. 6. Lincoln: University of Nebraska Press.

Munger, T. 1927. *Timber Growing Practices in the Douglas Fir Region.* Bulletin 1493. Washington, D.C.: U.S. Department of Agriculture.

Pacific Monthly. Vol. 11, no. 1 (January 1904).

Rajala, R. 1989. "Managerial crisis: The emergence and role of the West Coast logging engineer, 1900–1930." In P. Baskerville, ed., *Canadian Papers in Business History.* Victoria: Public History Group, University of Victoria.

Robbins, W. G. 1986a. "The Indian question in western Oregon: The making of a colonial people." In G. T. Edwards and C. A. Schwantes, eds., *Experiences in a Promised Land: Essays in Pacific Northwest History.* Seattle: University of Washington Press.

———. 1986b. *Hard Times in Paradise: Coos Bay, Oregon, 1850–1986.* Seattle: University of Washington Press.

Ross, A. 1986. *Adventures of the First Settlers on the Oregon or Columbia River, 1810–1813.* Lincoln: University of Nebraska Press.

Vancouver, G. 1960. *A Voyage of Discovery to the North Pacific Ocean, and Around the World.* 3 vols. London: British Museum. Originally published in 1798.

Wall, B. 1972. *Log Production in Washington and Oregon: An Historical Perspective.* Resource Bulletin PNW-42. Portland, Ore.: Pacific Northwest Forest and Range Experiment Station, USDA Forest Service.

Washington State Planning Council. 1941. *The Elma Survey.* Olympia, Wash.

West Shore. Vol. 1, no. 6 (January 1876); vol. 6, no. 3 (March 1880); vol. 7, no. 8 (August 1881).

White, R. 1980. *Land Use, Environment, and Social Change: The Shaping of Island County, Washington.* Seattle: University of Washington Press.

Worster, D. 1979. *Dust Bowl.* Oxford: Oxford University Press.

———. 1993. *The Wealth of Nature: Environmental History and the Ecological Imagination.* Oxford: Oxford University Press.

13. Economic and Demographic Transition on the Oregon Coast

Hans D. Radtke, Shannon W. Davis, Rebecca L. Johnson, and Kreg Lindberg

The communities of the coastal temperate rain forest are in the throes of a profound economic and social transformation. Long characterized by resource extraction, the bioregion's economy is undergoing structural changes caused by the decline of salmon runs in the southern part of the region, the displacement of people by machines in the timber industry, and the increasing export of raw logs. Commercial and recreational salmon fisheries, for example, which once employed half of the 900 residents of Ilwaco, Washington, in fishing boats and canneries, now provide a fraction of the jobs available only five years ago. In the Clayoquot Sound watershed on southern Vancouver Island, timber workers accounted for half of the direct employment in the region—at wage rates that made the town of Port Alberni one of the richest communities in Canada during the 1970s. Despite a continuing high timber cut, wood products jobs now make up only a quarter of direct employment, a reduction that is due primarily to mechanization.

In parts of the coastal temperate rain forest where U.S. federal agencies administer public land, the curtailment of logging to protect endangered species has had significant effects, at least in the short term. At the same time, population and wealth in the cities of the greater Pacific Northwest—Portland, Seattle, Vancouver, Victoria—have grown. Comparatively wealthy (and newly arrived) urban residents vacation in coastal communities, and many choose to retire there. Later in this chapter we explore some of the socioeconomic effects of the increasing numbers of tourists and retired people on coastal rain forest communities and examine the attitudes of residents toward tourism and its consequences.

Although communities of the coastal temperate rain forest share many socioeconomic and ecological characteristics, significant differences can be found as well. Some result from the different stages of evolution in which communities find themselves relative to a northward-moving "frontier" of resource extraction. From California to southern British Columbia, for example, most of the old-growth forest has been logged and few healthy stocks of wild salmon remain (Plate 3). In northern British Columbia, cutting of old-growth forests, primarily for pulp, still dominates economic activity. Further north, Alaska's fisheries remain healthy and, in fact, many fishers from the southern part of the bioregion have moved operations to Alaska, at least for part of the year. Alaska is unique as well in the presence and development of large oil reserves: the petroleum sector contributes approximately 70 percent of the gross state product (Larson 1990).

Our focus here is on communities of the Oregon coastal temperate rain forest. Many of the patterns we can discern in Oregon influence other parts of the coast, as well, but the differences should be kept in mind. The Oregon coast contains some of the most unfragmented old forests remaining on the West Coast of the United States. Changes taking place in Oregon with its mix of old-growth stands and industrial forests are similar to the changes taking place in other parts of the West Coast (Figure 13.1). Traditional resource-based industries like fishing and wood products have declined, while coastal economies have been supplemented by tourism and the arrival of retired people. These changes, particularly the development of tourism, have had both positive and negative effects. This chapter summarizes two of our studies that focus on these effects. The first study describes the demographic and economic trends taking place in Oregon's coastal communities up to 1991. In the second study, we evaluate certain impacts of tourism and economic development on Oregon coast residents in economic terms.

Demographic and Economic Changes

Although the Oregon coast population has grown in the last twenty years, its growth has generally lagged behind the state as a whole (Figure 13.2). Notable exceptions are Lincoln, Curry, and coastal Lane counties, which have received significant increases in retired people. Curry County also has experienced population increases in the Brookings area from jobs at a new large prison complex just south of the Oregon–California border. Generally, however, coastal counties are losing higher-paid jobs and young adults tend to leave the area to

Figure 13.1. Most coastal old growth, like this Douglas-fir photographed near Saddle Mountain in Clatsop County, Oregon, has long since been harvested. (Photo courtesy of Oregon Historical Society, #OrHi 27346.)

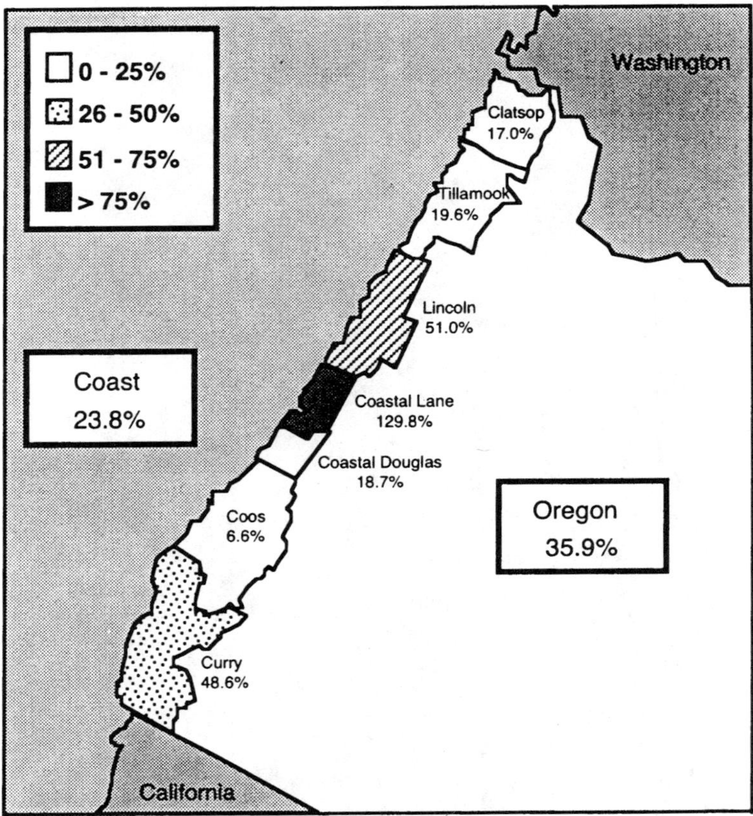

Figure 13.2. Changes in the population of coastal counties and state of Oregon: 1970–1990.

find education and work. With these population changes, coastal communities are experiencing significant shifts in their demographic structure. Certainly the national population is aging as large segments of the population move into older age groups. The trends are the same for Oregon, too, but more so in coastal counties (Figure 13.3). Buying habits, service needs, community involvement—all change with the changing age structure.

The demographic factors and the coast's reliance on uncertain natural-resource-based industries have resulted in low economic growth and greater than average unemployment (Figure 13.4). Real net earnings have decreased slightly in the last several years in some coastal counties, largely because of the decline in higher-wage jobs in the timber industry. Clatsop County, for example, experienced a drop of 6 percent in earned income between 1989 and 1991. Coos and Curry counties experienced decreases, too, but at smaller rates.

What is the source of personal income received by coastal communities?

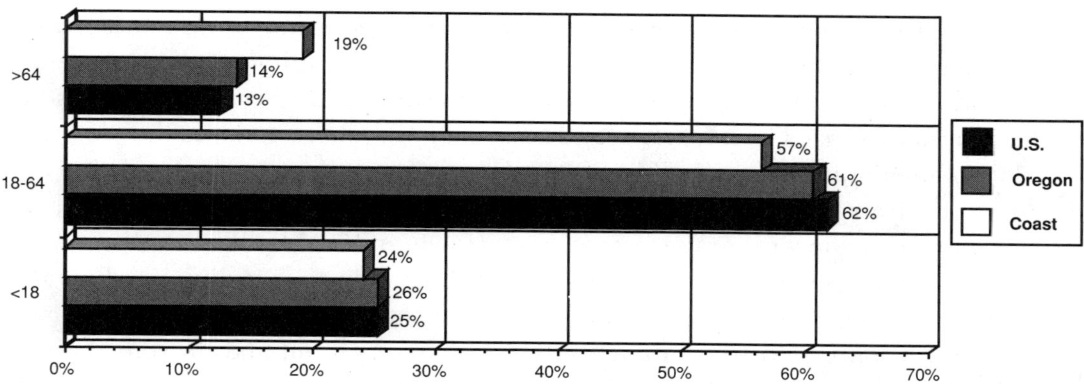

Figure 13.3. Relative population in selected age groups: 1990. *Source:* U.S. Bureau of the Census.

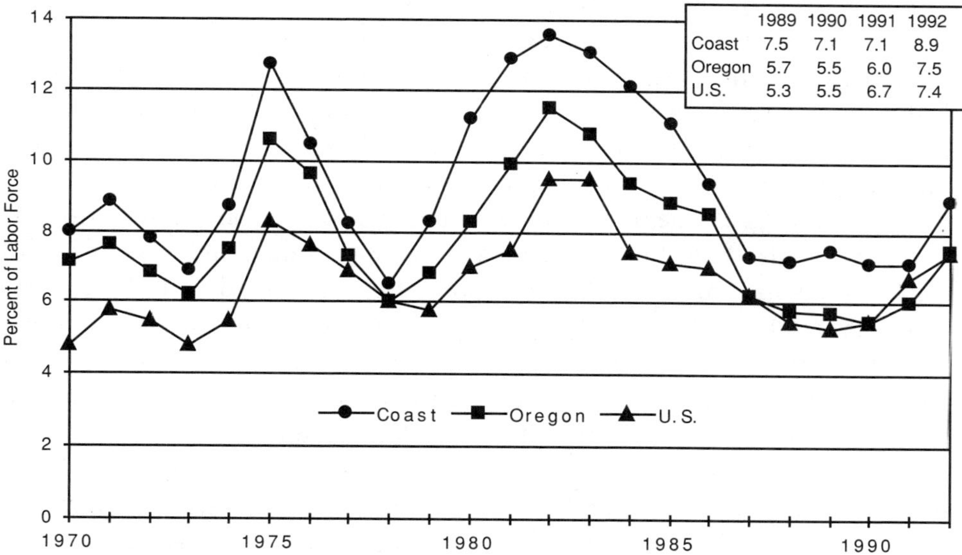

Figure 13.4. Unemployment rate: 1970–1992. *Source:* Oregon Employment Division.

Natural-resource-based industries such as timber, tourism, fishing, and agriculture accounted for 28 percent of income in 1991. Of the "exporting" natural-resource-based industries, the timber industry contributes about 12 percent of the total personal income of coastal residents. Tourism and fishing are major contributors, as well, making up about 8 percent and about 5 percent, respectively. Agriculture is a significant contributor to some communities—in Tillamook County it contributes 20 percent of total income—but coastwide it contributes only about 4 percent of total personal income. Both tourism and agriculture are experiencing growth on the coast. Although the growth of tourism tends to diversify the economic base of coastal counties, this industry is characterized by low wages and seasonal jobs—characteristics that do little to ameliorate seasonal effects from the other natural-resource-based industries. Transfer payments, investments, and other earnings make up the balance of personal income of the Oregon coast (Davis and Radtke 1994).

Income generated by the timber and fishing industries is declining for various reasons: decreasing availability of the resource, new demands to use natural resources for recreation and habitat preservation, and, in some cases, falling prices. Overall, natural-resource-based income has decreased from 1987 to 1991. For timber, stumpage prices have increased as final product prices have increased; therefore, transportation costs have become a smaller part of final manufacturing costs. As mills become more willing to expand the area from which they get timber, there has been a dramatic reduction in processing activity on the coast. Most timber is now shipped to the major processing centers of Roseburg, Eugene, Albany, and Portland.

For fisheries, three developments are affecting the contribution this industry can make to the coastal communities. First, declining consumer demand and increasing global supplies of all fish products have depressed the real per pound ex-vessel price for products such as salmon and shrimp that are also produced by aquaculture (Anderson 1994). The second development is the crisis facing the salmon industry. Because of unfavorable ocean conditions, inland habitat deterioration, and multiple demands for the harvest rights of the salmon resource, the availability of salmon for commercial ocean harvesting has declined steadily along the Oregon coast. Because salmon stocks are facing threatened or endangered status listing from the Sacramento River in California to the Queets River in Washington, regulations have reduced the ocean troll harvest to a fraction of historic levels. Because of reduced harvests, coastal communities have experienced a decrease of annual personal income attributable to salmon of 86 percent—from averages of about $28 million per year (1971 to 1993) to $3.5 million in 1993 (Pacific Fishery Management Council 1995). Especially affected are southern Oregon ports, where today there is no commercial salmon season at all. Small ports in this area have always relied on the salmon trolling industry to generate local income and support vital services such as local marinas and have used the local fishing industry to justify dredging operations by the U.S. Army Corps of Engineers. The third critical fishery issue is the allocation of the Pacific whiting fishery. The Pacific Fishery Management Council allocated about 70,000 metric tons of this species in 1995 for harvesting by vessels with home ports on the Oregon coast and processing by businesses located there. This means that coastal communities can expect to receive $40 million to $50 million in personal income from this fishery alone (Radtke 1995). This recently developed fishery contributes overall to the commercial fishing industry, but because of the need for special gear and holding facilities it is not a substitute fishery for most salmon troll vessels in Oregon.

Thus transfer payments and returns from investments (nonwage income) have become a major source of income for most coastal communities. Transfer payments (such as retirement program payments) and investment income (such as the sale of property elsewhere) made up about 45 percent of the total personal income on the Oregon coast in 1991. This compares with about 35 percent for Oregon and 33 percent for the United States. Spin-off jobs traced to these income sources, however, are lower-wage jobs and consumer-service oriented. The arrival of more retired people has helped to increase real transfer payments (mostly social security) from 20 percent of total personal income in 1987 to 24 percent in 1991. The returns to investments fell slightly between 1987 and 1991, from 21 to 17 percent, most likely due to falling interest rates. Growth in "other earnings" (27 percent of personal income in 1991) indicates an increase in small manufacturing and businesses. These small companies—plastic wedge and plaster water tank manufacturers, computer hardware and software developers, writers, artists—sell products outside the coastal area and return income to the coastal economy for spending. Overall the shifting demographic and economic factors have caused per capita incomes of coastal residents to lag ever further behind the rest of Oregon and the United States (Figure 13.5). The smaller earnings component of per capita income, largely comprising wages and salaries, is the dominant reason for the decline.

The social and health characteristics of the coastal population show that educational attainment is lower, access to health services is more difficult, the rate of poverty is higher, and there is a higher proportion of substandard housing relative to all of Oregon (Table 13.1). Housing costs as a percentage of household income are slightly less along the coast than in the state as a whole, but affordable housing for low-income households is in short supply. The median value of owner-occupied homes is less than the rest of the state, but the residential assessed value per capita is much higher. This demonstrates the existence of higher-valued second homes at the coast (13 percent of all vacant homes) versus 3 percent for the rest of the state. The combined crime rate for personal and property infractions is less than the state's.

Quality of life is frequently cited as a reason for population growth on the Oregon coast. Certainly the coast exemplifies several elements of high quality of life: recreational opportunities, moderate climate, lower crime rate, and lower consumer prices. But economic development requires other quality of

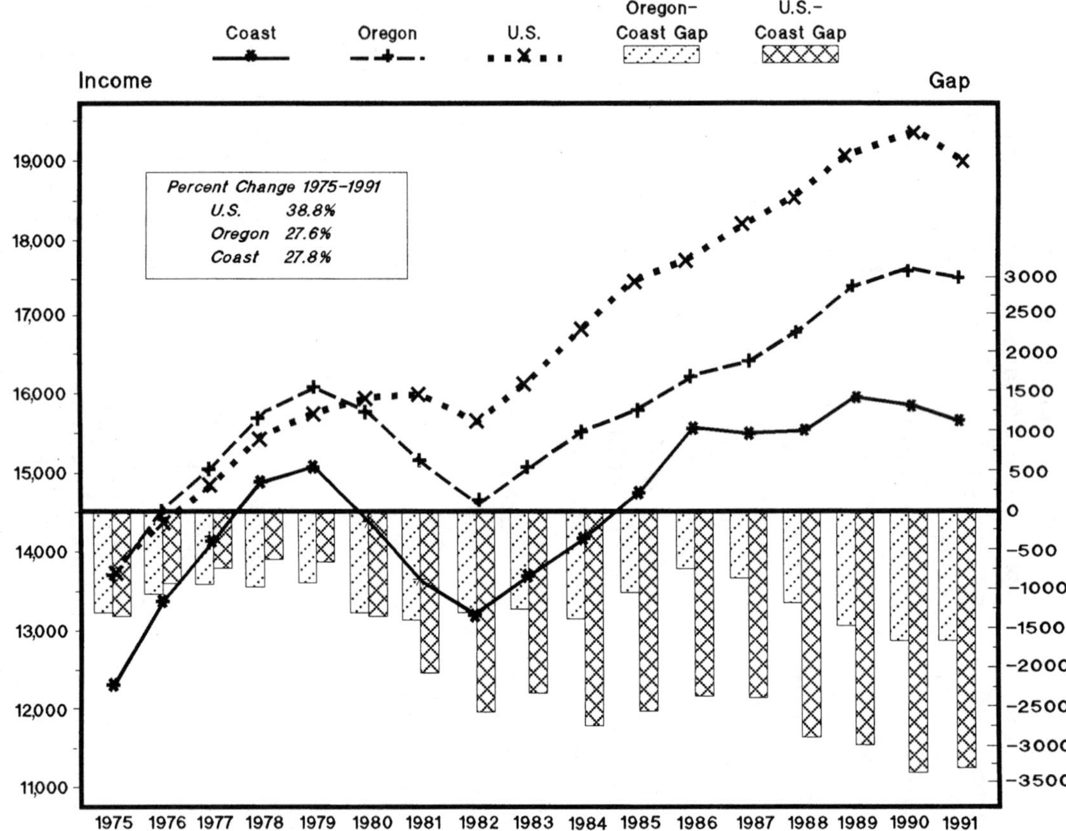

Figure 13.5. Real personal income per capita: 1975–1991. *Notes:* 1. Coast per capita real personal income is a weighted average based on population. 2. Dollar adjustment uses the U.S. GNP implicit price deflator developed by the U.S. Bureau of Economic Analysis. *Source:* Davis and Radtke (1994) and U.S. Bureau of Economic Analysis (1993).

Table 13.1. Health and Social Characteristics for the Coast and Oregon in 1991

Characteristic	Coast	Oregon
Completed high school (age 25 and over)[a]	78.2%	81.5%
Hospital beds (per 1000)[b]	2.1	2.6
Poverty status (% of population)[a]	15.4%	12.1%
Substandard housing (% housing with inadequate plumbing)[a]	1.1%	0.9%
Renter costs (costs as % of income)[a]	25.1%	25.5%
Owner costs (costs as % of income)[a]	17.4%	18.2%
Index crime rate (per 10,000 persons)[c]	474	586
Residential assessed property value (per capita)[d]	$31,400	$19,400
Vacancy rate (seasonal or recreational housing units)[a]	13.0%	2.5%
Owner-occupied median value [a]	$61,626	$66,800
Net property tax rate (per $1000 assessed value)[d]	$16.20	$20.66

[a] U.S. Bureau of Census (1992a).
[b] Oregon Health Division (1992).
[c] Oregon Criminal Justice Services Division (1993).
[d] Oregon Department of Revenue (1993).

life elements—such as health and education facilities—as well as the basic infrastructure for business, such as communication, transport, water and sewers, and zoning for diverse industries. Proximity to markets and major transport facilities, the presence of education and business links, the existence of technological support services, local control over investment capital—all are less available in the coastal area. These impediments are likely to cause dramatic changes in Oregon's settlement patterns. The coast's economic fortunes will continue to be tied to the prosperity of the Portland metropolitan area and other Willamette Valley urban centers. Vigorous efforts to package and offer investment incentives and promote quality of life variables will help to overcome impediments for new businesses.

Population and economic trends indicate that future employment growth will be in the consumer (health, retail) and producer (finance, real estate) service sectors. Manufacturing-sector employment will decline as a percentage of total employment, especially in timber and fisheries. The government, education, utilities, and transportation sectors will have moderate growth. The shift to service-sector employment and the rise of the information economy have modified the nation's as well as the coast's occupational employment structure. Generally, blue-collar middle-class jobs are disappearing and being replaced by either high-paying professional and technical jobs or low-paying service and clerical jobs. With the loss of middle-class jobs, the workforce is becoming increasingly stratified by skill and wage.

Because of the centralization of natural-resource-processing activities, the coast is losing industrial jobs to Oregon's urban areas. In turn, urban dwellers from the Portland metropolitan area and Willamette Valley will continue to travel to the coast for recreation. Private efforts funded by foundations, along with public programs (the Northwest Economic Adjustment Initiative from the Clinton administration's Forest Plan, for example, and the Oregon governor's Salmon Initiative), may also be helpful in promoting economic development and resource sustainability. But there is a side-effect from promoting economic development: there will be new environmental protection challenges. Demands on water, energy, waste treatment and disposal, and pressures to convert important natural habitats will diminish environmental quality unless there is careful land use planning and public facilities are properly designed. Most coastal counties are forecast to have population growth in the next twenty years. More important for land use planning than the number of new people is encouraging development into areas where services, facilities, and utilities already exist or where cost-effective services can be supplied. Growth should be directed away from natural hazard areas (such as unstable hillsides) and away from valuable agricultural and forest lands or other open-space areas where development would cause adverse social impacts or require the wasteful expenditure of public tax dollars.

In sum, then, population growth on the coast has generally lagged behind the rest of Oregon except for communities with large increases in retired people. The shifting demographic and economic factors have caused average per capita incomes of coastal residents to fall behind the rest of Oregon and the United States. The earnings component of per capita income, largely comprised of wages and salaries, is the dominant reason for the decline. Communities where residents rely on transfer payments, such as social security checks, have generally experienced a growth in real income.

The natural amenities that the coast offers are certain to attract increased population and economic growth. But these population increases and development must be anticipated and planned. If growth is not planned, the usual problems will result: overallocation of resources (such as water); increased dependence on energy, especially for transportation; and increased pressure to convert resource land to nonresource uses. The coast is expected to have a continued steady population increase and, as it occurs, the complexity of land use and natural resource management will intensify. If the growth is properly anticipated and planned, the increased needs of more people can be met with little sacrifice of the area's livability.

Tourism's Social Impacts

When population growth materializes on the coast, it will not come without negative elements, despite efforts for land use planning and natural resource management. Here we discuss the results of a survey of Oregon coast residents that evaluates one aspect of growth caused by the natural amenities: increased visitors and the new businesses that cater to them.

Significant contributions have been made recently to understanding tourism's social impacts, both actual and perceived, and the factors that affect residents' attitudes toward tourism (Ap 1992; Lankford and Howard 1994). Understanding these impacts means that we can now incorporate them into the policymaking process. Because the methods used in past research do not measure social impacts in a manner consistent with those used to measure economic impacts, economic benefits and costs often tend to dominate decisions on tourism planning and development (Choy 1991).

Tourism's social impacts can be either positive or negative (Bull 1991). On the whole they tend to be negative, however, so that their exclusion from consideration leads to overestimation of the net social benefits of tourism development—for example, the benefits of additional jobs may be emphasized and the problems of increased congestion overlooked. Because different tourism development paths produce different impacts, excluding the social impacts from consideration may also lead to selecting a path that is less socially desirable than others. Assuming that increasing social welfare is the goal of economic development programs, social impact values should be estimated and incorporated into the policymaking process.

Here we wish to introduce a technique—contingent valuation—for measuring the economic value of certain social impacts associated with tourism. By measuring these impacts in economic terms, researchers can integrate social and economic factors. Most goods and services are traded in well-developed markets. The behavior of consumers in these markets (the amount of each good they consume at different prices) provides the information needed by economists to determine the willingness of consumers to pay for each good. This willingness to pay is a measure of the economic value of the good (Peterson et al. 1990). Some goods and services, however, are not traded in markets; they are nonmarket goods. For example, many campsites on public land are free. Although a consumer's willingness to pay for these goods is generally positive, it cannot be measured by price because the price is zero.

Economists have developed techniques for measuring willingness to pay for nonmarket goods. One of these techniques is the contingent valuation method, which presents a hypothetical market to consumers. The behavioral intent reported in response to questions about the hypothetical market is used in lieu of actual behavior as a basis for estimating willingness to pay. Some economists

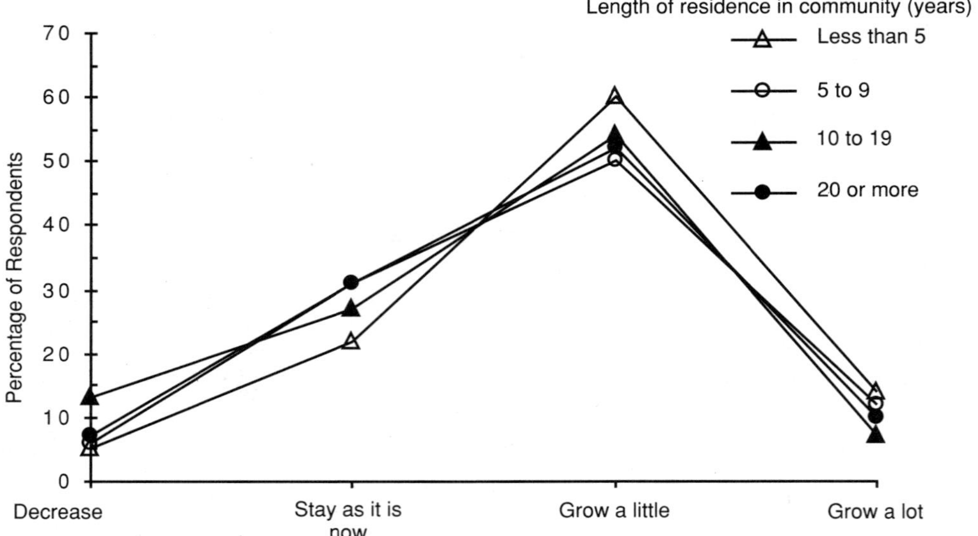

Figure 13.6. "Do new residents want less growth than long-term residents?" *Source:* Lindberg and Johnson (1995).

and psychologists question whether the contingent valuation (CV) method yields valid estimates (Harris et al. 1989; Hausman 1993). Nonetheless, the CV method has gained wide acceptance as at least a starting point for estimating willingness to pay for nonmarket goods—provided the survey methods are rigorous and certain standards are met (Arrow et al. 1993; Mitchell and Carson 1989).

Using the principles of Dillman's (1978) "total design method," we asked a variety of questions—related to growth generally and to individual industries in particular—of 945 randomly selected coastal residents. We found that a majority of these residents would like their community to grow a little; roughly 10 percent desire either a decrease in population or a lot of growth (Figures 13.6 and 13.7). Attitudes toward growth do not vary according to length of residence and income as much as one might think: newcomers have as much desire for community growth as long-term residents. The most notable difference across groups is that respondents living in the community from ten to nineteen years are more likely to favor a decrease in population. Respondents from low-income households (less than $20,000 per year) express less desire for growth than do respondents from higher-income households. Although most respondents from these low-income households appear to be retired people (54 percent are at least sixty years old), desires regarding growth are quite similar across all age groups of respondents living in low-income households. These results contradict to the beliefs that newcomers want to "lock up"

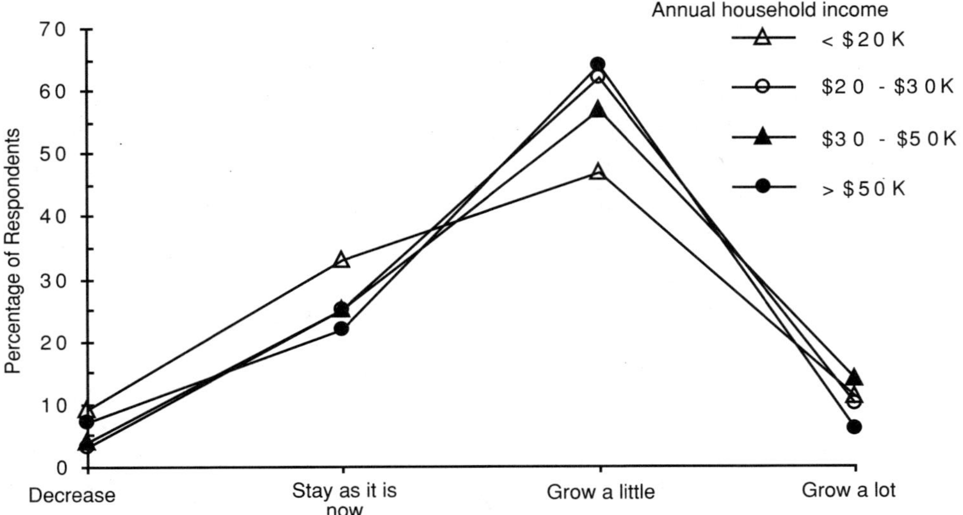

Figure 13.7. "Do high-income households want less growth than low-income households?" *Source:* Lindberg and Johnson (1995).

the community and that lower-income households, in particular, want growth in order to maintain or improve their job opportunities.

Additional questions concerned the role of newer residents. Seventy-eight percent of respondents agreed or strongly agreed with the statement, "Newcomers bring skills and business opportunities that contribute to the local economy." Given that newcomers may play an important role in future economic growth, we asked newer residents (those living in the community less than five years) why they moved to the community and whether an experience as a tourist contributed to their decision to move there. The most commonly cited motives for moving included the natural environment, a job opportunity, the small-town lifestyle, and family reasons. These responses illustrate one of the major dilemmas for many coastal communities: the natural environment and small-town lifestyle are two qualities that make communities attractive to both newer and long-term residents; yet by attracting newcomers these communities run the risk of tarnishing these attractive qualities.

A tourist experience can contribute to the decision to move to the coast, but that experience is not the only reason for moving. For example, we found that a tourist visit was the main reason for 18 percent, part of the reason for 43 percent, and not a reason at all for 40 percent of the residents moving to the coast. (Some 26 percent of these had moved within the last five years and 68 percent had visited the coast before moving there.) Residents express a strong desire for growth in local industries: fishing, timber and wood products, retail/tourism, agriculture, and high technology. Indeed, 87 percent of the respondents agreed or strongly agreed with the statement, "Creating jobs for residents should be a high priority for this community." But most options for creating jobs involve costs. These costs might be economic, social, or environmental. Given these trade-offs, residents were asked to evaluate the acceptability of general options for creating jobs. Despite the desire to create jobs, respondents opposed all the options presented in this survey, including allowing more pollution (opposed by 87 percent) or converting forest or farmland to industrial, commercial, or residential use (opposed by 70, 69, and 58 percent, respectively). Although the options were limited and were not community-specific, the results demonstrate the strong desire of residents to avoid sacrificing the local quality of life to create jobs. In short, residents want a vigorous economy while maintaining quality of life.

Residents were asked to list the most important perceived benefits and problems associated with tourism. Not surprisingly, the key benefits are economic, including the creation of jobs and local business opportunities. Some residents also noted that tourism development expands the facilities available to residents, too, and that tourists bring new ideas into the community.

An important issue is whether tourism development creates the "family wage" jobs necessary to replace the jobs lost as the timber and fishing industries decline. Tourism has provided a few jobs to residents who worked in fishing or timber, but most tourism workers have always worked in tourism or have come from the retail or service sectors (Figure 13.8). On the other hand, the majority (57 percent) of tourism employees earn at least as much as they did in their previous job; 15 percent earn about three-quarters as much; 23 percent earn one-half or less (5 percent responded "don't know"). These results suggest that tourism jobs have not directly replaced timber or fishing jobs, at least in the communities surveyed for this project. Tourism does, however, make an important contribution to communities. It helps diversify local economies, many of which have limited options for development. Indeed, 90 percent of the residents surveyed agreed or strongly agreed with the statement, "Tourism helps diversify the local economy, and therefore it is an important industry for this community."

Although the perceived problems of tourism development are similar to those found in many tourism-dependent communities, they are also due partly to local geography and the type of tourism development. Highway traffic, by far the most commonly cited problem, was noted by 47 percent of the respondents. U.S. Highway 101, constrained between the Pacific Ocean and the Coast Range mountains, is the primary north–south route followed by tourists and residents alike. Travel time for residents increases dramatically during the tourist season. Crime, reported as a problem by 14 percent of the respondents, consists primarily of minor violations such as disorderly conduct by visitors.

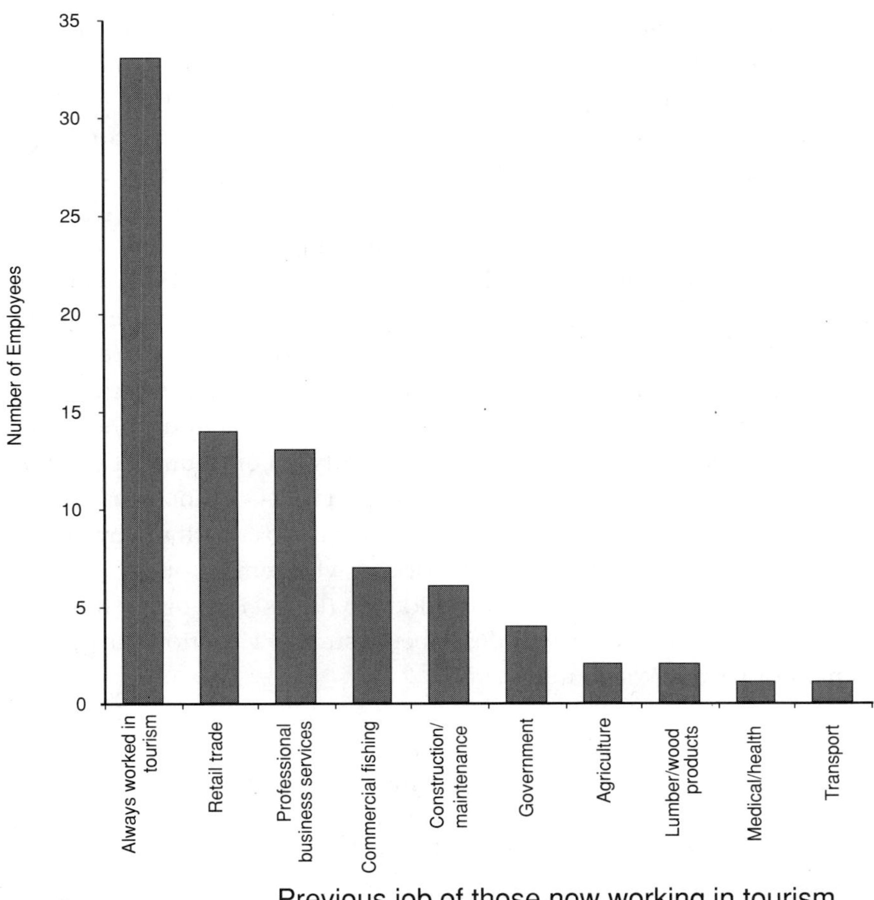

Figure 13.8. "Do tourism jobs replace jobs lost in other industries?" *Source:* Lindberg and Johnson (1995).

These infractions are particularly disruptive to residents because they often occur at vacation homes located in residential areas. Other reported problems include crowding in stores, bayfronts, and other areas, as well as competition for parking spaces. The linear nature of most coast communities exacerbates the problems of traffic and crowding and lack of parking; there is simply no place to put all the people and cars.

Despite the tourism-related problems, the majority of residents believe that tourism has been positive for them individually and for their community. Most residents agree or strongly agree with the statement, "Overall, for me personally, the benefits of tourism outweigh the costs of tourism" (Table 13.2). An even greater majority agree or strongly agree that the benefits to the community outweigh the costs to the community. These beliefs contribute to a desire by some for future increases in tourism development (Table 13.3). Although the majority prefer the status quo, increases are favored over decreases for all types of tourism, though short-term vacation rentals (less than one week) and day visitors are desired less than other types. Responses to other survey items suggest that this ranking is a result of the relatively low level of economic benefits and relatively high level of disruption associated with these two types of tourism.

The relationship between tourism and its impacts—such as a change in crime rates—remains subject to debate (Crotts and Holland 1993; Milman and Pizam 1988). Nonetheless, an indication of this relationship is presented here. Tourism and traffic, noise, and crime are highly correlated (Figure 13.9). This correlation, combined with corroborative evidence, indicates that tourism contributes significantly to traffic congestion and noise and minor crime in Newport. The relationships vary from community to community, but generally they show similar patterns. And although tourism is by no means the sole cause of the lack of low-income housing, it appears to contribute by increasing housing costs and attracting migrant workers who remain out of work or underemployed and are often eventually added to the list of those in need of low-income housing. Murphy (1985) describes a similar situation during the development of Disney World in Florida.

Table 13.2. Level of Agreement with the Statement: "Overall, for me personally/for my community, the benefits of tourism outweigh the costs"

Level of Agreement	For Me Personally	For My Community
Strongly agree	23%	33%
Somewhat agree	29%	43%
Neutral	24%	12%
Disagree	11%	6%
Strongly disagree	11%	4%

Source: Lindberg and Johnson (1995).

Table 13.3. Preferences for Change in Specific Types of Tourism Expressed as a Percentage of Respondents

Type of Tourism	Increase	Stay the Same	Decrease
Hotels/motels	39%	59%	2%
Short-term vacation rentals	35%	47%	18%
Destination resorts	49%	43%	8%
Day visitors	39%	45%	16%

Source: Lindberg and Johnson (1995).

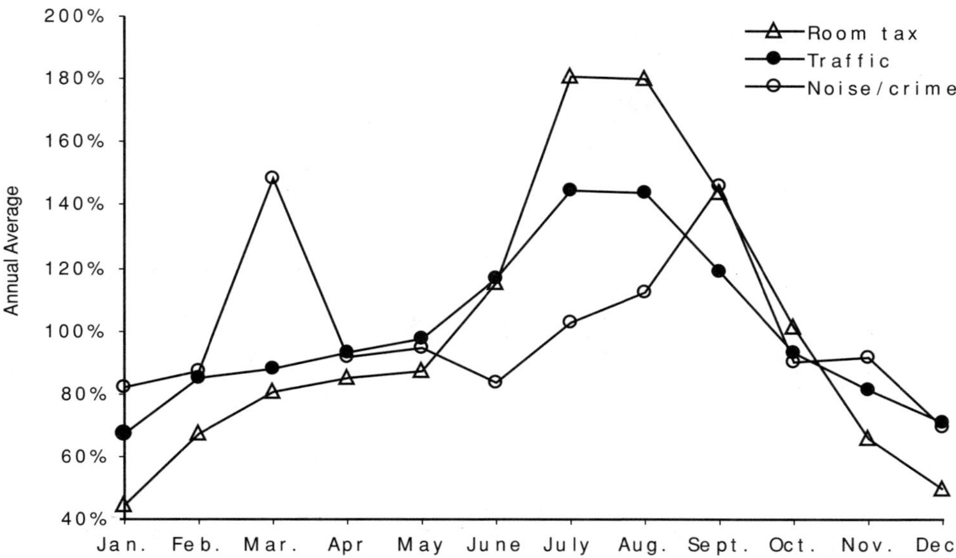

Figure 13.9. Link between tourism, traffic, and noise/crime (monthly percentage of 1991–1992 two-year average, Newport, Oregon). *Source:* Lindberg and Johnson (1995).

Based on the responses to our survey pretest, as well as discussion with community leaders, we created three contingent valuation scenarios:

- A program that would reduce traffic congestion on Highway 101 by 25 percent or 50 percent during busy periods
- A program that would reduce noise and minor crime by 30 percent during summer and holiday periods
- A program that would provide low-income housing for all qualifying families in the community

The first scenario was fashioned after a concurrent highway improvement process undertaken by the Oregon Department of Transportation. Because this process involved community meetings and received coverage in the local press, the contingent valuation scenario based on this process was likely to be understandable and plausible. We used the dichotomous-choice referendum method to elicit the respondent's willingness to pay (Mitchell and Carson 1989). Each survey contained all three scenarios in random order. In each scenario respondents were presented with a bid that was randomly selected from a group of sixteen values in the range of $5 to $1000 per household per year. Our results indicate that households would indeed be willing to pay to reduce problems of congestion, noise, and affordable housing—they indicated an average of $100, $95, and $103 per year, respectively.

Although there are no directly comparable analyses that can be used to evaluate the full validity of these results, similar analyses have produced generally similar results. Ahearn (1984), for example, found that households would be willing to pay $51 a year for a one-third reduction in the risk of burglary in Oregon communities. Adjusted for inflation, this equals $74 in 1993 dollars. Navarro and Carson (1991) used an election returns method to infer that the average San Diego household is willing to pay $138 per year to increase jail and court capacity in an effort to reduce crime. These survey responses support several general conclusions. First, the surveys and other research show that tourism has made an important contribution to coastal economies. Its significance is recognized by most residents in the communities studied. Most residents agree or strongly agree that, on the whole, tourism has been beneficial for them and for their communities. These perceptions have led many residents to support future increases in tourism.

Second, despite this general support, residents are sensitive to negative effects from tourism development. Responses to the contingent valuation scenarios indicate that these negative effects are significant. For example, tourism is likely to increase traffic congestion by more than 50 percent during busy periods. The average household's willingness to pay $100 a year is likely to be a conservative estimate of the loss in social welfare caused by traffic congestion. The significance of the negative impacts indicates that they should be mitigated whenever possible. If these impacts are not managed, they may lead to reactive regulations that are less effective, and less popular, than proactive measures.

Third, growth is occurring in many communities on the coast. This growth can threaten desirable qualities, among them the small-town lifestyle and high-quality natural environment, that are important to newcomers and long-term residents alike. Therefore, maintaining these qualities should be a continual priority of coastal residents and governments.

Finally, most respondents desire modest growth in their community, although some desire either more significant growth or a decrease in population. Our results show that some popular conceptions—that newcomers oppose growth, for example—are not always correct.

Acknowledgments

The authors wish to thank Christopher N. Carter and James Cornelius for their reviews of an early version of this chapter.

References

Ahearn, M. C. 1984. "An analysis of contingent valuation applied to air quality and public safety from crime." Ph.D. diss., Oregon State University, Corvallis.

Anderson, J. L. 1994. "The growth of salmon aquaculture and the emerging new world order of the salmon industry." Paper presented at the Fisheries Management—Global Trends conference at the University of Washington, Seattle, June 14–16, 1994.

Ap, J. 1992. "Residents' perceptions of tourism impacts." *Ann. Tour. Res.* 19:665–690.

Arrow, K., R. Solow, P. Portney, E. Leamer, R. Radner, and H. Schuman. 1993. "Report of the NOAA Panel on Contingent Valuation." *Fed. Reg.* 58: 4602–4614.

Bull, A. 1991. *The Economics of Travel and Tourism.* Melbourne: Longman Cheshire.

Choy, D.J.L. 1991. "Tourism planning: The case for 'market failure.'" *Tour. Mgmt.* 12(4):313–330.

Crotts, J. C., and S. M. Holland. 1993. "Objective indicators of the impact of rural tourism development in the state of Florida." *J. Sust. Tour.* 1:112–120.

Davis, S. W., and H. D. Radtke. 1994. *A Demographic and Economic Description of the Oregon Coast.* Newport: Oregon Coastal Zone Management Association.

Dillman, D. 1978. *Mail and Telephone Surveys.* New York: Wiley.

Harris, C. C., B. L. Driver, and W. J. McLaughlin. 1989. "Improving the contingent valuation method: A psychological perspective." *J. Envir. Econ. and Mgmt.* 17:213–229.

Hausman, J. A. 1993. *Contingent Valuation: A Critical Assessment.* Amsterdam: North Holland.

Lankford, S. V., and D. R. Howard. 1994. "Developing a tourism impact attitude scale." *Ann. Tour. Res.* 21:121–139.

Larson, E. 1990. *The Alaska Economy: An Overview.* Report prepared for the Alaska Department of Commerce and Economic Development and the Alaska Industrial Development and Export Authority. Anchorage: Institute of Social and Economic Research, University of Alaska.

Lindberg, K., and R. Johnson. 1995. "Estimating the economic value of tourism's social impacts." Unpublished manuscript. Department of Forest Resources, Oregon State University, Corvallis

Lindberg, K., R. Johnson, and B. Rettig. 1994. *Attitudes, Concerns, and Priorities of Oregon Coast Residents Regarding Tourism and Economic Development: Results from Survey of Residents in Eight Communities.* Report ORESU-T-94-001. Corvallis: Oregon Sea Grant.

Milman, A., and A. Pizam. 1988. "Social impacts of tourism in central Florida." *Ann. Tour. Res.* 15:191–204.

Mitchell, R. C., and R. T. Carson. 1989. *Using Surveys to Value Public Goods: The Contingent Valuation Method.* Washington, D.C.: Resources for the Future.

Murphy, P. E. 1985. *Tourism: A Community Approach.* New York: Methuen.

Navarro, P., and R. Carson. 1991. "Growth controls: Policy analysis for the second generation." *Policy Sci.* 24:127–152.

Oregon Criminal Justice Services Division. 1993. *The Report of Criminal Offenses and Arrests 1991–1992*. Law Enforcement Data System. Salem, Ore.

Oregon Department of Human Resources. 1993. *Oregon Coastal Employment and Payrolls by Industry and County*. Oregon Employment Division, Research and Statistics Section. Salem, Ore.

Oregon Department of Revenue. 1993. *Oregon Property Tax Statistics, Fiscal Year 1992–93*. Salem, Ore.

Oregon Department of Transportation. 1995. *Traffic Volume Tables*. Transportation Research Section. Salem, Ore.

Oregon Health Division. 1992. *Oregon Vital Statistics County Data 1991*. Salem, Ore.

Pacific Fishery Management Council. 1995. *Review of 1994 Ocean Salmon Fisheries*. Portland, Ore.

Peterson, G., B. Driver, and P. Brown. 1990. "The benefits and costs of recreation: Dollars and sense." In R. Johnson and G. Johnson, eds., *Economic Valuation of Natural Resources*. Boulder: Westview Press.

Radtke, H. 1995. *Windows on Pacific Whiting—An Economic Success Story for Oregon's Fishing Industry*. Newport: Oregon Coastal Zone Management Association.

U.S. Bureau of Economic Analysis. 1993. *Regional Economic Information System*. Washington, D.C.: Government Printing Office.

U.S. Bureau of the Census. 1992a. *Census of Population and Housing: Summary Tape File 3A*. Washington, D.C.

———. 1992b. *Summary Social, Economic, and Housing Characteristics*. 1990 Census of Population and Housing. Washington, D.C.: Government Printing Office.

WoodNet

TODD THOMAS

Communities on Washington's Olympic Peninsula had always relied on the plentiful timber that blanketed the coastal lowlands of the Olympic Mountains to support their core economy. Logs were exported from the peninsula for further processing elsewhere and used for high-volume, low-margin, commodity manufacturing, such as lumber, chips, shakes and shingles, pulp, veneer, and plywood.

In the late 1980s, however, the peninsula's timber harvest was abruptly reduced because of court action over endangered species, government regulations, overharvesting, and the beginnings of an ecosystem approach to forest management. The peninsula's primary wood products manufacturing sector was crippled, sending severe ripple effects through the peninsula's entire economy. And the future timber harvest is certain to remain lower than in the past.

WoodNet was created as a local effort to help the peninsula's struggling small and medium-sized wood products manufacturers face the economic challenges arising from declining timber harvest levels. WoodNet was one of five "flexible manufacturing network" experiments funded by the Northwest Area Foundation in 1991. Today's flexible manufacturing networks in Sweden, Norway, and Italy are patterned after centuries-old European guilds. Although the foundation sought to test the flexible manufacturing network model in a U.S. setting, the five networks, spread throughout the Northwest, were allowed to develop independently and today bear little resemblance to each other.

Adapting old and new models, WoodNet was designed to promote the creation of short- and long-term relationships between manufacturer networks to accommodate joint production, material procurement, and product marketing. The goal: to replace jobs lost due to reduced timber harvest levels while raising the overall standard of living on Washington's rural Olympic Peninsula.

The strategy: to develop a value-added, secondary wood products manufacturing industry on the Olympic Peninsula based on a sustainable timber harvest.

Moving from commodity production—logs—to the marketing and manufacture of value-added secondary wood products—cabinetry, furniture, boats, musical instruments—required fundamental changes in business management. New concepts such as market research and segmentation (niche markets), craftsmanship, quality design, short production runs, and computer-aided design and manufacture all played a part.

The peninsula is peopled by extremely independent folk used to a business culture characterized by cutthroat price competition—a pattern familiar to firms engaged in commodity product manufacture. Because networking often involves collaboration among direct competitors (working together to solve common problems and so forth), WoodNet's knack for getting firms to trust each other was an essential element of its success.

By adding more value to the primary natural resource, jobs lost in commodity manufacturing are being replaced. Working through WoodNet, manufacturers had a chance to pool resources, overcome common threats, and capitalize on previously unavailable market opportunities.

Gus Kostopulos, WoodNet's executive director, says: "Working together, member firms pursued business opportunities none could pursue alone." As a result, the network model is now an oft-cited strategy for industries in the midst of major restructuring.

Through WoodNet, small and medium-sized manufacturers were able to join forces and thus capitalize on market opportunities they might not have been able to undertake otherwise. Collaboration allowed smaller manufacturers to enjoy economies of scale in production through use of others' excess capacity and task specialization, as well as in purchasing and marketing. Thus they gained the competitive strength generally enjoyed only by large manufacturers. And they did so without incurring the capital investment or creating the bureaucracy typical of larger firms.

WoodNet was governed by a board of peninsula-based wood products manufacturers. WoodNet member firms range in size from one to fifty employees—from new enterprises to firms that have been in operation since 1909. Members produce everything from kitchen cabinetry to boats and musical instruments. Often referred to as an "information clearinghouse," a "dating service," and a "full-service center for peninsula wood products manufacturers," WoodNet evolved into much more than its founders initially envisioned.

WoodNet continually sought to satisfy its members' sales, marketing, management, financial, technology, and raw material needs through group training, special group projects, and confidential one-on-one management and technical assistance. WoodNet produced full-color catalogs featuring members' consumer product lines and has distributed these catalogs to over 10,000 wholesale buyers nationwide. WoodNet members attended domestic and foreign trade shows together. Recently they held a private trade show in Seattle, the nearest major metropolitan area, where scores of domestic and international buyers saw WoodNet products. WoodNet also hosted a series of evening sessions around the peninsula offering members the opportunity to meet their local bankers face to face.

WoodNet helped members gain access to raw materials by hosting delegations representing Russian timber interests and by acting as a clearinghouse for the transfer of manufacturing by-products from firms that generate them as waste to other members who use them as raw materials. In addition, WoodNet conducted training sessions on product pricing, promotion, distribution, market research, loan packaging, and financial management.

WoodNet's quarterly newsletter was packed with information on sales, joint manufacturing, marketing, raw material, financing, and equipment opportunities. Moreover, WoodNet published a member directory detailing many members' products, markets, raw material use, manufacturing equipment and capacities, and contact information.

WoodNet's most ambitious goal was to establish a wood products manufacturing technology center on the Olympic Peninsula. The center would have provided small, rural, wood products firms with previously unavailable access to advanced information, training, and production.

By working together and adding more value locally, wood workers on the peninsula have begun to address the region's need for jobs and the nation's need for wood products.

Epilogue: WoodNet closed its doors after five years in December 1995, unable to achieve financial self-sufficiency in its remote rural market. Its many accomplishments and unfortunate closure both attest to the difficulties facing sustainability initiatives in rural, resource-dependent communities along the rain forest coast.

Todd Thomas, formerly WoodNet's business development specialist, lives in Sequim, Washington.

Financing Small Business on the Olympic Peninsula

PATTY GROSSMAN

Seattle-based Cascadia Revolving Fund has become an unusual ally of economic development efforts on Washington's Olympic Peninsula. With money raised from socially and environmentally conscious investors, Cascadia provides loans and technical assistance to small businesses that in turn provide family-wage jobs.

Communities on the peninsula have depended for generations on the harvest and primary processing of timber and fish. Local sawmills provided family-wage jobs and an attractive lifestyle in a rural area. Stable housing and land values offered a safe haven for private investment and solid collateral value. Entrepreneurial activity was primarily concentrated in small private sawmills, logging companies, and service industries for the rural communities. But when unanticipated reductions in timber harvests from national forests resulted in mill closings and job losses in logging, trucking, and related businesses, this fabric of economic stability began to unravel.

Today fewer people work in the traditional natural resource industries, both on the peninsula and throughout the Northwest, but a "jobs versus the environment" fight bodes ill for both jobs and the environment. Innovative technologies and conservation-based industries are an essential part of creating a sustainable regional economy supported by healthy communities. Cascadia's support of environmentally sound enterprises is an important aspect of being a positive force in both the restoration of the Northwest environment and the creation of jobs. Cascadia has increased jobs and the quality of life for a variety of urban entrepreneurs and has extended its activities to small businesses in rural communities around the region.

Cascadia, a private nonprofit community development loan fund, is able to make higher-risk loans than banks and to provide the intensive support that new businesses require to succeed. Since Cascadia began lending in 1987, over 90 percent of its borrowers have remained in business after three years—compared with the 20 percent survival rate for new businesses nationwide.

The need to extend Cascadia's efforts to rural communities became critical as the economic crisis resulting from the loss of logging and primary manufacturing jobs intensified. Cascadia teamed up with the Olympic Peninsula wood products group WoodNet, the Clallam County Economic Development Council, and the Forks Economic Development Steering Committee to begin efforts to diversify the economic base and promote entrepreneurial activities. With support from Washington's Department of Community, Trade, and Economic Development, the partnership set to work. Its aims were to research the local business climate and the economy, to determine what types of business were likely to succeed in a setting of reduced timber availability, to contact potential borrowers, to work with businesses that receive loans, and to set up a network of technical assistance providers and volunteers to support businesses that became borrowers.

In the first three years, Cascadia disbursed over $1 million in loans to rural businesses. In fact, more than half of the fund's lending activity was rural-based. But Cascadia's officers became convinced that the availability of loans is not the only, or even the major, financial obstacle to the creation of new jobs in these areas. "We've taken debt places debt has never been before," says Cascadia president Rich Feldman. "It wasn't enough."

Cascadia discovered that promotion of new businesses in timber-dependent communities would require access not only to loans but to equity investment capital too. Traditional venture capital funds have been instrumental in starting new companies, primarily in electronics and biotechnology. Community development venture capital is a new and exciting application of venture-type investing.

Cascadia Revolving Fund proposes to establish and manage a rural development investment fund for the purpose of making equity investments in businesses for the timber communities of the region. Candidate businesses include new and existing value-added wood products manufacturers, value-added fish processors, manufacturers of products from recycled materials, and alternative agriculture. These businesses, often too new to borrow conventionally and too small to attract traditional venture capital investments, are the firms that can use existing skills in the woods. What is more, their products resolve the "jobs versus environment" dichotomy. The fund would complement Cascadia's lending activities and is an excellent way to establish a public/private partnership. Cascadia is seeking to match public risk capital with an equal or greater amount of private funds in order to use the power of private enterprise and entrepreneurship to help rural communities build sustainable economies.

Patty Grossman is the executive director of the Cascadia Revolving Loan Fund in Seattle, Washington.

□ □ □ □ □ □ □ □ □ □ □ □ □ □

ShoreTrust Trading Group

NANCY HAUTH

In the small fishing town of Ilwaco, Washington, a new nonprofit organization is redefining good business. ShoreTrust Trading Group is supporting small, ecologically sound businesses to reverse a declining economy. The approach that the ShoreTrust staff calls "conservation-based development" is helping set new goals for the Willapa watershed.

ShoreTrust is the result of a long-term development strategy created by Shorebank Corporation and Ecotrust. Shorebank Corporation is a bank in Chicago that invests in disadvantaged communities in the United States and overseas in support of community development. Ecotrust is a nonprofit organization that supports practical examples of conservation-based development in the rain forest communities of North America. For three years, Ecotrust and Shorebank met to develop a regional strategy aimed at invigorating business activity that would restore the environment in the Willapa region. The partners share a vision that ecological vitality is the basis for economic vitality—especially in natural-resource-dependent economies like the communities along Willapa Bay.

The Willapa watershed, which spans 680,000 acres, drains into one of the cleanest estuaries in the continental United States. This is a land of plenty: rich in cranberries, wild mushrooms, oysters, salmon, clams, crab, towering forests, and streams. Yet the human population of about 23,000 suffers higher than average unemployment, low wages, and out-migration of its young adults. Recent statistics show a 58 percent increase in the incidence of poverty in Pacific County (comprising most of the communities in the Willapa watershed) between 1989 and 1990, compared with an 11 percent increase for the rest of Washington State.

In 1992, Alana Probst, Ecotrust's director of economic development and later ShoreTrust's interim managing director, looked at a dozen small businesses in the region of southwestern Washington. She found rural coastal towns from Astoria, Oregon, to Grays Harbor, Washington, brimming with entrepreneurial energy and creativity.

Lacking, however, were resources, such as seed capital and marketing advice. Although banks in the Willapa area are energetic, flexible, and responsive to commercial lending opportunities, they are conservative in their lending to start-up businesses, product or market innovation, and natural resource sectors perceived to be in decline. Also, there was little information to be gathered about the growing "green market"—a $115 billion market driven by customers willing to pay a premium for products grown and manufactured responsibly in clean, lush environments like Willapa.

The Shorebank/Ecotrust team worked closely with the Willapa Alliance, a local group of farmers, oyster growers, fishermen, small business owners, and community leaders with interests in fisheries issues, education, and communication. Together Shorebank, Ecotrust, and the Alliance began to list a variety of services needed in the area to give local residents the business development tools that were lacking in remote coastal communities.

The Willapa Alliance appointed an economic development task force including bank executives, small business owners, and local leaders to review the ShoreTrust strategy and assist in the development of this new nonprofit organization. The task force helped customize the tools needed for Willapa Bay businesses to succeed. "Their involvement at every stage of the implementation process, from hiring to helping us get the word out to the community, has been critical to our success," says Probst.

Almost every day since ShoreTrust opened its doors in September 1994, someone has walked in with a product idea or a dream. What most lack is the capacity to develop and sell it, and that is where ShoreTrust comes in.

ShoreTrust's criteria for lending to small businesses differ from those of traditional banks. ShoreTrust's lending criteria are based on the principles of conservation-based development. "We're looking for financially stable businesses, of course, but we are swayed by businesses whose products add value to the region, enterprises that control and use industrial waste, and those that enhance the environment and community," says John Berdes, ShoreTrust's current managing director.

By offering business and marketing assistance and nonbank credit to environmentally responsible business, ShoreTrust creates self-interest in conservation within the business community. Yet ShoreTrust staff are determined that the lending services are not just for those few businesses already able to meet stringent environmental criteria. The staff is prepared for the long haul. "Conservation-based development takes a long time," says Ecotrust president Spencer Beebe. "Its constituency must be built house by house, farm by farm, oyster bed by oyster bed."

An example of conservation-based development is Willapa's oyster industry. Oysters are to the bay what canaries once were in the mine shaft. If there is any imbalance in the bay, oysters will show it—and oyster growers, in turn, will have to keep their products off the market. Some Willapa oyster growers have begun to use approved biocides on their oyster beds in the bay to control epidemic populations of burrowing shrimp that compete with oysters for food and lodging. Scientists suspect that overfishing the natural predators of shrimp—salmon and sturgeon—has contributed to the explosion of these native but noncommercial shellfish, though the question is difficult to study and results so far are inconclusive. Willapa remains one of the cleanest and most productive estuaries in the United States. Yet few market benefits accrue to Willapa seafood businesses because of this distinction.

Willapa products have never been differentiated in the marketplace from products grown and harvested elsewhere in worse environmental conditions. To strengthen the commitment of oystermen to the best environmental practices feasible in the Willapa estuary, the demand for Willapa products from quality and natural food stores and restaurants must be increased.

Three seafood businesses (oysters, clams, and crab/salmon) were selected to test-market products in natural foods supermarkets in Portland, Oregon. The executives of Nature's fresh Northwest! stores are committed to an urban/rural partnership to help match the demand from the green market, which recognizes the value of healthy, clean, and pure products, to entrepreneurs seeking solutions to environmental problems.

Nature's fresh Northwest! stores in Portland showcased Willapa seafood, telling the story of the bay and the growers with photographs and advertising. A special weekend event was held at which growers met Nature's customers. The interest shown by these customers suggested that the natural foods market on the West Coast could attract as much as 10 percent of the oysters and seafood from Willapa Bay. Growers agreed to donate 1 percent of their profits from this project to restoration activities around the bay.

This green marketing test run was successful: Willapa oyster growers directly benefited from the connection between environmental stewardship and economic returns. Yet the project also brought up contentious issues on water management that have, to some degree, divided Willapa Bay communities. The invasive, nonnative cordgrass spartina, for instance, is rapidly filling parts of the bay and turning gently sloping mudflats into elevated salt marsh. If spartina's spread continues unchecked, three-quarters of the tidelands—and the oysters they support—are at risk. Add to that the challenge of controlling burrowing shrimp and the Willapa story is a complex one, possibly too complex to communicate easily to the average green consumer.

The difficulty in marketing is to develop the story in a way that engages green market consumers while honestly disclosing the issues that local oyster growers face. ShoreTrust's philosophy is far from purist, yet the oyster growers' position on controlling pests—both native and exotic—must be thoughtful and carefully managed if they hope to hold the commitment of the green market. Is it ShoreTrust's role to monitor the balancing act between the health of the bay and the health of the business? How much information do consumers expect before buying their oysters or other products? How should a long-term marketing plan address these issues?

These are questions ShoreTrust continues to confront as it encourages Willapa suppliers to grow and harvest their products responsibly while they enter the lucrative green market.

Epilogue: In late 1995, the strategy progressed to a new phase, the development of a commercial bank. From headquarters in Southwest Washington, ShoreTrust Bank will make commercial loans to environmentally responsible businesses in targeted communities of the coastal temperate rain forest.

Nancy Hauth, former coordinator of the economic development team at Ecotrust, now runs a small business with her husband Dennis in Portland, Oregon.

□ □ □ □ □ □ □ □ □ □ □ □ □ □

Rural Development Initiatives

JENNIFER PRATT

In the 1980s, Oregon's traditional social and economic fabric began to rip. A series of technological, biological, and philosophical shifts fell into alignment and wrought tremendous changes, particularly on the rural populations of the state. Oregon faced crises in the timber industry, declining salmon runs, and restrictions on the use of public grazing lands.

The state had to respond or run the risk of sliding into the kind of structural poverty and economic dislocation associated with the decline of the coal economy in Appalachia. In 1989, the Oregon Economic Development Department established Community Initiatives to help rural communities cope with these changes. Oregon's rural communities needed to diversify their economies to avoid repeating the mistake of single-industry dependency, but they needed to do so in a manner that would preserve their high quality of life. After two years of trial-and-error effort with Communities Initiatives, a nonprofit organization called Rural Development Initiatives (RDI) was founded with a $500,000 annual budget financed by state lottery dollars.

The state's Community Initiatives effort had initially tried to address specific economic problems, such as plant closures. This project-by-project response quickly proved to be too narrowly focused: the program needed to address the wider economy to understand sustainable development in all its complex dimensions. So a different approach took shape. This new approach was based on four building blocks of a healthy, well-developed community: workforce development, business development, infrastructure, and quality of life.

Although created with a very specific objective, RDI has evolved and continues to incorporate learning into its program activities. Central to the organization's approach from the beginning, however, has been the notion that development is sustainable only if it is informed and if it is community-driven. RDI's efforts to organize community coalitions who will work together to achieve common goals have expanded over the years. The breadth of representation among community teams is now much greater. Equally important is the effort to gather information, which has become more rigorous and comprehensive. To ensure that decision making is grounded in analysis, RDI undertakes an assessment of each community's strengths, weaknesses, opportunities, and threats. The results are then presented to the community in a public forum. This analysis, like all efforts taken during the planning process, relates explicitly to the "building blocks."

RDI strives for flexibility in community planning while remaining focused on the objectives of a well-thought-out strategic plan and the challenge of creating broad-based and effective local development organizations. Understanding that every community is unique, RDI simply offers a framework upon which communities can create plans. RDI facilitates community efforts to focus on development for the long term using achievable short-term objectives. The goals, strategies, and action items that result are as individualized as the people involved.

Through trial and error, RDI has learned that a plan is only as effective as the community members who will implement it. Not all community members possess the skills needed to initiate and manage meaningful community and economic development activities at the outset of the process. Thus RDI now emphasizes skill development among community team members. Among these skills are building partnerships, locating and using financial resources, engaging new players, and building long-term commitment to the community's future.

RDI sees Reedsport as one of its successes. Within the last decade, Reedsport has twice found itself facing the unimaginable: first its timber industry, then its salmon industry, faced severe decline. Reedsport began taking steps to diversify its economic base following the timber decline, and it worked even harder after the curtailment of salmon fishing. With RDI's assistance, the community developed a long-term economic development strategy. Building on its waterfront and connection to the sea, Reedsport has made tangible progress in diversifying its economy and stabilizing its community. Located along its scenic waterfront, run predominantly by community volunteers, and funded by a broad cross section of public and private partners, Reedsport's Umpqua Discovery Center is considered a success. The center has only been open since 1994, but already it has drawn enough visitor traffic to spark investment in the town's previously neglected but historic Old Town. Along with the increase in foot traffic and community activities, several waterfront restaurants are now open and boat rides are available.

Originally conceived as a business development strategy to increase tourism, the Discovery Center offers an educational experience as well. The center's interpretive displays introduce sea life, the Umpqua estuary, and surrounding sand dunes. The retired antarctic research vessel *Hero* is a further educational attraction. The city is applying for a National Science Foundation grant to further develop the center's antarctic displays and this floating classroom. Various other enterprises have opened their doors nearby.

Many lessons have been learned over the last five years. Some of these lessons have to do with the communities themselves. Others have more to do with RDI. The essential lessons about RDI are:

- Corporate support is harder to raise than initially expected.
- We are fortunate that RDI's executive director came from the state's Economic Development Department and already had a great many contacts and an established reputation. Without this advantage, it is doubtful whether RDI could have gained such legitimacy across the state and its access to state institutions in such a short time.
- External pressures sometimes force rapid organizational growth. Growth, however, is not necessarily compatible with the effort to refine work processes. RDI is working to integrate its commitment to experimentation.
- Standards are essential—you cannot afford to invent every step from scratch—but flexibility is important too.

The essential lessons about community are:

- Communities have to experience *need* before this kind of effort will work. It appears to be nearly impossible to organize communities beforehand.

- As initial pressures are released through the planning process and communities begin to organize and take control of their futures, their needs change.
- Communities need quick, tangible evidence of success. There is nothing like planning and more planning to guarantee burnout.
- Whether locally based or available through an outside organization, communities need full-time staff to do the work of long-term development. Part-time volunteers cannot do it alone.
- Trust and strong social networks are indispensable.
- Sustainable development takes time—there are no quick fixes.

Jennifer Pratt is RDI's community development coordinator for the mid-Willamette Valley, based in Lowell, Oregon.

14. From Ecosystem Dynamics to Ecosystem Management

KEN LERTZMAN, TOM SPIES, AND FRED SWANSON

Understanding how natural disturbance regimes shape forest ecosystems has become a key element of new approaches to forest management. Indeed, many of the conservation and management problems in the temperate rain forests of western North America relate directly to the differences between the management regimes we have imposed on the forests and the natural disturbance regimes that dominated the forests before the arrival of European settlers and industrial development. A variety of recent initiatives in forest policy in both the United States and Canada have emphasized the processes and structural consequences of natural disturbance as models for management (Swanson et al. 1993; Scientific Panel for Sustainable Forest Practices in Clayoquot Sound 1995a). Implementing this approach, however, is limited by our understanding of natural disturbance regimes, by our ability to make direct management recommendations from what we do know, and by constraints imposed by the social context in which forest management takes place.

The current focus on using natural disturbance patterns as models for management in temperate rain forests is part of a much broader shift toward ecosystem-based management regimes in both Canada and the United States. "Ecosystem management" includes more elements of landscapes than forests, however, and many more considerations than natural disturbance regimes. Key dimensions of ecosystem management include the integration of social and ecological data and values and the application of the principles of adaptive management. Nonetheless, understanding the *processes* responsible for shaping temperate rain forests and maintaining their biodiversity in the past is fundamental to successful ecosystem management.

Ecosystem Dynamics

Over the past few decades, the disturbance regimes to which forests are subjected have emerged as key features distinguishing different types of forests around the world. A forest's natural disturbance regime can be defined as the long-term pattern of the frequency, intensity, spatial extent, and heterogeneity of disturbances (Pickett and White 1985). Some forests, such as those of most boreal regions, are characterized by relatively frequent large-scale disturbances that destroy forest stands, resulting in a coarse-grained mosaic in which relatively young, even-aged stands are distributed in large patches. Other forests experience large-scale disturbances only infrequently, resulting in a finer-grained mosaic of older, often uneven-aged, forest patches. Temperate rain forests generally follow this pattern. When major disturbances are separated by long intervals, the forest's structure, the kinds of habitats it provides, and the suite of animals living in it are all affected (Figure 14.1) Temperate rain forests are especially distinguished by their tremendous accumulations of biomass, by the great longevity of canopy dominants (Table 14.1), and by the widespread distribution of late-successional (old-growth) elements over forest landscapes.

A number of general conclusions about the dynamics of temperate rain forests have been drawn from recent research (Table 14.2). Most of these ideas arise from research on a variety of forest types well distributed over the temperate rain forest region and, at least qualitatively, should apply to all forest

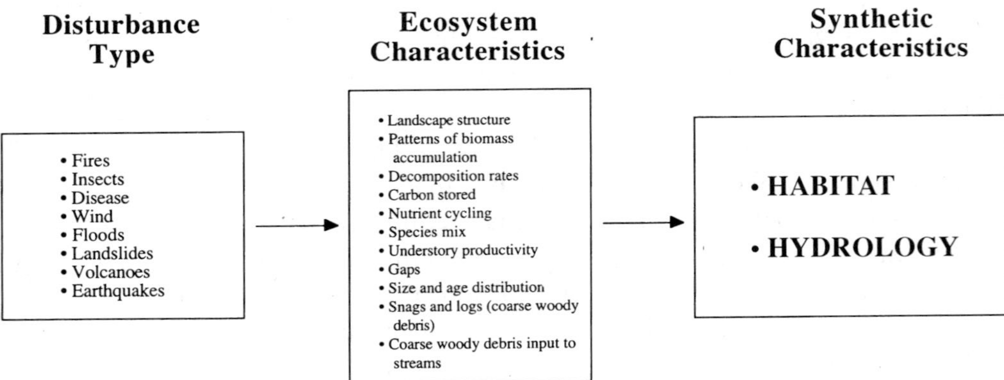

Figure 14.1. The influence of disturbance on temperate rain forests. The forest's hydrological character and its ability to provide habitat are synthetic variables that arise from the combined effects of a number of specific attributes.

Table 14.1. Characteristics of Northwest Trees

	Typical (of productive very old forests)			Maximum		
Species	Age(years)	Diameter(cm)	Height(m)	Age(years)	Diameter(cm)	Height(m)
Pacific silver fir (*Abies amabilis*)	>400	90–110	45–55	750	237	72
grand fir (*Abies grandis*)	>400	90–150	45–60	>500?	202	81
noble fir (*Abies procera*)	>400	100–150	45–70	>500	275	90
Port-Orford cedar (*Chamaecyparis lawsoniana*)	>500	120–180	60	—	359	—
yellow-cedar (*Chamaecyparis nootkatensis*)	>1000	100–150	30–40	1824 (>2000?)	365	62
western redcedar (*Thuja plicata*)	>1000	150–300	40–50	1400 (>2000?)	631	71
Sitka spruce (*Picea sitchensis*)	>500	180–230	60–75	1350	525	95
Douglas-fir (*Pseudotsuga menziesii*)	>750	150–220	70–80	1300	440	100
coast redwood (*Sequoia sempervirens*)	>1250	150–380	75–100	2200	501	—
western hemlock (*Tsuga heterophylla*)	>500	90–120	50–65	1238	275	75
mountain hemlock(*Tsuga mertensiana*)	>500	75–100	35+	>1000	221	59

Sources: Modified from Waring and Franklin (1979) and Pojar and MacKinnon (1994).

Table 14.2. What Key Lessons Have We Learned About Ecosystem Dynamics in Coastal Temperate Rain Forests?

- Disturbances have had major impacts on ecosystem dynamics, and disturbances play varying roles in forests.
- Disturbances vary in their importance at several spatial scales.
- General patterns of structural development have emerged, but there are diverse pathways of ecosystem change in response to disturbances.
- Ecosystem structure is strongly linked to biodiversity at multiple spatial scales.
- Late-successional elements are significant in temperate rain forest landscapes.
- Terrestrial and aquatic ecosystems are strongly interconnected.
- Considering large scales of time and space is critical.
- Intensive plantation forestry does not maintain late-successional elements of ecosystems.

ecosystems in the region. Logs and snags, for instance, are important habitat elements in all forest types, irrespective of variations in tree species, soils, climate, or natural disturbance regimes. Some forest types are less well studied than others, however, and much more research is required before quantitative statements can be made about many of the ideas discussed here. Several of these conclusions (especially the first four in Table 14.2) apply not only to temperate rain forests but to a great variety of other ecosystems.

Ecosystem Management

The emerging cluster of concepts known as ecosystem management carries with it a gestalt of holism rather than reductionism, a subordination of human desires to ecosystem health, and recognition of a broader range of values in ecosystems than past practices have acknowledged (Grumbine 1994). The goal of ecosystem management is to manage for the long-term integrity of whole ecosystems, not for the production of single resources. This goal is easier to state than to implement, of course, and is tied to a broad range of social and institutional issues. As nations have begun to articulate commitments to "sustainable development" (Scientific Panel for Sustainable Forest Practices in Clayoquot Sound 1995b), ecosystem management provides scientific, social, and institutional concepts that set a context for thinking more broadly about sustainability in land use planning and management. A hallmark of current thinking about ecosystem management is that, much more than past approaches, it recognizes that people and their values are part of the system to be managed.

The application of ecosystem management is much broader than just forest management or conservation. Indeed, it incorporates many ideas other than

Table 14.3. Dominant Themes of Ecosystem Management

- *Hierarchical context:* Ecosystem management requires consideration of all levels in the biodiversity hierarchy—genes, species, populations, ecosystems, landscapes—and managers must seek connections between all levels.
- *Ecological boundaries:* Ecosystem management requires working across administrative and political boundaries and defining ecological boundaries at appropriate scales.
- *Ecological integrity:* Managing for ecological integrity means protecting all elements of native diversity and the ecological processes and patterns that maintain it.
- *Data collection:* Ecosystem management requires more information about natural systems and better use of existing information in management.
- *Monitoring:* The consequences of decisions and actions must be tracked in order to evaluate success or failure quantitatively.
- *Adaptive management:* Ecosystem management recognizes our uncertainty about the dynamics of natural systems and acknowledges that our management actions are experiments that must be designed, monitored, and used to change future management.
- *Interagency cooperation:* No single agency or interest group has a lead role in management. Managers and others must work together to integrate conflicting mandates and management goals.
- *Organizational change:* Making ecosystem management a reality will require diverse changes in institutional structure and behavior that range from minor to fundamental.
- *Humans as part of nature:* People and their actions cannot be separated from nature. Their mutual influences on each other must be recognized.
- *The importance of values:* Regardless of scientific knowledge, human values play a fundamental role in determining our goals for managing ecosystems.

Source: Modified from Grumbine (1994).

ecosystem dynamics (Grumbine 1994) (Table 14.3). Despite the eclecticism, ecosystem management has become the central concept around which new approaches to forest management and conservation are being organized. (See FEMAT 1993; Scientific Panel for Sustainable Forest Practices in Clayoquot Sound 1994, 1995a.) Though visions of ecosystem management vary substantially (Franklin 1993b; Swanson et al. 1993; Carpenter 1995; Malone 1995), the primacy of maintaining ecological integrity is a consistent component. There is also general agreement on some of the requirements of "ecological integrity" (Table 14.4). Understanding ecosystem dynamics and incorporating that understanding in management is essential for meeting these requirements for ecological integrity in coastal temperate rain forests. Managing forest ecosystems based on their natural dynamics means considering not just silvicultural or harvesting systems but landscape planning and design, rates and types of disturbance to the hydrological system, and the system of riparian and other

Table 14.4. Ecological Integrity as a Goal of Ecosystem Management

Objectives Within the Overall Goal of Sustaining Ecological Integrity
• Maintain viable populations of all native species in situ.
• Represent, within a system of protected areas, all native ecosystem types across their natural range of variation.
• Maintain evolutionary and ecological processes (disturbance regimes, hydrological processes, nutrient cycles, and the like).
• Manage over periods of time long enough to maintain the evolutionary potential of species and ecosystems.
• Accommodate human use and occupancy within these constraints.

reserves to be maintained within a planning unit. We have indentified, and will discuss, six key ideas about natural disturbance regimes and forest dynamics that set the context for ecosystem management.

Ecosystem dynamics and patterns that maintained biological diversity and ecological function in the past are our best model for doing so in the future. We have inherited forest landscapes produced not by centuries of stability but by a long and variable history of ecosystem change. These same landscapes provide ecological services and habitats for diverse communities of plants and animals that vary over time in any given location and vary geographically within the region. The life forms present today have obviously been able to survive the range of past variation in environmental conditions. We tend to take ecological services for granted because they have been present throughout the history of our interaction with coastal rain forests. The more forests diverge from their historical range of ecosystem states, the less certain we can be that they will continue to provide the habitats and services they have conferred in the past. In theory, ecosystem management should maintain forests within the range of variability they have experienced over the preceding centuries and even millennia. For landscapes and ecosystems that have already been substantially modified by human activities, however, this goal may no longer be attainable at large spatial scales. Our knowledge of the dynamics of temperate rain forest ecosystems is so limited that a precautionary approach to management is imperative. This approach should have two components: an effort to emulate conditions that we know did not compromise biodiversity or ecological function in the past, and a commitment to adapt our practices as we learn more about the behavior of temperate rain forest ecosystems.

We already know enough about natural disturbance regimes to design silvicultural disturbance practices that incorporate key aspects of natural disturbance better than past practices have done. This is especially true at the forest stand level. Many researchers have proposed forest practices informed by natural ecosystem dynamics. (See Franklin et al. 1986; Hansen et al. 1991; Hopwood 1991; Swanson and Franklin 1992; Franklin 1993a, 1993b; McComb et al. 1993.) These proposals, however, have mostly been based on research in the Douglas-fir forests of the southern part of the bioregion. Although the factors of concern are generally similar across the coastal rain forest region—planning for biological legacies, for example, and maintaining the integrity of riparian networks—differences in management are indicated by regional gradients in disturbance regimes (Figure 14.2).

Three caveats are important. First, while these new approaches to forestry are based on a substantial body of research and are being tested widely through the temperate rain forest region, they remain highly experimental and have not been practiced long enough to demonstrate that they will indeed maintain biodiversity. Second, given our rudimentary knowledge of linkages between forest dynamics and biological diversity, formal and informal reserves are a key component of an overall conservation strategy. Third, natural processes, both known and unknown, should be allowed to operate in such reserves—so they can serve both as refugia for life forms that do not find suitable habitat in the more heavily managed matrix and as models for management in matrix landscapes.

The idea that new approaches to silviculture based on natural disturbance will introduce and maintain late-successional elements of biological diversity in managed forests remains a largely untested hypothesis. Novel silvicultural practices complement, rather than replace, the need for reserves of a variety of sizes. Unfortunately, though most proposals for maintaining biological diversity in managed landscapes emphasize the interdependence of stand-level and landscape-level level strategies (Franklin 1993c), the background science and detailed management recommendations for new stand-level approaches are substantially better developed than our ability to understand and manage at the landscape scale.

At the stand level, forest harvesting occurs on a gradient of the removal or retention of trees: clearcuts are at one end of the gradient and nearly undisturbed forest is at the other (Scientific Panel for Sustainable Forest Practices in Clayoquot Sound 1995a). The greater the proportion of trees removed, the more other components of the forest (such as understory vegetation and forest floor biota) will be disturbed. Silvicultural prescriptions for Douglas-fir forests with a history of stand-initiating disturbances emphasize retaining certain trees and patches of forest as biological legacies (McComb et al. 1993). Such prescriptions have generally proposed that a fairly low level of retention (perhaps 5 to 15 percent) is sufficient to produce structurally complex stands made up of older, multiaged remnants in an even-aged matrix. As they develop,

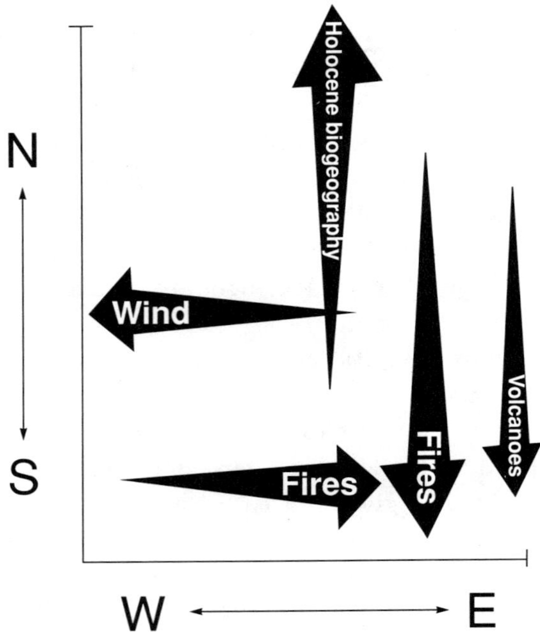

Figure 14.2. The variation of disturbance influences on North American coastal temperate rain forests along gradients of latitude and longitude. Broadly influential, but expressed differently over gradients, are Pleistocene glaciation, erosion, landslides, floods, and stream channel geomorphology.

stands of this type will resemble unmanaged old-growth forest more than intensively managed plantations. Recent research comparing such retention harvests with traditional clearcutting supports this expectation. Retained trees and forest patches—and the structurally complex forests they will become—are likely to retain a greater variety of mid- to late-successional species of various taxa than are clearcut areas (Hansen et al. 1995; Schowalter 1995).

In wetter types of forest, a higher level of retention is more likely to be appropriate. After studying patterns of forest structure in wet temperate rain forest on the west coast of Vancouver Island, Lertzman et al. (1996) suggested a management regime that would maintain late-successional character in managed stands by creating small gaps within a matrix of old forest. This approach should better maintain conditions similar to the continuous, uneven-aged, late-successional forest characteristic of wetter types of coastal temperate rain forest and, moreover, would provide settings for regeneration similar to those responsible for many of the current canopy trees.

The proposal by the Scientific Panel for Sustainable Forest Practices in

Clayoquot Sound (1995a) to implement a "variable retention" silvicultural system can incorporate both the lower levels of retention appropriate in the southern part of the coastal temperate rain forest region and the higher levels of retention that may be required in the northern parts of the region. Similarly, McComb et al. (1993) describe an approach to developing desired future conditions for managed stands based on stand-initiating disturbances that produce a one-story canopy with old-growth remnants, stand-maintaining disturbances that produce a multistoried, multiaged stand with patchy gaps, and disturbances that are intermediate in intensity, producing stands with two to three distinct age classes.

A truly ecosystem-based management regime will require detailed information on forest structure and disturbance history at several scales. It will also require a substantial commitment to monitoring in order to determine if its objectives are being met (FEMAT 1993; Scientific Panel for Sustainable Forest Practices in Clayoquot Sound 1995a). In designing silvicultural practices that incorporate natural ecosystem dynamics, ecosystem-specific information on the following variables should be considered:

- The range of disturbances and causes of tree mortality (windthrow, fire, physical damage)
- The range of intensitiy of disturbances
- The range of sizes and shapes of disturbed patches
- The internal heterogeneity of disturbed patches
- The spatial relationships of disturbed patches to one another
- The temporal frequency of disturbances
- The spatial variability in disturbance intensity and frequency

Knowledge of historical dynamics is a fundamental aspect of implementing an ecosystem dynamics approach to management. Ecosystem-based silvicultural disturbance regimes cannot be designed without understanding the natural disturbance regime for the ecosystems in question. While certain ideas are likely to apply to most ecosystem types—such as the importance of biological legacies at all scales of analysis—even nearby watersheds experiencing similar climatic conditions may exhibit substantial variation in disturbance history. For instance, Morrison and Swanson (1990) concluded that topography was important in how the natural disturbance regime was expressed in the two watersheds they compared. Local information in such detail is unavailable, however, for much of the coastal temperate rain forest region.

Managers should not refrain from applying ideas about ecosystem dynamics to the design of management regimes simply because their information about local disturbance histories is limited. The art of management consists of combining the known, however incomplete, with professional judgment to achieve management goals. This requires a willingness to proceed from general principles to specific practices in the absence of hard rules—a willingness that is rare among managers in both the United States and Canada. Nevertheless, acting with incomplete information, while acknowledging its limitations, will be an essential component of professionalism in future forest management. A challenge to ecosystem management is to strike a balance between giving managers the freedom to adapt practices to complex ecosystems and creating policies that provide for basic ecological protection.

We know enough to begin structuring management around various types of natural disturbance regimes. The *Biodiversity Guidebook* of the Forest Practices Code of British Columbia provides a good example (British Columbia Ministry of Forests 1995). In this guidebook, management recommendations for maintaining biological diversity are stratified by the "natural disturbance type" (NDT) of the forest being considered: NDT 1 refers to ecosystems in which stand-initiating events are rare; NDT 2 refers to ecosystems in which stand-initiating events are infrequent; NDT 3 refers to ecosystems in which stand-initiating events are frequent; NDT 4 refers to ecosystems in which stands are maintained by frequent fires. For instance, the distribution of seral stages to be maintained in a landscape planning unit varies from more emphasis on early seral stages in NDT 3 to more emphasis on mid to later seral stages in NDT 1. This framework is a significant step toward a more ecologically based approach to management. Even under such guidelines, of course, both the stand-level and landscape-level characteristics of the forest are substantially modified from their natural state. Natural disturbances may kill trees, but they do not remove the bodies—which is, after all, a major objective of forestry.

The approach we describe here is intended to maintain key ecological processes in managed forests, not to maintain completely unmodified ecosystems. A forest managed for timber production will always differ in significant ways from a forest subject solely to natural disturbances. Our objective is not merely to "mimic" natural disturbances, but to incorporate the attributes of natural disturbance that allowed species and ecological processes to persist through or recover from disturbances in the past. Managers must always be aware of the degree to which they are generalizing from other ecosystems. The problem of limited local information on disturbance histories is exacerbated by the substantial variation in disturbance regimes over the coastal temperate rain forest region. Disturbance ecologists, managers, and landscape modelers must work together to design and implement management regimes that reflect both local ecological conditions and general concepts of forest dynamics.

A focus on ecosystem dynamics forces management to consider not only ecosystem states but trajectories. One important consequence of the focus on ecosystem dynamics is its shift in emphasis from the current state of a forest to the trajectory the forest will follow over the course of time. Because forests change so slowly compared with human time scales, it is easy to ignore the substantial changes they undergo over several centuries. Placing forests in their historical context allows us to better assess their current state and encourages us to look beyond current conditions to projected future trends. Consider, for example, an area where a stand of old-growth forest was recently cut but some snags were retained. These snags will provide habitat for cavity-nesting birds over the short to medium term. But if no large living trees were left and the stand is scheduled to be cut again in eighty years or less, it will be on a trajectory of declining habitat for species that depend on large snags. Once the initial cohort of snags has decayed, they will not be replaced. It is much more important to plan for the structural characteristics that will develop over the full rotation (and more) than it is to produce some desired condition immediately after harvest.

The dispersion of logged patches in space is a good example of the difference between states and trajectories at the landscape scale. Distributing logging in small dispersed clearcuts has been a common response to concerns about the cumulative effects of logging on both sides of the international border. This policy has created the checkerboard patterns now so familiar on federal lands in the U.S. Pacific Northwest (Spies et al. 1994). Dispersed clearcuts have often been implemented, however, without any change in the overall rate at which the forest is cut. Yet the consequences of an excessive rate of logging in a watershed will be the same in the long term, whether the cut is dispersed in many small patches or in a few large ones, and may be worse in the short term under a dispersed cut scenario because more roads must be built and the remaining forest is fragmented more rapidly. Eventually, cut patches will coalesce, residual forest patches will be small, and continuity among forest patches will be lost (Franklin and Forman 1987; Spies et al. 1994). The trend (historically in the U.S. Pacific Northwest and currently in British Columbia) to disperse cut patches masks these consequences, but only temporarily. In general, forest landscape planning has been plagued by too little analysis of the long-term and large-scale consequences of planning rules conceived with small-scale, short-term variables in mind.

Natural states and processes are sometimes undesirable. Should we emulate all naturally occurring processes or patterns? No. This is both a scientific question and a matter of social values. Natural events, such as large landslides, can be extreme; life forms and ecosystems are unlikely to have well-developed adaptive responses to events that occur only rarely in their evolutionary or ecological histories. Moreover, the extremes of natural disturbance intensity, frequency, or spatial extent will also often have undesirable social or ecological consequences—disruption to hydrological regimes with resulting impacts on fish stocks, for example, or private property put at risk by large wildfires. The effects of natural and human-induced disturbances are cumulative, and many components of temperate rain forest ecosystems are already stressed by human activity (for instance, by overfishing of anadromous fish). The current state of such systems constrains future trajectories that we may consider acceptable. In general, desired states or trajectories for management should be bounded by the range of conditions resulting from natural disturbance regimes, but they need not represent all the states or trajectories possible under those disturbance regimes (Figure 14.2).

A focus on ecosystem dynamics leads to a consideration of whole landscapes. With the shift in focus from managing timber to managing whole ecosystems comes an explicit need to assess and plan for whole landscapes. The emphasis on planning for those parts of the landscape where logging does not occur is a keystone of ecosystem management in both British Columbia (Scientific Panel for Sustainable Forest Practices in Clayoquot Sound 1995a) and the northwestern United States (FEMAT 1993). In both countries, this focus has led to increased emphasis on the integrity of riparian networks and the importance of riparian ecosystems for both aquatic and terrestrial life forms (Chapters 5 and 6).

Landscape elements—riparian zones, mid-slope forests destined for harvest, late-successional reserves—cannot be treated independently. These elements are functionally linked by numerous physical and ecological processes (Swanson et al. 1988). Conservation, once primarily concerned with reserves, now must address the whole landscape. From a conservation standpoint, a landscape can be thought of as reserved areas plus a surrounding matrix of managed forest. The matrix plays at least three key roles in conserving biological diversity (Franklin 1993c; FEMAT 1993):

- The matrix can provide habitat elements (such as logs and snags) well distributed in space.
- Management of the matrix can increase the effectiveness of reserved areas.

- The matrix controls landscape connectivity, which influences such processes as the movement of animals between reserves.

Conservation biology initially treated reserves as islands in a sea of inhospitable terrain, functionally the equivalent of oceanic islands (Simberloff 1988). From this perspective, concerns about continuity across the landscape take on an almost artificial air—like building bridges between islands that were never linked ecologically (Simberloff et al. 1992). But in temperate rain forests, habitat islands are remnants of what was once more-or-less continuous forest. Matrix management in such landscapes should emphasize maintaining significant elements of continuity by reducing the contrast between the reserves and the surrounding matrix, or "softening the matrix" (Franklin 1993c). This approach has tremendous appeal and is a cornerstone of ecosystem management on federal lands in the U.S. Pacific Northwest (FEMAT 1993). The trade-offs between investing in matrix management versus reserves, however, remain largely unexamined and need substantial research.

Ecosystem management provides a context for ecosystem-focused as well as species-focused conservation. Traditional approaches to conservation have focused on species or populations of particular interest. Four key problems with the species-based approach have emerged:

- Species cannot be maintained in situ without their habitat or the ecosystems that provide it.
- Species-specific plans are too expensive, time-consuming, and labor-intensive to implement for more than a very small fraction of the species known to inhabit temperate rain forests.
- The vast majority of species in temperate rain forests are little known, as are their ecological relationships.
- Because many species have conflicting needs, a management regime designed for one species is likely to have negative impacts on others.

If our objective is to conserve biological diversity, adopting a conservation strategy that places more emphasis on ecosystems and landscapes is the only feasible approach (Franklin 1993c). The *Biodiversity Guidebook* of the British Columbia Forest Practices Code (BC Ministry of Forests 1995) is predicated on this idea: one of its key elements is the delineation of "forest ecosystem networks" (FENs). FENs contain specified distributions of seral stages of forest that represent the distribution of ecosystems in an unmanaged landscape and maintain some of the connectivity inherent in the landscape before logging. This idea is, in general, consistent with the approach we describe here, though the quantitative prescriptions for the amount of area in FENs have more to do with policy objectives than with disturbance ecology. FENs are intended to provide habitat or refugia for the majority of species that require natural forest conditions—without adopting species-focused management plans. Species with special requirements beyond what FENs provide are treated separately, not as part of the strategy for biological diversity in general. Although society will continue to demand special treatment for some species of special concern, such attention will continue to be limited to a very small portion of the temperate rain forest biota. Such an approach is unsustainable in isolation from a broader conservation strategy.

Key Problems

We have learned a tremendous amount about the dynamics of temperate rain forests and are beginning to apply that knowledge to management and conservation. Predictably, our understanding highlights a number of unresolved issues. We know enough to identify problems, and sometimes enough to propose solutions, but rarely can we assess the long-term consequences of specific management actions.

Ecological dogmas and assumptions, including those of ecosystem management, need continous testing and revision. But much of this testing can be done through management. Forest management in temperate rain forests has always been a large-scale experiment. We now have to design the experiment so that it tests our hypotheses more effectively. In the near future, new tools will exert a strong influence on research on ecosystem dynamics and on the management prescriptions that evolve from that research. By combining remotely sensed data, geographic information systems (GIS), and spatially explicit simulation models, we can address fundamental questions at larger scales than ever before.

We expect ten problems in temperate rain forest ecology and management to dominate the research agenda over the next decade (Table 14.5). As with the lessons outlined earlier in Table 14.2, many of these problems are not restricted to temperate rain forests, or even to forests in general.

Key Linkages Between Ecosystem Structure, Biodiversity, and Ecosystem Function

We understand well the biology of habitat dependence for a few significant temperate rain forest species, but even basic habitat relationships remain poorly described for many groups. Furthermore, we rarely know the extent to which the few well-known species are representative of a fauna or flora as a whole. Issues such as functional redundancy (Walker 1995) or keystone taxa and processes (Willson and Halupka 1995) in temperate rain forests remain

Table 14.5. What Are the Outstanding Problems in Ecosystem Dynamics and Ecosystem Management in Temperate Rain Forests?

- Key linkages between ecosystem structure, biodiversity, and ecosystem function
- Understanding the structure and dynamics of late-successional ecosystems
- Historical contingency
- Structure and dynamics of large ecosystems
- Active management for conservation objectives
- Whole landscapes = reserves + matrix
- Adaptive management
- Dynamics of recovery in dysfunctional landscapes
- Natural dynamics + anthropogenic changes + global change = ?
- Ecosystem management: more than science

largely unexamined. Although we know that the structure of forest ecosystems relates to their ability to provide habitat and maintain various ecological processes, we are far from fully understanding the nature of this relationship.

Understanding Late-Successional Ecosystems

Late-successional (old-growth) ecosystems have received a lot of attention, but much of it has focused narrowly on individual species, such as the spotted owl, or on policy-motivated efforts such as ecosystem-specific definitions of "old growth." We have just begun to understand the dynamics *within* late-successional ecosystems. Forests are routinely labeled old-growth whether they are 200 or 1000 years old, yet we know that substantial structural and compositional change occurs between these ages. Relatively little is known about changes in soil biology that occur late in succession, for instance, or about changes in the forest canopy structure and canopy biodiversity. Some general ideas have been proposed about how the dynamics of late-successional forests vary across landscapes and over the temperate rain forest region as a whole, but the details of this variation have not been described. The role of late-successional remnants in the recovery of disturbed landscapes is a problem of obvious importance, but one about which we know almost nothing.

Historical Contingency

The analysis of natural disturbance regimes is generally limited to a description of what happened in a particular place during a particular period. Many aspects of natural disturbance processes are stochastic; that is, they have probabilistic components, such as the coincidence of particular wind patterns with a lightning storm. Many patterns of events *could* have happened; the landscape we see, however, is contingent on a particular history. The extent to which alternative histories were possible remains largely unexplored. For instance, a single, intense stand-initiating fire may shift stands from one trajectory of structural development or another. How contingent are our conclusions on a particular history of prior events? Our ability to answer this question is limited. Yet the ability to answer such questions is important in our quest to design landscape management policies: we want those policies to reflect fundamental aspects of ecological process, not historical accidents. These problems of historical contingency cannot be resolved without models of landscape dynamics more refined than the ones we currently have.

Structure and Dynamics of Large Ecosystems

Research on patterns and processes at scales from landscapes to regions is becoming more common (Spies et al. 1994; Wallin et al. 1994), but the science of large ecosystems remains in its infancy. Although the need to characterize, measure, and predict the cumulative ecosystem consequences of different land use practices is widely acknowledged, we have done so in only a few cases. We are not yet able to say how management regimes interact with other ecological processes to determine large-scale patterns of dynamics.

Future research in both managed and unmanaged ecosystems must focus on processes that integrate ecosystems across large scales. The large, relatively pristine ecosystems still found in some parts of the coastal temperate rain forest region therefore represent an internationally significant scientific resource (Ecotrust et al. 1995). They represent, for instance, some of the few undisturbed river systems in the world where it is still possible to study interactions between the drainage network and the surrounding forests.

Active Management for Conservation Objectives

Traditional approaches to ecosystem conservation have, for the most part, been passive. In large landscapes little modified by human activity, merely refraining from intervention is sometimes sufficient to conserve species or features of interest. But such landscapes are rare. In many cases, active manipulation may be necessary to push ecosystem change in desired directions. Active management may be particularly necessary where past human intervention has caused ecosystems to follow a trajectory substantially different than they would have in the absence of human activity. Circumstances likely to require active intervention include the reintroduction of fire to fire-dependent forests, the restoration of degraded riparian zones, and vegetation control in ecological reserves that have been invaded by exotic weeds. This approach demands caution because, in the past, such interventions have caused problems as often as they have solved them.

Whole Landscapes = Reserves + Matrix

The idea of whole landscapes or watersheds as units for study or planning is now a central focus for both research and management (FEMAT 1993; Scientific Panel for Sustainable Forest Practices in Clayoquot Sound 1995a). At present, however, we have little basis for quantitative projections of the consequences of alternative management scenarios. Evaluating the consequences of varying degrees of emphasis on reserves versus matrix management is a good example. Ecosystem-based landscape planning at present relies on a set of working hypotheses that, by necessity, will be tested in their application. We will learn the most from these applications if they are formally designed as hypothesis tests in an adaptive management approach. Taylor (1995) shows how experimental design can be incorporated into the planning of FENs in the southern interior of British Columbia.

Adaptive Management

Adaptive management is the implementation of policy or planning decisions as experiments intended to test hypotheses about the system being managed (Walters and Holling 1990; McAllister and Peterman 1992). This form of management is a particularly important approach to decision making when there is substantial uncertainty regarding the dynamic behavior of the system and its responses to management: information is as much a product of adaptive management as are the commodities or resources that are the more familiar focus of management activities. Adaptive management has become a critical component of ecosystem management because we do not yet understand the dynamics of diverse ecosystem components at larger scales of space and time (FEMAT 1993; Scientific Panel for Sustainable Forest Practices in Clayoquot Sound 1995a). While formal adaptive approaches to management have a substantial history in fisheries (McAllister and Peterman 1992) and forest managers have long implemented silvicultural practices as experiments at the stand level, there have been few attempts to apply formal experimental management approaches to the larger scales and longer time frames we describe here.

Dynamics of Recovery in Dysfunctional Landscapes

Throughout the coastal temperate rain forest, numerous landscapes can be described as dysfunctional with regard to various ecological processes. Though public attention has focused more on the protection of less disturbed landscapes, the restoration of disturbed landscapes will command substantial public and scientific focus in the future. Restoration has been attempted at the scale of forest stands or stream reaches, but neither the dynamics of landscape recovery nor practical efforts to restore whole landscapes have received much attention. Just as the response times of landscapes are slow (Wallin et al. 1994), managing their recovery will be commensurately slow and challenging. The role that late-successional legacies may play in the recovery of disturbed landscapes deserves particular emphasis in research.

Natural Dynamics + Anthropogenic Changes + Global Change = ?

Natural disturbance regimes and forest dynamics are driven by changing climate. The many possible interactions and feedbacks between climate, patterns of land use, and various agents of disturbance in forests (such as fires and insects) create substantial uncertainty about the ecological changes to expect over the next two to four decades (Kasischke et al. 1995). Research on the interactions among processes that cause change in ecosystems (disturbances, biological invasions, habitat loss, disease, and physiological stresses) and processes that buffer change (community "inertia" and competition, the self-stabilizing microclimatic feedbacks in massive forests, active management efforts) can reduce this uncertainty. Forest management policy and planning processes have yet to deal seriously with the degree of uncertainty they face due to the combination of long time scales, rapid climate change, and many biological feedbacks, both positive and negative.

Ecosystem Management: More Than Science

Ecological science is only one aspect of the design of management practices in coastal temperate rain forests. Social responsibility, economic feasibility, political acceptability—all will shape the management paradigm that leads to ecological sustainability. Land management is not a scientific process. Though it should incorporate scientific ideas and information, it inevitably reflects substantial elements of consensus and compromise achieved in political and social settings. Ecological science has often played a smaller role in natural resource decision making than we would like, sometimes with disastrous consequences.

If ecosystem management is to fulfill its promise, ecosystem scientists must be prepared to create and accept roles in the management process. These roles should reflect both our knowledge and our uncertainty about ecosystem dynamics. We should use all the tools we have to project and understand the consequences of alternative management actions. Ecosystem scientists must also be willing to step beyond the confines of ecosystem science to work with social scientists and managers to build management options. Ecological, social, and economic criteria for managing ecosystems have nowhere been effectively combined. Success could come in the coastal temperate rain forest. But it will require unprecedented cooperation, and humility, from scientists, managers, and citizens alike.

Acknowledgments

The authors are grateful to James R. Karr, Ellen W. Chu, Peter Schoonmaker, and Dana Lepofsky for their reviews of an early draft of this chapter. The ideas presented here arose from research supported by the Natural Sciences and Engineering Research Council (Canada), Forest Renewal British Columbia, the Canadian Forest Service (Natural Resources Canada), and by the H. J. Andrews Experimental Forest Long Term Ecological Research Program supported by National Science Foundation grant BSR 90-11663.

World Wide Web Home Page Addresses

Andrews Forest LTER:
http://www.fsl.orst.edu/lterhome.html

B.C. Ministry of Forests. For Forest Practices Code:
http://www.for.gov.bc.ca/

Reports of the Scientific Panel for Sustainable Forest Practices in Clayoquot Sound:
http://conservation.forestry.ubc.ca:8080/panel/clayhome.html

References

Agee, J. K. 1993. *Fire Ecology of Pacific Northwest Forests.* Washington, D.C: Island Press.

British Columbia Ministry of Forests. 1995. *Biodiversity Guidebook.* Forest Practices Code of British Columbia. Victoria, B.C.

Bunnell, F. J. 1995. "Forest-dwelling vertebrate faunas and natural fire regimes in British Columbia: Patterns and implications for conservation." *Cons. Biol.* 9:636–644.

Carey, A. B., and M. L. Johnson. 1995. "Small mammals in managed, naturally young, and old-growth forests." *Ecol. Appl.* 5:336–352.

Carpenter, R. A. 1995. "A consensus among ecologists for ecosystem management." *Bull. Ecol. Soc. Am.* 76:161–162.

Ecotrust, Pacific GIS, and Conservation International. 1995. *The Rain Forests of Home: An Atlas of People and Place.* Part 1: *Natural Forests and Native Languages of the Coastal Temperate Rain Forest.* Portland, Ore.

Forest Ecosystem Management Assessment Team (FEMAT). 1993. *Forest Ecosystem Management: An Ecological, Economic, and Social Assessment.* Washington, D.C.: U.S. Departments of Agriculture, Commerce, and the Interior, and the Environmental Protection Agency.

Franklin, J. F. 1993a. "Lessons from old-growth." *J. Forestry* 91:10–13.

———. 1993b. "The fundamentals of ecosystem management with applications in the Pacific Northwest." In G. Aplet, N. Johnson, J. Olson, and V. Sample, eds., *Defining Sustainable Forestry.* Washington, D.C.: Wilderness Society and Island Press.

———. 1993c. "Preserving biodiversity: Species, ecosystems, or landscapes?" *Ecol. Appl.* 3:202–205.

Franklin, J. F., and R.T.T. Forman. 1987. "Creating landscape patterns by forest cutting: Ecological consequences and principles." *Landscape Ecol.* 1:5–18.

Franklin, J. F., and T. Spies. 1991. "Composition, function, and structure of old-growth Douglas-fir forests." In L. F. Ruggiero, K. B. Aubry, A. B. Carey, and M. H. Huff, tech. coords., *Wildlife and Vegetation of Unmanaged Douglas-fir Forests.* Gen. Tech. Rep. PNW-GTR-285. Portland: USDA Forest Service, Pacific Northwest Research Station.

Franklin, J. F., T. Spies, D. Perry, M. Harmon, and A. McKee. 1986. "Modifying Douglas-fir management regimes for non-timber objectives." In C. D. Oliver, D. P. Hanley, and J. A. Johnson, eds., *Douglas-fir: Stand Management for the Future.* Symposium proceedings. Seattle: College of Forest Resources, University of Washington.

Gregory, S. V., F. J. Swanson, W. A. McKee, and K. W. Cummins. 1991. "An ecosystem perspective of riparian zones." *BioScience* 41(8):540–551.

Grumbine, R. E. 1994. "What is ecosystem management?" *Cons. Biol.* 8:27–38.

Hansen, A. J., T. A. Spies, F. J. Swanson, and J. L. Ohmann. 1991. "Conserving biodiversity in managed forests: Lessons from natural forests." *BioScience* 41(6): 382–392.

Hansen, A. J., W. McComb, R. Vega, M. Raphael, and M. Hunter. 1995. "Bird habitat relationships in natural and managed forests in the west Cascades of Oregon." *Ecol. Appl.* 5:555-569.

Harmon, M. E., J. F. Franklin, F. J. Swanson, P. Sollins, S. V. Gregory, J. D. Lattin, N. H. Anderson, S. P. Cline, N. G. Aumen, J. R. Sedell, G. W. Lienkaemper, K. Cromack, Jr., and K. W. Cummins. 1986. "Ecology of coarse woody debris in temperate ecosystems." *Adv. in Ecol. Res.* 15:133–302.

Hopwood, D. 1991. *Principles and Practices of New Forestry.* Land Management Report 71. Victoria: British Columbia Ministry of Forests.

Huff, M. H. 1995. "Forest age structure and development following wildfires in the western Olympic Mountains, Washington." *Ecol. Appl.* 5: 471–483.

Kasischke, E., N. L. Christensen, Jr., and B. Stocks. 1995. "Fire, global warming, and the carbon balance of boreal forests." *Ecol. Appl.* 5:437–451.

Kusel, J. and L. Fortmann. 1991. *Well-Being in Forest-Dependent Communities.* Berkeley: Department of Forestry and Resource Management, University of California.

Lertzman, K. P. 1992. "Patterns of gap-phase replacement in a subalpine, old-growth forest." *Ecology* 73:657-669.

———. 1995. "Forest dynamics, differential mortality rates and variable transition probabilities." *J. Veg. Sci.* 6:191–204.

Lertzman, K. P., and C. J. Krebs. 1991. "Gap-phase structure of a subalpine old-growth forest." *Can. J. For. Res.* 12:1730-1741.

Lertzman, K. P., G. Sutherland, A. Inselberg, and S. Saunders. 1996. "Canopy gaps and the landscape mosaic in a temperate rain forest." *Ecology* 77:1254–1270.

Malone, C. R. 1995. "Ecosystem management: Status of the federal initiative." *Bull. Ecol. Soc. Am.* 76:158–161.

McAllister, M. K., and R. M. Peterman. 1992. "Experimental design in the management of fisheries: A review." *North Am. J. Fish. Mgmt.* 12:1–18.

McComb, W. C., T. A. Spies, and W. H. Emmingham. 1993. "Douglas-fir forests: Managing for timber and mature-forest habitat." *J. Forestry* 91:31–42.

Morrison, P., and F. J. Swanson. 1990. *Fire History and Pattern in a Cascade Range Landscape.* Gen. Tech. Rep. PNW-GTR-254. Portland: USDA Forest Service.

Pickett, S.T.A., and P. S. White. 1985. *The Ecology of Natural Disturbance and Patch Dynamics.* New York: Academic Press.

Pojar, J., and A. MacKinnon. 1994. *Plants of Coastal British Columbia.* Vancouver, B.C.: Lone Pine Publishing.

Scientific Panel for Sustainable Forest Practices in Clayoquot Sound. 1994. *Progress Report 2: Review of Current Forest Practice Standards in Clayoquot Sound.* Victoria, B.C.

———. 1995a. *Report 5: Sustainable Ecosystem Management in Clayoquot Sound: Planning and Practices.* Victoria, B.C.

———. 1995b. *A Vision and Its Context: Global Context for Forest Practices in Clayoquot Sound.* Victoria, B.C.

Schowalter, T. D. 1995. "Canopy arthropod communities in relation to forest age and alternative harvesting practices in western Oregon." *For. Ecol. and Man.* 78:115–125.

Simberloff, D. 1988. "The contribution of population and community biology to conservation science." *Ann. Rev. Ecol. and Syst.* 19:473–511.

Simberloff, D., J. A. Farr, J. Cox, and D. Mehlman. 1992. "Movement corridors: Conservation bargains or poor investments?" *Cons. Biol.* 6:493–504.

Spies, T. A., J. F. Franklin, and T. B. Thomas. 1988. "Coarse woody debris in Douglas-fir forests of western Oregon and Washington." *Ecology* 69:1689–1702.

Spies, T. A., W. J. Ripple, and G. A. Bradshaw. 1994. "Dynamics and pattern of a managed coniferous forest landscape in Oregon." *Ecol. Appl.* 4:555–568.

Swanson, F. J., and J. F. Franklin. 1992. "New forestry principles from ecosystem analysis of Pacific Northwest forests." *Ecol. Appl.* 2:262–274.

Swanson, F. J., J. A. Jones, D. Wallin, and J. Cissel. 1993. "Natural variability: implications for ecosystem management." In M. Jensen and P. Bourgeron, eds., *Eastside Forest Ecosystem Health Assessment II: Ecosystem Management Principles and Applications.* Report PNW 89-103. Portland: USDA Forest Service, Pacific Northwest Research Station.

Swanson, F. J., T. K. Kranz, N. Caine, and R. G. Woodmansee. 1988. "Landform effects on ecosystem patterns and processes." *BioScience* 38(2):92–98.

Taylor, B. 1995. "Forest ecosystem networks: A decision-making framework and its application in the Nehalliston landscape unit." Master's thesis, Simon Fraser University, Burnaby, B.C.

Vannote, R. L., G. W. Minshall, K. W. Cummins, et al. 1980. "The river continuum concept." *Can. J. Fish. and Aqua. Sci.* 37(1):130–137.

Walker, B. 1995. "Conserving biological diversity through ecosystem resilience." *Cons. Biol.* 9:747–752.

Wallin, D. O., F. J. Swanson, and B. Marks. 1994. "Landscape pattern response to changes in pattern generation rules: Land-use legacies in forestry." *Ecol. Appl.* 4:569–580.

Walters, C. J., and C. S. Holling. 1990. "Large-scale management experiments and learning by doing." *Ecology* 71:2060–2068.

Waring, R. H., and J. F. Franklin. 1979. "Evergreen coniferous forests of the Pacific Northwest." *Science* 204:1380–1386.

Willson, M. F., and K. C. Halupka. 1995. "Anadromous fish as keystone species in vertebrate communities." *Cons. Biol.* 9:489–497.

15. A Vision for Conservation-Based Development in the Rain Forests of Home

PETER K. SCHOONMAKER, BETTINA VON HAGEN, AND ERIN L. KELLOGG

The coastal temperate rain forest as it exists today is the sum of myriad colonizations during the past 10,000 years. Plants—especially the long-lived, coniferous fir, spruce, hemlock, and cedar—and animals—salmon and humans in particular—arrived at their pre-European distribution less than 4000 years ago. The coastal rain forest of 200 years ago, when Europeans sailed the coast, was a biologically and culturally rich and diverse bioregion. Some of this legacy has been lost, but much remains.

Forest historians have yet to estimate the standing biomass of the pre-European rain forest, and scientists cannot yet present a comprehensive picture of presettlement salmon distribution and abundance. We do know that many river systems supported an order of magnitude more salmon than at present; the same mammoth logs that choked hundreds of miles of coastal rivers, turning away early explorers and settlers, also provided ideal salmon habitat.

First Nations peoples numbered in the hundreds of thousands, occupying the entire coast at a density surpassed only by their contemporaries in Central America. Indeed the pre-European population density rivaled that of today, if major metropolitan areas are excluded (Plate 8). More than sixty separate languages, representing diverse cultures, were spoken (Plate 4).

The changes wrought throughout the bioregion (Figure 15.1) during the last

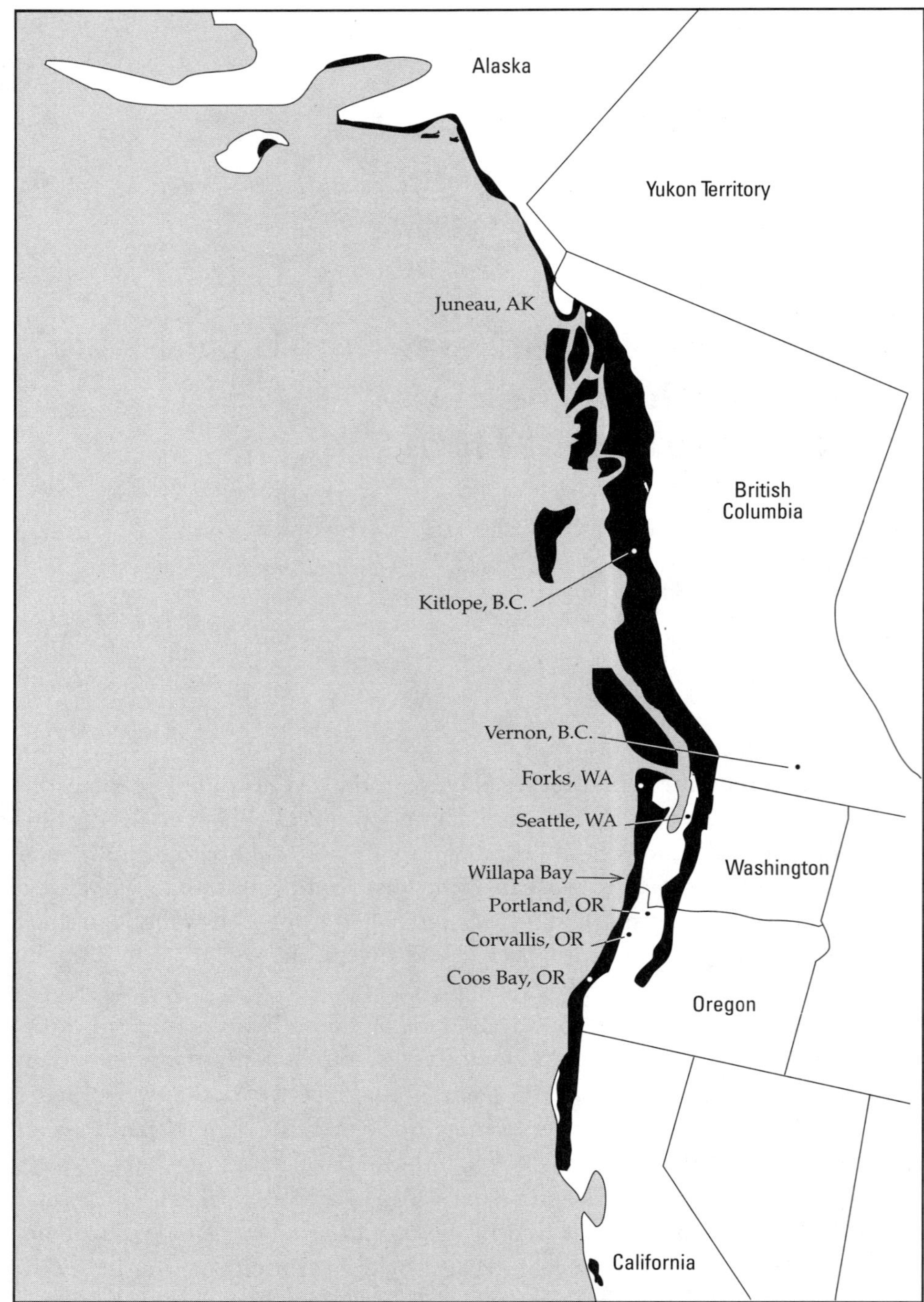

Figure 15.1. Selected locations mentioned in Chapter 15.

200 years can be seen as the final chapter of Euro-American expansion across North America. Europeans arrived in small numbers at first—explorers, traders, trappers, then settlers—always driven by a quest for untapped resources. By the mid-nineteenth century, it had become their "manifest destiny" to rush westward and reap the abundance of a bioregion that seemed to be lying fallow. Following the rush of settlers came the salmon canneries and the timber industry. The lumbermen, moving ever westward, had depleted the forests of New England and the upper midwest. During the late nineteenth and early twentieth centuries, they began liquidating coastal old growth at ever-increasing rates.

The natural resource trail ended at the shores of the Pacific. By the mid-twentieth century, large trees and fat salmon had grown scarce from California to Washington State. So the westward rush turned right—northward to British Columbia and Alaska. After the heyday of old-growth logging, many timber companies bounced back—back to the southeastern United States' productive pineries. Timber towns like Coos Bay, Oregon, or Grays River and Forks, Washington, have yet to bounce back. Many, like Valsetz, Oregon, have disappeared, as have the fishing villages that dotted the Columbia River during the early twentieth century. Gone, too, are many First Nations people and their languages (Plate 4), replaced in large part by English-speaking people of European descent along with a diverse mix of people from Asia, Latin America, and Africa.

The timber industry has arrived at an equilibrium of sorts in the bioregion. Where second-growth forests dominate, 50-year cutting rotations have somewhat smoothed the cycles of economic prosperity and recession (Lemaster and Beuter 1989). The old boom-and-bust pattern still prevails, however, where old-growth timber is extracted: a one-time wave of record harvests continues to sweep northward in British Columbia and Alaska.

The region's salmon fishery reflects the history of our forests for good reason. To complete their life cycle, salmon depend on habitat provided by mature functioning forest ecosystems. We mined our old-growth forests and dammed our rivers, trading salmon habitat for low-cost lumber and subsidized electricity, while simultaneously over-harvesting the fishery. The high-seas fishery offers another example of the westward search for resources, and, concurrently, an eastward incursion by large Asian fishing conglomerates. Once again, as the resource has been depleted, we have moved northward.

Timber and salmon. Forests and fish. These dualities reflect in a sense our still simplistic view of the bioregion. This simplistic outlook has resulted in a legacy of environmental diminution closely coupled to contemporary economic distress. Outside ownership of resources, large-scale extractive strategies employed by these outside owners, and a general disinvestment in the region as extraction moves from south to north have fueled this economic decline. Absentee landownership, concentrated in large corporate and government hands, strongly correlates with the incidence of poverty in rural areas (Kusel and Fortmann 1991). Traditional corporate timber and overcapitalized fisheries communities on the coast generally experience lower average incomes and higher unemployment than the state or province in which they are located (Chapter 13; Labor Market and Eonomic Analysis Branch 1995). As resource extraction washes over the region, an economic culture based on tourism and retirement follows in its wake. Newcomers bring jobs and innovation, they diversify economies, yet tourism and retirement industries often place stress on communities, produce low-wage service-sector jobs, and yield a dual economy of the older rich and the younger poor (Chapter 13).

Given these uncertain conditions, coastal residents need to chart their course by examining their past, taking stock of the present, and envisioning their future—a future that links long time residents (with their rich knowledge of local patterns and processes) and new residents (who bring fresh approaches, capital, and links to outside markets and resources). Such a future demands that those who create it be both outward-looking and locally focused. Such a future reconnects environment and people. Such a future fuses economic development and environmental conservation into conservation-based development.

What might this conservation-based development future look like? The possible pathways are many, but the key ingredient is a revitalization of community and an reintegration of environment, economy, and culture. We could begin by reconciling political boundaries with environmental ones: counties would follow watersheds, natural resource areas, or geomorphic provinces where possible. These new entities would be represented not just by elected officials but by nonprofit coalitions dedicated to a long-term vision of environmental, economic, and cultural health. Natural resource stewardship would be vested in local businesses accountable to their community—but overseen by regional and national guidelines rather than managed for shareholder benefits by multinational corporations and other absentee landlords. Land trusts would help ensure a long-term balance of production and biodiversity maintenance. Salmon fisheries would focus on natural production fisheries rather than hatcheries. Land use planning would emphasize compact communities; it would halt or reverse suburban sprawl; and it would ensure a land base for natural-resource based businesses. Local entrepreneurs would be encouraged through public and private incentives to add value to natural resources. Communities would embrace environmentally benign industries that diversify and stabilize their economy.

Learning would not stop with a high school diploma or college degree. It would be extended through formal and informal community organizations and through expanded educational and research institutions to a lifelong process that would both deepen one's local knowledge and expand one's horizons to the bioregion and beyond. Indeed, not only would there be meaningful connections between communities; there would also be a renewed engagement between urban and rural citizens as new forms of communication build a web of dialogue throughout the coastal rain forest. Such a dialogue between individuals and groups dedicated to conservation-based development would help residents to choose qualitative rather than quantitative development: fewer children, better cared for; moderating harvest, adding value; refocusing from consumption to creation; and, perhaps most importantly, moving from demanding rights to taking personal responsibility.

To realize this vision of conservation-based development in the coastal temperate rain forest bioregion, local, regional, and national policymakers must play a role. National and state or provincial governments must undo misguided incentives that allow individuals and organizations to externalize environmental and social costs: residential, municipal, and industrial pollution; depletion of mineral, timber, and wildlife resources, often at a loss to national or regional government accounts; rapid suburban development that encourages inefficient, isolated lifestyles and leaves urban centers destitute; international trade policies that favor multinational corporations at the expense of communities and the environment (Ryan 1995).

Is it naive to envision politicians at the national level reversing course on issues that are so closely watched by powerful special interest groups? Perhaps. But many historically meaningful processes were viewed as naive by contemporary pundits: the American revolution, abolition in the mid-nineteenth century, the labor movement in the early twentieth century, the civil rights and environmental movements in the mid-twentieth century. And while politicians may operate in the national forum, they are elected and reelected at home.

Many would call this vision radical. In fact it is mainstream. What we have just described is what residents in many communities—Corvallis, Sherwood, and Portland in Oregon; Seattle, the Willapa watershed, and Thurston County in Washington; Juneau in Alaska—envision when they come together to chart their future (e.g., Sustainable Seattle 1993; Powell 1994; Willapa Alliance and Ecotrust 1995). The radical part is the implementation. In the following pages we discuss the barriers to this vision and then present strategies to overcome them. Because the intersections between human needs and ecosystem integrity vary from one community to the next, they require different kinds of understanding, conservation, economic development, and policy reform—the four cardinal points of conservation-based development.

Barriers to Sustainability

Creating the conditions for conservation-based development is no small task. There are formidable barriers to its acceptance: lack of understanding, uncertain access to resources, fear of change, an infrastructure based on large-scale production and cheap fossil fuels, control of land by "absentee landlords" (e.g., government and large corporations), and a plethora of interests vested in the status quo. The challenge is to create the conditions under which residents, working in their own self-interest, will also contribute to ecosystem health and community vitality. This strategy harnesses the forces of the market and the creativity of people for the benefit of the ecosystem. The conditions that will lead to conservation-based development include access to information, local organizational capacity, enhancement of entrepreneurial capacity, access to resources, and a supportive policy climate for appropriate development.

Understanding: Local Access to Information

Developing long-term strategies for resource stewardship is possible only if there is accurate information on the distribution and condition of these resources and their patterns of use. In many communities (particularly those dependent on natural resources), large corporations and government agencies own or control both the resources and the information about their condition. A growing number of initiatives, such as the local indicators, bioregional atlas, and on-line services described in this chapter, focus on brokering information about resources to local communities, thereby providing the perspective and capacity needed for effective local stewardship. Clearly the resources exist to address these issues—dozens of annual conferences, hundreds of extension agents and libraries, thousands of federal and state land managers, and millions of dollars in research—yet these resources are not being used to maximum effect. Critical information and expertise are not being communicated to people in temperate rain forest communities, and worse yet, the people in these communities—and their knowledge and experience—are not part of this information resource.

Local Organizational Capacity

Many rural communities in the bioregion developed as "company towns" in which companies provided the economic activity and also developed and controlled key infrastructure and civic activities. As forest products companies have abandoned communities because of timber depletion and the economic imperatives of mechanization, and as fisheries have collapsed, the social infrastructure has often collapsed as well.

Lack of organizational capacity has left communities ill equipped to capture external resources and vulnerable to undesirable development. One approach is to create permanent, locally controlled, self-sustaining organizations for conservation-based development that can articulate a shared vision, champion local initiatives, and broker outside resources. When issues are centered in a place, and when the decision makers share a common commitment and context, solutions are more likely to emerge than when the issues are abstract and the debate ideological. The most innovative and practical solutions to restoring watersheds and jobs have emerged from these local community initiatives.

Entrepreneurial Capacity

Conservation-based entrepreneurs must understand the environmental and community implications of their business activities while meeting the usual challenges of finding markets, locating consistent sources of supply, developing appropriate technical and managerial skills, and finding the necessary capital. These challenges are greater in rural communities where infrastructure is often underdeveloped or deteriorated, markets are remote or inaccessible, and resources and technical expertise are limited. With few options to turn to, local communities continue to participate (often unwillingly) in the depletion of their natural capital by outside owners and even take part in exporting the benefits of that resource extraction out of the region. Providing key resources such as credit, access to markets, and technical assistance to environmentally restorative businesses can be invaluable in spurring entrepreneurial activity.

Local Ownership

Landownership is heavily concentrated in the coastal temperate rain forest region. In California, over half the land is held by a single owner in twelve forest counties, and about half of the total acreage of forestland is owned by a person or corporation with an address outside the county (Kusel and Fortmann 1991). The pattern is similar in Oregon where the ten largest owners control 59 percent of total private forest acres, ranging from a low of 55 percent in Yamhill County to a high of 94 percent in Clatsop County (Willer 1995). In British Columbia, 95 percent of forestland is owned by the provincial government, which awards tree farm licenses based on an annual allowable cut. Control of the allocated cut in B.C. has been increasingly concentrated in fewer hands: the share of committed harvesting rights held by the ten largest companies increased from 37 percent in 1954 to 69 percent in 1990, with the largest company controlling about 17 percent (M'Gonigle and Parfitt 1994; Forest Resources Commission 1991).

Concentration of landownership matters, according to Kusel and Fortmann (1991), because it is correlated with poverty. Concentration of landownership also reduces the diversity of management strategies, making communities less resilient to economic recessions and increasingly dependent on policies set in distant board rooms. Finally, outside ownership reduces opportunities for local innovation and entrepreneurship and for responsible behavior: as resources are depleted, outside owners can and do move away and are not obliged to live with the consequences of their actions.

Regulatory Uncertainty

Resource management is unlikely to be sustainable unless resource owners receive the full benefits of their investments and stewardship. Moreover, they must be assured a degree of certainty about regulations and tax structures. Some recent cutting of forests on private land, especially of mature forests that provide valuable wildlife habitat, has been prompted by fear of additional regulation to protect the spotted owl, marbled murrelet, and other potentially threatened and endangered species. In some cases, this response has been self-serving, but in others, it reflects legitimate concerns. For many rural residents, whose land is both a source of income and a legacy to their children, regulatory changes have significant consequences. As counterproductive as the apparent short-term perspective of business can be, it is reinforced by an unpredictable regulatory and political environment. A more predictable regulatory climate and the ability to reap the benefits of stewardship, however, must be balanced by the recognition that land and resource ownership brings responsibilities as well as rights.

Government Structure and Subsidies

Much of the support for community renewal and ecosystem restoration must come from private initiative and through market forces. A more favorable policy climate, however, can expedite and extend the reach of private resources and initiative. The current tax and regulatory system offers powerful incentives for unsustainable business practices—and thus presents a formidable barrier to conservation-based development. For example, fossil fuel subsidies contribute to the competitive advantage of imported over locally grown foods. Government policies themselves, the structure of government, the way in which these policies are developed—all can be realigned to provide incentives for stewardship, and to engage local communities in a more democratic process of revitalization.

Beyond Barriers: Fusing Conservation and Development

At the heart of conservation-based development is the integration of conservation with economic development so that development activities, on balance, maintain and nurture healthy, functioning ecosystems. In conservation-based development, preservation of key intact ecosystems is still critical, but the purpose of these natural areas is to provide a template and a hedge rather than to be the full sum of conservation activity on the landscape. The conservation impact of protected areas is complemented by the conservation activity that occurs in nonprotected areas, where appropriate development maintains biodiversity and ecosystem services such as clean air and water.

Conservation-based development implies not only a different approach to conservation but, more acutely, a different approach to development. Economic activity has done more than cause significant and permanent environmental impoverishment, such as the loss of biodiversity and global climate change. It has failed in its fundamental purpose of enhancing social welfare. Conservation-based development—a synthesis of community revitalization and ecosystem conservation—arose to address some of the failures of traditional economic development and environmental advocacy.

For conservation to succeed, development must succeed: local communities must have a vested interest in maintaining protected areas through sharing in the benefits of conservation. And for development to succeed, conservation must succeed: no development activity can be sustained without a healthy, productive ecosystem that provides both the raw materials for development and the assimilative capacity for waste products. The task of conservation-based development, then, is simultaneously to revitalize communities and restore ecosystems in a way that is economically viable and self-perpetuating. Traditional funding sources for conservation and ecosystem restoration—namely taxation and private donations—are necessary but grossly insufficient for the level of activity needed. A deeper, more catalytic source of energy and creativity is needed: the entrepreneurial spirit of local residents and their commitment to place. Small-scale entrepreneurs, grounded in a place and community through ties of family and history, have the capacity to develop the community's resources for their best and most restorative use.

Other social movements have captured this source of energy and creativity. Community development, which grew out of the civil rights movement, has attempted to address the needs of poor and disadvantaged communities. Permanent, self-sustaining development institutions have been created to foster community renewal through entrepreneurship. This focus on individual residents as agents of greater social change is an important lesson for ecosystem restoration. Community development combines social capital (creativity and labor) and financial capital to address social problems. In turn, ecosystem restoration can help community development become truly self-sustaining by incorporating natural capital (environmental goods and services such as forests, clean air, and clean water) into long-term reinvestment strategies.

While seemingly worlds apart in intent and approach, ecosystem restoration and community reinvestment share many core principles and values, making their union particularly compelling (Table 15.1). The mission of each, for example, is to enhance productivity, diversity, and resilience: in the case of ecosystem restoration, in watersheds; in the case of community reinvestment, in neighborhoods. In both cases, activity is rooted in the local context. Both movements draw their strength from developing local knowledge; success hinges on an intimate understanding of underlying structures and processes. Ecosystem restoration aims to restore natural processes, using intact, functioning ecosystems as a guide. Community revitalization aims to restore market forces, including access to credit and markets for local products, using the discipline of the market to ensure efficient and enduring solutions. In both cases subsidies, if they are employed, are temporary and used to reestablish the defining ecosystem patterns and processes or market forces rather than artificially creating success that must be permanently subsidized. For example, ecosystem restoration might restore a healthy, commercially productive forest on a cutover area by allowing the forest to undergo natural succession, including a period in which nitrogen-fixing alder is the dominant species. The alternative "artificial" approach is intensive use of herbicides, which leads more quickly to the desired conifer species but does so at a high economic and environmental cost. In the case of community revitalization, working with market forces means creating local businesses that meet real market demand, rather than subsidizing the purchase of substandard local goods.

Ecosystem restoration employs native species, which are well adapted to local conditions and consequently need minimal care and attention once they are established. Similarly, community revitalization draws on local resources and historical strengths that have demonstrated adaptation to local conditions. In contrast, many conventional economic development efforts rely on "imported" economic activities in which the scale of activity and the type of technology may require continuous subsidies. The key in creating self-perpetuating systems is to use locally adapted species and resources to maximize the natural and competitive advantages of a place.

Finally, conservation-based development draws on the experience of community revitalization in creating permanent institutional capacity, anchored in the watershed and neighborhood. As the racial mix of urban neighborhoods changed in the United States in the 1950s and 1960s, many banks and other service providers "redlined" these neighborhoods, declining to provide any credit to neighborhood residents, thereby hastening urban decay (Taub 1994). Resources, particularly labor, continued to be exported, but few resources were imported. To reverse this trend, revitalization efforts in the 1970s and beyond focused on "greenlining" the neighborhoods, capturing external resources and targeting them exclusively to greenlined areas. The resource flow

Table 15.1. Principles of Conservation-Based Development

	Community Development (People)	+	**Ecosystem Restoration** (Nature)	=	**Conservation-Based Development** (People and Nature)
Goal	Community revitalization		Ecosystem integrity		Social and environmental resilience: *Foster vibrant communities in healthy, productive ecosystems.*
Approach	Restore market forces		Restore natural processes		Restore natural and market forces: *Focus on community's unique assets, using nature and the market as guides for enduring solutions.*
Place	Neighborhood		Watershed		Place-based community: *Ground strategy in local context and conform to natural and social contours of the local landscape.*
Institution	Community development corporation		Conservation organization		Permanent conservation-based development institution: *Create new, adaptive learning organizations to foster permanent process of renewal.*
Development Focus	Social capital		Natural capital		Social and natural capital: *Capture local energy, creativity, and vested interest in ecosystem and community well-being.*
Conditions for Replication	Individual initiative		Natural productivity		Entrepreneurial stewardship: *Foster commitment and capacity for change by creating government support for conditions of renewal, but rely on individual initiative.*

in watersheds containing natural resources and resource-based communities has been similar. Resources—timber, fish, agricultural products—have been extracted, but significantly less value has been returned, judging by relative income levels and environmental degradation. Permanent institutions, committed to the well-being of the watershed and buffered from the short, unpredictable cycles of politics, are needed to foster the long-term health of ecosystems and their residents.

A New Path: Beginning with Understanding

Any realistic hope for sustainable economic development must begin with an effort to understand the social, cultural, economic, and political forces, in addition to the ecological processes, that shape the coastal region. The channels to knowledge and information must be open regionwide, so that ecologically sound economic and community development can be enhanced by the ability to access, analyze, and apply scientific, economic, and cultural information. Access to information should be interactive: local communities should enter into dialogues among themselves and with academic and government institutions.

Indeed, there is a historical precedent for a shared ecological, economic, and cultural information system within the coastal temperate rain forest. For thousands of years First Nations people exchanged information through common languages (such as the Chinook trading language) and ritual ceremony (such as the potlatch) so that they could trade, coexist, and better manage the ample resources of their coastal environment (Chapter 10). What we need now is a modern analog to this indigenous information and communication system, one that connects ecological understanding with economic and community development at local and regional scales.

Our first task in developing a strategy for conservation and economic development is to gather, synthesize, and share ecological and socioeconomic information about the region along a spectrum of temporal and spatial scales. A focus on bioregional and watershed scales allows us to see long-term, large-scale patterns while simultaneously working on more tractable local issues that fit into the bioregional picture. The dissemination of this information can take many forms: databases, geographic information system (GIS)-based atlases, books, conferences, and workshops, and on-line information systems, to name a few. The important point is that the media must be appropriate for and accessible to the information user (Davis 1995).

While a bioregional view provides the perspective to see key trends and opportunities, our understanding of the bioregion and our progress toward sustainability must be grounded in local knowledge. But how do we evaluate whether or not the activities undertaken by local residents are leading toward sustainability? Locally generated indicators or "vital signs" of community and ecosystem health have emerged as one mechanism to measure progress toward sustainability while giving the community the tools to direct, influence, and evaluate that progress. Such community-based assessment is essential for any effort to preserve, restore, or enhance community amenities.

There are numerous examples of monitoring environmental, economic, and social conditions by tracking a set of indicators through time. In many respects, communities along the Pacific Coast have led the way in developing readily measurable community indicators as bellwether tests of sustainability (Oregon Progress Board 1992; Sustainable Community Roundtable 1993; Sustainable Seattle 1993; Powell 1994). Few, if any, of these community indicator systems have been established in rural, natural-resource-dependent areas.

One example of a rural, community-based indicator system is a project jointly undertaken in southwestern Washington by the Willapa Alliance, a local nonprofit organization, and the authors. The resulting community report, "Willapa Indicators for a Sustainable Community" (Willapa Alliance and Ecotrust 1995), organized a few key indicators of health into three broad categories—environment, economy, community—deemed relevant for the coastal temperate rain forest region of North America in general and the Willapa watershed in particular (Table 15.2). Such indicator systems are successful when communities own them and use them, and when they are transferred to (and transformed by) other communities and incorporated into a bioregional database.

Although numerous locally based projects and data sets help to paint a bioregional picture in mosaic form, a more comprehensive overview results when we begin with a bioregional perspective that tracks large-scale ecological and socioeconomic changes in the coastal rain forest. An example of such a large-scale perspective is a recent atlas of natural forests and Native languages of the coastal temperate rain forest that describes pre-European and present forest cover, watershed condition, and the distribution and status of indigenous languages (Ecotrust et al. 1995) (Plates 3, 4, and 5). This coarse-grained approach has revealed:

- Over half of the world's coastal temperate rain forest is in North America.
- Most of that forest south of Vancouver Island has been recently altered.
- Most watersheds are in less than pristine condition.
- Very little coastal rain forest is set aside in protected areas south of Canada.

A series of atlases describing long-term changes in fisheries resources, wildlife, cultural diversity, and economic conditions is planned. The patterns

Table 15.2. Key Indicators of Health for the Willapa Watershed

Category	Primary Indicator
Environment	
• Water resource quality	Oyster condition index (1963–1993)
• Vegetation cover	Land use change (1950–1991)
• Species populations	Salmon escapement and catch (1980–1993)
Economy	
• Productivity	Key natural resources harvests (1976–1993)
• Opportunity	Number of businesses (1987–1993)
• Diversity	Sectoral employment (1970–1992)
• Equity	Personal income per capita (1969–1992)
Community	
• Lifelong learning	High school graduation rate (1982–1989)
• Health	Percentage of healthy-birthweight babies (1953–1993)
• Citizenship	Voter participation (1940–1993)
• Stewardship	Recycling and waste stream generation (1989–1992)

Source: The Willapa Alliance and Ecotrust (1995).

and trends that are uncovered can inform us about opportunities for protecting natural areas of special concern and for developing sustainable businesses that add value to raw resources, that use overlooked resources (such as nontimber forest products), or that find more environmentally benign practices. Indeed this has happened with the greater Kitlope ecosystem—the last large pristine temperate rain forest watershed in North America (more than 300,000 hectares, or 741,000 acres). Slated for timber harvest in 1991, this unique resource has now been protected by an unprecedented joint management agreement between the province of British Columbia and the Haisla Nation, whose traditional territory includes the Kitlope.

What else might we discover when we look at coastal estuaries, marine resources, transportation infrastructure, or forest-product stocks and flows at a bioregional level? Understanding patterns of change at local and bioregional scales provides the basis for developing a conservation template, a sort of "temperate rain forest wildlands project" (Noss 1992a). Several conservation plans have been developed for subzones of the bioregion (the Oregon Coast Range plan (Noss 1992b); President Clinton's forest plan (FEMAT 1993), and British Columbia's Protected Areas Strategy, for example. A large-scale, long-term conservation plan that explicitly integrates socioeconomic considerations has yet to be developed for the entire coastal temperate rain forest.

Yet the planning of core, buffer, matrix, and corridor areas, which results in setting aside land exclusively for conservation, is by itself an insufficient strategy (Chapter 14). The vast areas needed to maintain viable populations and functioning ecosystems over the long term may not be available as human populations continue to grow. Furthermore, conservation biology and ecosystem management are conceptually young, with relatively little empirical grounding, and subject to unknown future contingencies (such as climate change and catastrophic disturbances). Future conservation strategies must reconcile ecological health with human needs and therefore move beyond a "conservation plan" to a shared community vision.

Creating Local Organizational Capacity

Conservation-based development requires the creation of new kinds of civic organizations. A growing number of sociologists, political scientists, and ordinary citizens are calling not only for a reinvention of government, but for a reinvigoration of civic democracy—a reinvention of the governed (Kemmis 1990; Putnam 1993; Lee 1993). They argue that the current system discourages meaningful public participation and that people need ample opportunities to practice citizenship—the routine, face-to-face engagement in civic issues that encourages all of us to rise above our own self-interest and see how our interests can actually be served in the context of the community's needs (Kemmis 1990). Conservation-based development seeks to build new organizational capacity at the community level. These new institutions will be:

- Patient: They will have the commitment and tenacity to work in very long time frames.
- Values-based: A commitment to environmental quality, economic vitality, and equity will underlie their philosophy.
- Entrepreneurial: They will be flexible and opportunistic in their work.
- Empowering: An interest in enabling people will motivate their actions.
- Resourceful: They will be bent on assembling significant resources and accessing information and services from the public and private sectors.
- Autonomous: They will be community controlled and self-sustaining in the long term.
- Learning organizations: They will be willing to redefine problems, reassess tools, and reformulate approaches to solutions.

These organizations in essence become the community's permanent, locally controlled, self-sustaining conservation-based development institutions. They serve as a community's means of not only adapting to change but channeling the forces of change to its advantage.

People constitute organizations. Too often, the human development aspects of organizational and community capacity building are overlooked. In many parts of the bioregion, as in the United States and Canada as a whole, political leadership and citizenship have become sharply stratified. Career politicians with the financial resources to mount large-scale political campaigns are becoming more and more isolated from communities and community development. We complain about the lack of strong national leadership without recognizing that it begins at the community level with support for local leaders. The proliferation of informal groups and coalitions in the region—like the Northwest Sustainability Working Group, which arose from people's interest in having more regular opportunities to discuss the key issues they faced in their work—attests to the value of linking and supporting those who play leadership roles in their communities.

Creating Entrepreneurial Capacity

Access to entrepreneurial resources—such as credit for start-up businesses, market research, brokering, and product distribution—as well as creating an entrepreneurial culture are critical for small business formation. Profit incentives can be aligned with ecosystem and community health by serving local markets. If local markets are limited in size and scope, rural entrepreneurs can be connected to external "green" markets to forge some of the same incentives for business stewardship. This relationship between business self-interest and the promotion of ecosystem and community health is best illustrated by example. In Willapa Bay in southwestern Washington, two groups of businesses are of particular interest: those that rely directly on a healthy, productive, and intact ecosystem for successful production and those that use waste products or underutilized resources for a higher and better use. Examples of enterprises reliant on a healthy ecosystem include the oyster and wild edible mushroom industries. Oysters are the "canaries of the mudflats"; they are extremely sensitive to deterioration in water quality and, as a consequence, oyster growers have been reliable and persuasive guardians of the bay's water quality. Wild edible mushrooms are generally harvested from intact forests, and the value of recreational and commercial harvests provides an economic incentive for maintaining diverse and mature forests. By helping businesses in these sectors succeed, the constituency for a healthy and diverse ecosystem is strengthened.

Businesses that use waste products provide tremendous opportunities for effective development because their activity both reduces financial and ecological disposal costs and produces revenues. One company in Astoria, Oregon, for example, uses fish wastes from fish processing plants to produce an array of products, including organic fertilizers and fish meal. The fertilizer is sold to cranberry growers and replaces chemical fertilizers that would otherwise be used. This not only reduces chemical runoff: it also substitutes for imports into the region and maintains local wealth. Another example involves *Spartina alterniflora*, an invasive cordgrass that is spreading rapidly and threatening oyster beds and fish nurseries in Willapa Bay. Entrepreneurs are exploring uses for spartina ranging from clothing to paper; if they are successful, economic incentives may be provided to assist other efforts to eradicate it or control its spread. Case studies in the "Concepts in Action: Conservation-Based Development" section (pages 349–360) outline additional approaches and examples.

These examples are individually modest yet collectively powerful. Providing incentives to individual entrepreneurs will in time create a growing group of restorative businesses and socially and environmentally responsible investors and consumers—with commensurate influence. With a critical mass of businesses exploring new ways to operate in a more socially and environmentally responsible manner, opportunities arise for innovation on a larger scale. These businesses will eventually prompt a shift to more sustainable practices "to create a remarkably different economy, one that can restore ecosystems and protect the environment while bringing forth innovation, prosperity, meaningful work, and true security" (Hawken 1993:2).

Fostering Local Ownership

Enhancing capacity and creating the conditions for stewardship will foster increasing opportunities for local ownership of resources. As costs are internalized, the changing economics of resource extraction will cause landowners who currently bear a lower proportion of the environmental and social costs of their business activities to shift their ownership to those who can best internalize these costs. For example, some mechanisms are emerging to help forest owners recapture their investment in stewardship. These include third-party certification of good forest management that creates new market opportunities and can lead to higher profits (Mater 1995). Certification requires not only responsible environmental management but the protection of local rights and a reflection of true costs (Cabarle et al. 1995). Conservation easements that protect the ecological character and function of the land and compensate forest owners for the increased carbon storage associated with long-rotation forestry, are two other emerging mechanisms. ("The Pacific Forest Trust," pages 197–201). Adding value locally can also create incentives for local ownership of resources as secure access to raw materials and quality control become more important.

In British Columbia, where 95 percent of forestland is owned by the government and leased to corporations, change depends on government policies as well as market forces. In Vernon, the Ministry of Forest's Small Business

Forest Enterprise program is combining ecologically responsible silvicultural approaches with a sorting yard that grades logs individually—allowing wood to be sold for its best end-product use—and makes the logs available to all bidders, even on a single log basis (Smith 1995). Public opinion can be persuasive: the deep rifts over forest management and land tenure in British Columbia caused the president of Fletcher Challenge Canada Ltd., the second-largest forest company in the province, to publicly recommend the wider distribution of timber tenures: "There doesn't have to be a net loss of jobs. . . . This could stimulate the forest sector by bringing in new players with new ideas" (quoted in M'Gonigle and Parfitt 1994:95). The timber companies may have no choice: negotiations between the provincial and federal government and First Nations may result in the transfer of 20 percent of land in British Columbia to Native control.

Reducing Regulatory Uncertainty

As fish and old-growth forests become more scarce, as the makeup of coastal communities changes, as global market forces increasingly reverberate in local and regional economies, the region's political climate echoes individuals' reactions to these unsettling shifts. Fear of an uncertain future has led many communities and their elected representatives to call for a rollback of environmental and other regulations and, as well, for greater local autonomy over public resource decision-making. Private property rights groups up and down the coast—Wise Use in the United States and Share in British Columbia—reflect a growing disaffection with government nationwide. In many parts of the region, rural residents are voting to turn control of federal lands over to the county or state, to overturn environmental protection laws, and to ignore policy recommendations rooted in science, even though these are precisely the people who have the most at stake in healthy, functioning ecosystems. A fundamental premise of conservation-based development is that local residents—given the right incentives, with access to information, and a secure sense of options—will make choices that are in their own, and their ecosystem's, best interest.

Herein lies the difference between the conservation-based development view of local control and the view that underlies the Wise Use movement. Both advocate local control with the premise that local people have the best understanding of local conditions and have the most at stake in managing their resources. But the conservation-based development perspective is that this greater understanding and vested interest will promote more responsible behavior—especially when reinforced by policies that internalize costs. The Wise Use perspective is that property owners have a fundamental right to extract the maximum economic value possible from their property, if they wish, and any limitations on this extraction must be compensated. Another difference between the two lies in their definition of "local." The Wise Use movement's definition includes the large corporate interests that provide the bulk of the movement's funding. The conservation-based development definition is that local owners are those who live locally and are primarily responsible to their neighbors, rather than to distant shareholders. The shared future of the community and local property owner provides a powerful counterforce to the incentives to disregard environmental and social costs.

Finding ways to incorporate local knowledge and information in decision-making processes, to build on economic self-interest for conservation, and to provide a more predictable regulatory climate will go a long way toward reversing the uncertainty, fear, and alienation felt in too many coastal communities. Regulatory uncertainty, however, must not be confused with ecological uncertainty. We cannot legislate the effects of weather and other catastrophic natural events out of resource-based industries. If they are to endure, resource owners must devise ways in which to maintain resilience in the face of ecological change. A more predictable regulatory climate helps to buffer the ups and downs of natural patterns only in the sense that changing regulations do not exacerbate their effects.

Restructuring Government

Much of what needs to be done to ensure that conservation-based development flourishes in the region depends on individual initiative and the private sector. As our economic, social, and political structures have become increasingly complex, government's capacity to address local economic and environmental problems has diminished. Yet we have come to expect that government has a limitless capacity and responsibility to step in and provide financial assistance in the aftermath of economic crises brought about by ecological collapse, such as the extensive fishing closures along the coast and drastic reductions in timber harvests. Government should indeed play a role in environmental and economic policy. But there are different areas where federal, state, and local governments each have distinct advantages. Rather than abandoning environmental regulation altogether, as some propose, the federal government must set standards but then allow state and local governments flexibility in the ways they achieve them (John 1994). Government policies can go a long way toward incorporating values that are invisible to the market.

Governments can foster local responsibility for problem solving by supporting local initiatives. As this is a reversal of the role many government agencies are accustomed to playing, it may be difficult for them to embrace. The Willapa Fisheries Recovery Strategy in southwestern Washington provides a good example of this shift. In an effort to address the local economic upheaval caused by fishing closures, the Willapa Alliance convened local commercial and sport fishers, as well as the state agencies responsible for fisheries management in the Willapa watershed, and is facilitating the development of a comprehensive fisheries recovery strategy for the entire basin. State agencies play a supporting role in the strategy, and one representative said that participating in an advisory rather than a leadership role is "a fairly significant change in role for fisheries managers." Nonetheless, managers see the value in "giving away some control of the process, in exchange for higher local knowledge and buy-in" (Rose 1995).

Another way in which government can aid conservation-based development is to restructure agencies along ecological rather than political or administrative lines and take a longer-term view in their work than recurrent political cycles. Washington's Department of Ecology has reorganized its staff around a few key watersheds in the state to test this approach. The state of Oregon is fostering watershed councils along the coast and elsewhere composed of local residents and government agency staff. The program is now endangered by budget cuts, but it has successfully assembled stakeholders within watershed boundaries. (See "Oregon's Watershed Health Program," pages 304–306.) The Commission on Resources and the Environment (CORE) in British Columbia sponsored a series of land use planning processes in each ecoregion of the province (CORE 1993). The Scientific Panel for Sustainable Forest Practices in Clayoquot Sound changed the tenor of the relationship between science and politics in the region by providing the first real example of the political adoption of a scientifically based recommendation for resource management. It also illustrated the growing trend toward bioregionalism. Dr. Jerry Franklin, a leading U.S. forest ecologist who played a key role in the Forest Ecosystem Management Assessment Team (FEMAT 1993), served on the Clayoquot Scientific Panel at the invitation of British Columbia's government, demonstrating the potential links between forests and forest management policy in British Columbia and the U.S. Pacific Northwest.

The tax and incentive structure in the region could be a powerful force for conservation-based development if it were realigned. Economic policy should encourage activities consistent with conservation-based development, including long-term investments, conservation, and the efficient and sustainable use of natural resources. Subsidies that foster the depletion of natural capital, overconsumption, pollution, and waste should be discontinued (Repetto et al. 1992; Ryan 1995). Creating mechanisms that provide near-term economic benefits to landowners who adopt long-term resource management practices would go a long way toward realizing this shift. For example, tax-free bonds, carbon credits, or other financial mechanisms could be used for forestland owners who adopt lengthier rotations that produce high-quality wood, reduce the number of harvests, and increase the amount of older forest on private lands (Johnson 1995).

There are already several promising initiatives at the policy level in the coastal temperate rain forest region. In perhaps the most far-reaching and significant effort, British Columbia's CORE went beyond its mandate to create and implement a land use planning strategy for the province. Indeed, it proposed an overarching Sustainability Act for British Columbia (CORE 1993). These examples offer hope for the region and illustrate ways in which a favorable policy and institutional climate can promote conservation-based development. Nonetheless, the fundamental basis of conservation-based development is that a single community which supports its economy and quality of life, while restoring and maintaining its ecosystem, will be the most powerful symbol of hope and catalyst for change.

An Exceptional Place to Begin

Clearly the coastal temperate rain forest is among the most productive bioregions in the temperate zone, if not the world. The region's location straddling the North American, Pacific, and Juan de Fuca plates, and its position at the intersection of cyclonic weather patterns and ocean currents, ensure plentiful rainfall, mild temperatures, and geologically young, productive soils. Forest trees grow like weeds, streams and estuaries teem with life, and the diverse landforms and verdant flora give rise to a multitude of terrestrial and aquatic habitats from headwaters to nearshore. The physical setting of the bioregion also drives biologically rich and diverse offshore and ocean ecosystems.

Ocean and land are linked by more than physical forces. Of the thousands of vertebrate species that use this terrestrial/marine system, salmon in particular unite ocean and land. Like salmon, First Nations peoples have been tied to the rain forest coast for thousands of years in a relationship that their more nomadic European neighbors are just begining to understand. Now these more recent arrivals must strengthen their bonds to this relatively new home.

While the bioregion has suffered significant diminution of some of its attributes, the legacy of environmental simplification need not constrain economic and cultural development. Rather, we have the opportunity to maintain and restore this inherently productive corner of the earth to support a healthy economy and vibrant, diverse communities.

The same productivity and diversity that supported First Nations peoples and attracted newcomers make this bioregion the logical birthplace of conservation-based development in which small-scale, locally adapted development fosters healthy ecosystems and healthy communities. This is the place to explore the relationship between conservation and development and to innovate.

This is the place to capture and transform the commitment of local communities to their ecosystem into innovative and appropriate development.

The coastal temperate rain forest bioregion of North America is particularly well suited to conservation-based development for a number of reasons: it is a region of high biological productivity, diversity, and resilience; it contains numerous globally rare and unique ecosystems deserving our stewardship; environmental conditions occupy the full spectrum from pristine to highly degraded; small communities have multi-generational ties to their natural resource base; and residents tend to have strong commitments to place (Ryan 1994). These conditions invite thoughtful, enduring solutions.

Yet significant barriers to long-term sustainability exist. The ideas we have outlined in this chapter are the components of a sort of "skunk works" for overcoming these barriers. They are offered, not as solutions, but as working hypotheses designed to be tested in the real world and adapted as we learn. Sustainable development to date has shown a distinct disjunction between theory and practice. We begin with a vision, point out barriers to realizing that vision, and offer methods to overcome them. Our methods may change; our principles will not.

If the development and implementation of such a vision is possible at all, it is possible here, where the coastal temperate rain forest joins two of the wealthiest countries in the world—the United States and Canada. The United States has been the world's most successful exporter of ideology, technology, products, and culture. Canada has been a global leader in promoting concern for the environment and sustainable development. These two countries have both the resources and the moral obligation to develop a way of living that will result in true prosperity—vibrant communities in healthy, productive ecosystems—for us, and for those who follow.

Acknowledgments

The authors thank Spencer B. Beebe, Alan Thein Durning, Ian Gill, and Kai N. Lee for reviews of an early draft of this chapter.

References

Cabarle, B., R. Hrubes, C. Elliot, and T. Synnott. 1995. "Certification accreditation: The need for credible claims." *J. Forestry* 93(4):12–17.

Commission on Resources and Environment (CORE). 1993. *The Provincial Land Use Strategy.* Victoria: Provincial Government of British Columbia.

Davis, F. 1995. "Information systems for conserving diversity." *BioScience* 45(3): 416–423.

Ecotrust, Pacific GIS, and Conservation International. 1995. *The Rain Forests of Home: An Atlas of People and Place,* Part 1: *Natural Forests and Native Languages of the Coastal Temperate Rain Forest.* Portland, Ore.

Forest Ecosystem Management Assessment Team (FEMAT). 1993. *Forest Ecosystem Management: An Ecological, Economic, and Social Assessment*. Portland: USDA Forest Service, Pacific Northwest Research Station.

Hawken, P. 1993. *The Ecology of Commerce: A Declaration of Sustainability*. New York: HarperCollins.

John, D. 1994. *Civic Environmentalism: Alternatives to Regulation in States and Communities.* Washington, D.C.: Congressional Quarterly Press.

Johnson, K. 1995. *Building Forest Wealth: Incentives for Biodiversity, Landowner Profitability, and Value-Added Manufacturing.* Seattle: Northwest Policy Center.

Kemmis, D. 1990. *Community and the Politics of Place*. Norman: University of Oklahoma Press.

Kusel, J., and L. Fortmann. 1991. *Well-Being in Forest-Dependent Communities.* Sacramento: Forest and Rangeland Resources Assessment Program, California Department of Forestry and Fire Protection.

Labor Market and Economic Analysis Branch, Washington State Employment Security. 1995. *Labor Force and Employment in Washington State*. Olympia, Wash.

Lee, K. N. 1993. *Compass and Gyroscope: Integrating Science and Politics for the Environment*. Washington, D.C.: Island Press.

Lemaster, D. C., and J. H. Beuter. 1989. *Community Stability in Forest-based Communities.* Beaverton, Ore.: Timber Press:

Mater, J. 1995. "Certified forest products: Building tomorrow's market today." *J. Forestry* 93(4):36–37.

M'Gonigle, M., and B. Parfitt. 1994. *Forestopia: A Practical Guide to the New Forest Economy*. Madeira Park, B.C.: Harbour Publishing.

Noss, R. 1992a. "The Wildlands Project: Land conservation strategy." *Wild Earth* (Special Issue):10–25.

———. 1992b. *A Preliminary Conservation Plan for the Oregon Coast Range*. Newport, Ore.: Coast Range Association.

Noss, R., and A. Y. Cooperrider. 1994. *Saving Nature's Legacy: Protecting and Restoring Biodiversity*. Washington, D.C.: Island Press.

Oregon Progress Board. 1992. *Oregon Benchmarks: Standards for Measuring Statewide Progress and Government Performance.* Salem: Oregon Progress Board.

Parsons, L. S. 1994. "Management of marine fisheries in Canada." *Can. Bull. Fish. and Aquat. Sci.* 225.

Peel, A. L. 1991. *The Future of Our Forests*. Victoria, B.C.: Forest Resources Commission.

Powell, J. 1994. "Juneau indicators for sustainability." Paper presented at the 1994 Western Regional Environmental Indicators Conference, Sacramento, Calif., January 6–7.

Putnam, R. D. 1993. *Making Democracy Work: New Traditions in Modern Italy.* Princeton: Princeton University Press.

Repetto, R., R. C. Dower, R. Jenkins, and J. Geoghegan. 1992. *Green Fees: How a Shift Can Work for the Environment and the Economy.* Washington, D.C.: World Resources Institute.

Rose, B. 1995. "The Willapa Alliance: An assessment of the sustainability of the organization." Unpublished report prepared for the Willapa Alliance, South Bend, Wash.

Ryan, J. 1994. *State of the Northwest.* Seattle: Northwest Environment Watch.

———. 1995. *Hazardous Handouts.* Seattle: Northwest Environment Watch.

Smith, J. 1995. "The business of good forestry." Paper presented at the Sustainable Forestry Symposium, Simon Fraser University, Vancouver, B.C., November 16.

Sustainable Community Roundtable. 1993. *State of the Community: A Sustainable Community Roundtable Report on Progress Toward a Sustainable Society in the South Puget Sound Region.* Olympia, Wash.

Sustainable Seattle. 1993. *Sustainable Seattle: Indicators of Sustainable Community.* Seattle, Wash.

Taub, R. P. 1994. *Community Capitalism.* Boston: Harvard Business School Press.

Waring, R. H. and J. F. Franklin. 1979. "Evergreen coniferous forests of the Pacific Northwest." *Science* 204:1380–1386.

Willapa Alliance and Ecotrust. 1995. "Willapa Indicators for a Sustainable Community." South Bend, Wash.

Willer, C. 1995. *Gated Lands: A Report on the Ownership of Oregon's Private Coast Range Forests.* Newport, Ore.: Coast Range Association.

Contributors

Paul Alaback is assistant professor of ecology at the University of Montana's School of Forestry in Missoula.

Kenneth Ames is a professor at Portland State University's Department of Anthropology.

Eric C. Anderson is a graduate student in the School of Fisheries at the University of Washington in Seattle.

Spencer B. Beebe is founding president of Ecotrust, a private, nonprofit conservation-based development organization with headquarters in Portland, Oregon.

Fred L. Bunnell is director of the Centre for Applied Conservation Biology at the University of British Columbia in Vancouver.

Ann C. Chan-McLeod is a faculty member of the Centre for Applied Conservation Biology at the University of British Columbia in Vancouver.

Shannon W. Davis is a consulting community planner specializing in the response of local government to natural resource management changes.

Megan Dethier is a research assistant professor at the University of Washington's Friday Harbor Laboratories in Friday Harbor, Washington.

Jerry F. Franklin is a professor of ecosystem analysis in the College of Forest Resources at the University of Washington in Seattle.

Douglas Hay is a research scientist in the Science Branch of the Canadian Department of Fisheries and Oceans, Pacific Biological Station, in Nanaimo, British Columbia.

Richard J. Hebda is head botanist and earth historian at the Royal B.C. Museum. He is also adjunct associate professor of biology in the School of Earth and Ocean Sciences at the University of Victoria in Victoria, British Columbia.

Rebecca L. Johnson is an associate professor in the Department of Forest Resources, Oregon State University in Corvallis.

Erin L. Kellogg is director of policy at Ecotrust in Portland, Oregon.

Ken Lertzman is an assistant professor in the School of Resource and Environmental Management at Simon Fraser University in Burnaby, British Columbia.

Colin Levings is a research scientist in the Science Branch of the Canadian Department of Fisheries and Oceans, West Vancouver Laboratory, in West Vancouver, British Columbia.

James A. Lichatowich is principal fishery biologist with Alder Fork Consulting in Sequim, Washington.

Kreg Lindberg, formerly at Oregon State University, is a lecturer at Charles Sturt University in Albury, Australia.

M. Patricia Marchak is dean of the Faculty of Arts and a professor in the Department of Anthropology and Sociology at the University of British Columbia.

David R. Montgomery is a geomorphologist in the Department of Geological Sciences at the University of Washington in Seattle.

Robert J. Naiman is director of the Center for Streamside Studies and professor of fisheries and forestry at the University of Washington in Seattle.

Willa Nehlsen is a fish and wildlife biologist with the U.S. Fish and Wildlife Service in Portland, Oregon.

Jim Pojar is a forest ecologist with the B.C. Forest Service in Smithers, British Columbia.

Hans D. Radtke is a freelance economist in Yachats, Oregon, specializing in the relationship between resource-based industries of the Pacific Northwest and regional economies.

Kelly Redmond is regional climatologist and deputy director of the Western Regional Climate Center at the Desert Research Institute in Reno, Nevada.

William G. Robbins is a professor of history and associate dean of the College of Liberal Arts at Oregon State University. He is a recipient of a 1994–1995 National Endowment for the Humanities fellowship.

David K. Salmon is a faculty member at the Institute of Marine Science at the University of Alaska, Fairbanks.

Peter K. Schoonmaker is vice president of Interrain Pacific in Portland, Oregon, and a research associate at Oregon State University in Corvallis.

Charles A. Simenstad is coordinator of the Wetland Ecosystems Team at the School of Fisheries at the University of Washington in Seattle.

Tom Spies is a research ecologist with the USDA Forest Service's Pacific Northwest Research Station in Corvallis, Oregon.

Wayne Suttles is emeritus professor of anthropology at Portland State University. He now resides on San Juan Island, Washington.

Fred Swanson is a research geologist and ecosystem team leader with the USDA Forest Service's Pacific Northwest Research Station in Corvallis, Oregon.

George Taylor is the state climatologist for the state of Oregon and a faculty member at the College of Oceanic and Atmospheric Sciences at Oregon State University in Corvallis.

Nancy J. Turner is an ethnobotanist and professor in the Environmental Studies Program at the University of Victoria, British Columbia. She is also adjunct professor in the University of British Columbia's Department of Botany.

Bettina von Hagen is director of the Natural Capital Fund at Ecotrust in Portland, Oregon. She is currently working on her dissertation in nontimber forest products at the School of Urban and Public Affairs and is a lecturer for the Geography Department at Portland State University.

Cathy Whitlock is a professor in the Department of Geography at the University of Oregon in Eugene.

Edward C. Wolf is director of communications at Ecotrust in Portland, Oregon.

Key Organizations

Cascadia Revolving Loan Fund
157 Yesler Way
Suite 414
Seattle, WA 98104
206-447-9226

Clatsop County Economic Development Council Fisheries Project
250 36th Street
Astoria, OR 97103
503-325-6452

Coastal Studies and Technology Center/Center for Science Education
5107 Hwy. 101 N.
Seaside, OR 97138
503-738-4021

Ecoforestry Institute
785 Barton Road
Glendale, OR 97442
503-832-2785

Ecotrust
1200 NW Front Avenue, Suite 470
Portland, OR 97209
503-277-6225
503-222-1517 fax
info@ecotrust.org
http://www.ecotrust.org

Institute of Sustainable Forestry
P.O. Box 1580
Redway, CA 95560
707-923-4719

Interrain Pacific
1200 NW Front Avenue
Suite 470-A
Portland, OR 97209
503-226-8108

Kennedy Lake Technical Working Group
Ecotrust Canada
1122 Mainland Street
Suite 420
Vancouver, BC V6B 5L1
604-682-4141

The Florence R. Kluckhohn Center for the Study of Values
119 North Commercial Street
Suite 820
Bellingham, WA 98225
360-733-5648

Mattole Restoration Council
P.O. Box 160
Petrolia, CA 95558
707-629-3514

Nanakila Institute
P.O. Box 1101
Kitamaat Village, BC V0T 2B0
604-632-3308

Oregon Trout
117 SW Front Avenue
Portland, OR 97204
503-222-9091

The Oregon Watershed Health Program (Governor's Watershed Enhancement Board)
255 Capitol Street NE
Salem, OR 97310
503-378-3589 x831

Pacific County Noxious Weed Control Board
P.O. Box 88
South Bend, WA 98586
360-875-9300 x226

Pacific Forest Trust
P.O. Box 858
Boonville, CA 95415
707-895-2090

Pacific Rivers Council
P.O. Box 10798
Eugene, OR 97440
541-345-0119

Prince William Sound Science Center
P.O. Box 705
Cordova, AK 99574
907-424-5800

Rural Development Initiatives, Inc.
P.O. Box 265
Lowell, OR 97452
541-937-8344

ShoreTrust Trading Group
P.O. Box 826
Ilwaco, WA 98624
360-642-4265

Skeena Watershed Committee
P.O. Box 1056
Prince Rupert, BC V8J 4H6

Tillamook Bay National Estuary Project
613 Commercial Street
P.O. Box 493
Garibaldi, OR 97118
503-322-2222

Willapa Alliance
P.O. Box 278
South Bend, WA 98586
360-875-5195

WoodNet
P.O. Box 1028
Port Angeles, WA 98362
360-452-2134

Index

Abiotic factors affecting coastal communities, 164
Aboriginal Affairs and Forests, British Columbia Ministries of, xv
Aboriginal burning, 76
  see also History, pre-European; Traditional ecological knowledge
Aboriginal Fisheries Strategy, 202, 203
Active management for conservation objectives, 376
Adaptive management, 178–81, 365, 377
Adaptive strategies and traditional ecological knowledge, 284–85, 287
Agee, J. K., 73, 84, 85, 110
Aging population, 332, 333
Agriculture, 333–34
Agriculture, U.S. Department of (USDA), xv
Ahearn, M. C., 346
Ahousaht GIS project, 299–301
Alaback, P. B., 74, 79, 85, 115
Alaska, 107
  biogeochron sequences, 242
  economic health of, 330
  estuarine habitats, 163
  freshwater tidal habitats, 161
  Laurentide ice sheet, 236
  runoff from coastal rain forest rivers, 138
Alaska Current, 11, 12–13
Albano, M., 169
Alder stands, 246
Aleutian Low pressure system
  California and Alaska Currents, 12–13
  El Niño–Southern Oscillation, 14
  phytoplankton and zooplankton, 17–18
  ridges, blocking, 15
  storms, 8, 11
Alexander Archipelago, 54
Algae, 163, 165, 166
Allaye Chan, A., 115
Allen, G. B., 245, 246
Alley, N. F., 235
Alluvium, 48, 50–51
Alpine glaciation, 52, 64
Alpine rocklands, 80
Alsean peoples, 257
Ames, K. M., 265
Amos, Gerald, 90
Amphibians, 108, 109, 110, 112, 113
Anadromous fish, 84, 169, 214–15
Anderson, J. L., 334
Anderson, N. H., 142
Anderson, P. M., 230
Andrews, E. D., 141
Anoxia in the water column, 163
Anthony, G., 167
Antunez de Maylo, S. E., 15
Applegate Partnership, xi
Archaeological research on the Northern coast, 264–72
Arlecho Creek basin, 95–96
Armstrong, D. A., 174
Armstrong, D. D., 174
Armstrong, J. E., 236
Armstrong, R. J., 238
Arrow, K., 340
Art traditions, pre-European, 261, 262
Athapaskan-Eyak peoples, 257
Atleo, R., 278
Atwater, B., 150
Atwater, T., 46
Ausubel, K., 208
Avalanche disturbances, 64
Avalanche tracks, 80
Avanzino, R. J., 135

Babine Lake, 202
Backus, E., 145, 299, 301
Backyard biodiversity, 208–9
Baines, G., 275
Baker, J., 113
Banfield, A. W. F., 108
Banner, A., 74, 79, 245, 246
Barbeau, M., 277
Barman, J., 319, 324
Barnes, C. A., 155
Barnes, M., 204, 206
Barnosky, C. W., 230, 238, 240, 245, 247
Bartina, V. F., 17
Bartlein, P. J., 229, 230, 236
Baskets and traditional ecological knowledge, 285
Batten, A. R., 233
Baumgartner, T., 17
Bawden, C. A., 168
Beaches
  erosion, 154

Beaches (*continued*)
reflective gravel, 164
rocky, 150, 153, 164, 165
sandy, 78, 164–65
shingle, 77–78
Beamish, R. J., 15, 17, 18, 221
Bears, 89–91, 108, 113
Beck, M., 49
Bedrock, 50–51, 56–57, 62, 100–101, 164
Beebe, Spencer B., xiv
Beechie, T. J., 132–35, 142, 174, 177
Behling, P. J., 230, 236
Bell, M. A. M., 281
Bencala, K. E., 135
Benda, L. B., 52, 132–35, 142, 174, 177
Berg, D. R., 132–35, 142, 174, 177
Berger, T. R., 275
Berkes, F., 275
Berlin, B., 288
Berris, S. N., 174
Beschta, R. L., 174, 175
Best, C., 201
Beuter, J. H., 385
Bilby, R. E., 84, 134, 142, 214
Biodiversity Guidebook, 370
Biogeochrons, 228–29, 239, 242
Biogeoclimatic zones, 229
Biogeographic patterns, 72
perhumid temperate rain forest, 74–75
seasonal rain forest, 73–74
warm temperate rain forest, 73
Biological diversity
backyard biodiversity, 208–9
ecosystem-focused conservation, 373
past models used for the future, 366–69
patterns of, 80–84
uniqueness of, 227
vertebrates, protecting, 115–16
see also Latitudinal patterns of biological diversity
Bioregions, xiii–xiv
Biotic factors structuring coastal communities, 164–65
Biotic processing and transformation, 135
Bird, N., 211
Birds
endemic to coastal temperate rain forests, 109
feeding grounds, 166–68
forest cover, 110
habitat occurrence and use, 120–27
riparian forests, 114
salmon carcasses, 214
shorebirds/waterfowl, migrating, 106–7
species richness, 107
terrestrial, 168
Willapa Bay, local science in, 302
Birkeland, P. W., 247
Bisson, P. A., 84, 110, 132–35, 142, 143, 174, 177, 214, 220
Biswell, B. L., 113
Bjornn, T. C., 213, 216
Black bears, 113
Black brant, 168
Bliss, L. C., 74
Blocking ridges, 15
Blowdowns, 110–12
Boas, F., 277, 278
Bog ecosystems, 78, 246
Bond, C. E., 53
Booth, D. B., 51
Boreal species, 76
Bormann, F. H., 45
Bosman, A. L., 167
Bottom, D. L., 163
Bouchard, R., 277, 279, 282
Bouillon, D. R., 15, 17, 18, 221
Boyd, R., 263
Boyle, D. E., 161
Boyle, E. A., 229
Branson, B. A., 53
British Columbia Coast Range, 12, 54
Aboriginal Affairs and Forests, British Columbia Ministries of, xv
biogeochron sequences, 242
birds, 107
Carnation Creek estuary, 166
cavity nesters, 112
Commission on Resources and Environment, 402, 403
Cordilleran ice sheet, 238
ecoforestry, 206
economic health of, 330
employment and mechanization, 329
estuarine habitats, 163
fisheries, 202
flooding, 140
freshwater tidal habitats, 161
herring, 173
landownership, 389–90, 399–400
Laurentide ice sheet, 236
lumber production, 324
maritime habitats, 77
Nuu-Chah-Nulth peoples, 276
Protected Areas Strategy, 397
runoff from coastal rain forest rivers, 138

species richness, 103–4, 105
transitional forests, 245
vegetation, regional, 246
wind disturbance, 110
Brodeur, R. D., 17, 20
Brooke, R. C., 74
Brown, D. A., 168
Brown, E. R., 113
Brown, M., 311
Brown, P., 339
Brown, R., 279
Brubaker, L. B., 230, 250
Bryophytes, 71
Buffington, J. M., 60
Bull, A., 339
Bullrush marshes, 161
Bunnell, F. L., 103, 104, 105, 107, 109–15
Burke, R. M., 247
Bury, R. B., 113
Buskirk, S. W., 113, 115
Butler, R. W., 168

Cabarle, B., 400
Caine, N., 45, 57, 372
California, 107, 138, 140
California Coast Range, 54
California Current, 8, 10, 12–13
California Department of Fish and Game, 93
California Forest Practices Act, 200
Callenbach, E., 313
Canopy dominants, 362, 363
Cantwell, R., 318
Canyon winds, 37
Caplan, J. A., 181
Carbon cycles, 214
Carbon storage, forest, 200–201
Carbyn, L. N., 275
Carey, A. B., 113
Carleton, J. T., 179
Carlson, B. F., 282, 285, 289
Carlson, R. L., 265
Carmody, G. R., 115
Carnation Creek, 166, 178
Carpenter, R. A., 365
Carson, R., 346
Carson, R. T., 340, 345
Cartwright, C. W., Jr., 181
Cascade Mountains, 54, 142
Cascadia Revolving Fund, 352–53
Cascadia subduction zone, 46, 48, 150
Castilla, J. C., 168
Cattle, 168–69
Cavity nesters, 112
Cayan, D. R., 17
Cedar trees, 74, 75, 81, 94–96, 111, 160, 233, 235, 249
Cederholm, C. J., 143, 214
Center for Streamside Studies, xv
Ceremonies, pre-European, 261
Certification of forest products, 194, 206
Champion International Timber Company, 99
Change, why do ecosystems, 229–30
Channels, stream, *see* Rivers and streams
Chasan, D. J., 167
Chatel, K. W., 168
Chehalis River, 160
Chelton, D. B., 150
Chimakuan peoples, 257, 258
Chinookan peoples, 257
Chinook salmon, 213, 215, 218
see also Salmon
Chitons, 165
Chlorophyll in central North Pacific, 17
Choy, D. J. L., 339
Christensen, N. L., Jr., 378
Chum salmon, 213, 216
see also Salmon
Cissel, J., 361, 365
Clague, J. J., 236, 238
Clark, N., 319
Clark, W. C., 45
Clark, William, 315–16
Clarkson, L., 275
Classification/nomenclature of cultural resources/phenomena, 288–89
Clatsop Economic Development Council's (CEDC) Salmon Program, 195–97
Claxton, E., 278
Clayoquot Sound, 108, 276, 300
Clean Water Act in 1987, 307
Clearcutting, 58, 85, 174
Clemons, S. C., 229
Cliffs, coastal, 154
Climate of the coastal temperate rain forest
cloud cover and solar radiation, 37
decision making, 39–40
Early period (11,000–5500 B.P.), 266
ecotones, terrestrial/marine, 154–55
large-scale causes of change in, 229–30
location of rain forests, 25, 26, 27
moist forest systems, 250
ocean oscillations, 7
precipitation and humidity, 27–32

Climate (*continued*)
salmon populations, 222, 223
smelts, 172
snow, 37–38
streamflow, 32–33, 34
teleconnections, 38–39
temperature, 33–35
temporal variability, 15–17
transition to modern vegetation and climate, 245–47
winds, 35–37
Clinton, Bill, 396
Cloud cover and solar radiation, 37
Coastal Studies and Technology Center (CSTC), 309–11
Coast Range Mountains, 154
Coho salmon, 213, 215–16
see also Salmon
Cole, D. L., 214
Coleman, Elizabeth, xv
Collins, D., 276
Collinson, M. E., 233
Colluvium, 58
Colorado, P., 276
Columbia River, 33
discharge, annual, 155
ecotones, altering terrestrial/marine, 174–75
pollutants in, 177
postglacial connections, paucity of, 133
salmon, 213
tidal fluctuations, 160
turbidity maxima, 163
Columbus Day storm of 1962, 36
Commission on Resources and Environment (CORE), 276, 402, 403
Communication and exchange of traditional ecological knowledge, 277, 288–92
Community coalitions, x, 358, 359–60
see also Conservation-based development
Comprehensive Employment and Training Act (CETA), 195
Compton, B. D., 282, 289
Concepts in action, *see* Conservation-based development; Forestry and fisheries initiatives; Learning and decision making; Restoring and managing ecosystems
Coney, P., 49
Conflict resolution in multiparty comanagement settings, 203–4
Conifers, Pacific Northwest, 82–85, 232
Connaughey, R. A., 174
Connectivity among populations, forest practices disrupting, 115, 219–20
see also Human activities and land-shaping processes
Conservation-based development, xiii, 5, 383
barriers to sustainability, 388, 404
civic organizations, 397–98
economic development integrated with, 391–92
entrepreneurial capacity, 389, 398–99
financing small business on the Olympic Peninsula, 352–53
government involvement, 390–91, 401–3
information, local access to, 388
landownership, concentration of, 389–90, 399–400
organizational capacity, 388–89, 397–98
policymakers and, 387
principles of, 393
regulatory uncertainty, 390, 400–401
revitalization of community, 386
rural development initiatives, 357–60
ShoreTrust Trading Group, 354–57
understanding, beginning with, 394–97
vegetation from ridgetop to seashore, 85–86
WoodNet, 349–51
Conservation easements, 198–201
Consumer and products service sectors, 337–38
Contingent valuation (CV), 339–40
Convergent plate margins, 46, 48
Cooperrider, A. Y., 82
Coosan peoples, 257
Cordell, J. R., 179
Cordgrass, eastern Atlantic smooth, 179
Cordilleran ice sheet, 51, 229, 238, 241
Corn, P. S., 113
Corporate domination of timberland, 324, 326, 386
Coupland, G., 214, 268
Cox, A. L., 49
Cox, R., 316
Crabs, 174
Cretaceous and tertiary granitic batholiths, 48
Crime and tourism, 342–43, 345
Croes, D. R., 269
Cronon, W., 323
Crotts, J C., 344
Crown Pacific Timber Company, 95, 96
Crown Zellerbach Corporation, 195

Crows, 168
Culture/economy and environment, links between, 6, 263–64
Cummins, K. W., 138, 142
Currents
  Alaska Current, 11, 12–13
  California Current, 8, 10, 12–13
  Kuroshio Current, 8, 10
  North Equatorial Current, 8
  North Pacific Current, 8, 10, 11
  spring transition, 27
  winter current, 27
Cushing, C. E., 138
Cwynar, L. C., 243
Cybulski, J. S., 256, 269

Daly, C., 31
Danks, H. V., 141
Daust, D. K., 109, 112–14
Davis, A., 284
Davis, F., 395
Davis, R. E., 150
Davis, S. D., 247
Davis, S. W., 334, 336
Davis, W. M., 44
Dayton, P. K., 165
Dead and decaying organic matter, 166, 167
Debris flows, 60, 61, 100, 134, 163–64
Deciduous forest cover, 64
Decision making, weather/climate and, 39–40
  see also Learning and decision making
Deer, black-tailed, 113
DeFerrari, R. J., 135
Delgam Uukw, 280
del Moral, R., 175
Deltas, 150, 152, 160, 175
Demographic changes, *see* Oregon Coast, economic/demographic transition on
DeRiso, R. B., 18
Dethier, M. N., 153, 164
Detrital decomposition, 166, 177
Devil's club, 71
Dewberry, Thomas, 102
Dietrich, W. E., 52, 58
Diking, 175, 176
Dillard, A., 321
Dillman, D., 340
Dissection, fine-scale, 52
Disturbances, natural
  Aboriginal burning, 76
  avalanches, 64
  biodiversity, 83–84
  debris into estuarine habitats, 163–64
  as a dominant feature of the coastal temperate rain forest, 110–12
  historical contingency, 375–76
  plant and animal population dynamics, 53
  process domains, general, 64–65
  rivers and streams, 135
  seasonal rain forests, 74
  tectonic settings, 49
  understanding, 5
  winds, 36
  see also Ecosystem dynamics to ecosystem management
Diversity, *see* Biological diversity
Dodimead, A. J. F., 10
Doubleday, N. C., 275
Douglas-fir, 74, 76, 200, 235
Dower, R. C., 403
Downwelling, coastal, 12
Driver, B. L., 339, 340
Drought-tolerant species, 73, 84
Drucker, P., 258
Dryness, C. T., 73
Duffy, A., 229
Duncan, S. H., 174
Dunne, T., 138
Dupuis, L. A., 110, 112–15
Durning, A. T., 275
Duxbury, A. C., 155
Dynamic regions, rain forests as, 5
Dyrness, C. T., 53, 230, 233
Dysfunctional landscapes, dynamics of recovery in, 377

Earthquakes, 49, 58, 150
Easements, conservation, 198–201
Eastern North Pacific, oceanography of
  climatic changes, 7
  El Niño–Southern Oscillation, 13–17
  fisheries, 17–20
  overview of the region, 8, 9–10, 11–13
  weather systems connected to terrestrial environments, 19, 20, 21
Ebbesmeyer, C. C., 33
Ecoforestry, 204–6
Ecological integrity, 365–66
Ecological systems, *see* Ecosystem dynamics to ecosystem management; Geological processes influencing ecological systems; Restoring and managing ecosystems
Economic culture, xi, 313–14, 326–27
  see also Oregon Coast, economic/demographic transition on the

Economic development integrated with conservation-based development, 391–92, 393
Economy/environment and culture, links between, 6, 263–64
Ecosystem dynamics to ecosystem management
  disturbance regimes, understanding, 361
  dominant themes of ecosystem management, 364–66
  ecosystem dynamics, 362, 363, 364
  ecosystem-focused conservation, 373–74
  historical dynamics, knowledge of, 369–71
  past models used for the future, 366–69
  problems, key, 374–78
  time and, 228–29
  trajectory of forest over time, focus on, 371
  whole landscapes, 372–73, 376–77
Ecotones, terrestrial/marine, 76, 149
  adapting management to natural variability, 178–81
  altering, understanding consequences of, 174–77
  climate, 154–55
  estuarine habitats, 161, 162, 163–64
  exposed coast habitats, 164–65
  foraging habitats and food web interactions, 165–69
  freshwater tidal habitats, 157, 160–61
  geology, 150, 154
  gradients and habitats, ecological, 157, 158–59, 160–61, 162, 163–65
  life history transitions, 169, 170–71, 172–74
  nature of, 150, 152–53
  rivers and their inputs, 155–56
  tidal ranges, 156–57, 158–59
  uncertainty, coexisting with, 181–82
Ecotrust, xiii, 115, 189, 354
Educating the public, 211, 309–11, 386
Eelgrass, Japanese, 179–80
Eel River, 144
Efrat, B. S., 289, 290
Elevation
  habitats at high, 79–80
  plant communities, 53
  precipitation, 30–31
Elitism and ethnocentrism in western science, 275
Elliot, C., 400
Elliott, W. P., 31
El Niño–Southern Oscillation
  El Niño phase, 13–14
  La Niña phase, 14–15
  phytoplankton and zooplankton, 17–20
  teleconnections, 38–39
  temporal variability, 15–17
  terrestrial environments, modifying, 21
Emerson, G. H., 322
Emerson, George, 321–22
Emery, W. J., 18
Emigrants' Guide to Oregon and California (Hastings), 318
Emminghamn, W. H., 367, 369
Employment issues
  consumer and producer service sectors, 337–38
  mechanization, 329
  natural-resource-based industries, 332, 333
  tourism, 342, 343
Endangered Species Act of 1973, 214
Endemism to the rain forest, 108, 109, 110
Energy transference between ocean/atmosphere and landmasses
  El Niño–Southern Oscillation and Pacific mackerel, 14
  precipitation and elevation, 31
  shorelines, 84
  types of, 7
  weather systems, 19, 20, 21
Entrapment zones in estuaries, 166
Entrepreneurial capacity, 389, 398–99
Environmental Studies Program, xv
Environment/economy and culture, links between, 6, 263–64
Epiphytes, 71
Equator-to-pole temperature difference, 27
Ericaceae family, 71–72
Erman, D. C., 141
Erosion, 44–45
ESSA Technologies, 189
Estate taxes and older forests, 198
Estes, J. A., 167
Estuarine habitats, 175
  ecotones, terrestrial/marine, 161, 162, 163–64
  exotic species, 179
  rearing, variations in, 169
  restoration of, 176
  shorebirds/waterfowl, migrating, 106
  Tillamook Bay Estuary Project, 306–8
Ethnobotanical studies, 247
Ethnocentrism and elitism in western science, 275

Eulachons, 172–73
European settlement, legacy of, 248, 385
  agricultural model, 205
  boundary between U.S. and England expanded, 317–18
  corporate domination of timberland, 324, 326
  economic culture, 313–14, 326–27
  Lewis and Clark, 315–16
  logging operations, beginning of, 319, 320, 321
  magazines fueling speculation/expansion, 323–24
  railroads, 322–23
  steam power, 321–22
  technological innovations, 326
  see also History, pre-European
Eutrophication, 163
Evolution, 228
Excess Salmon to Spawning Requirements (ESSR), 203
Exotic species
  disproportionate effect of, 86
  geological age/disturbances and vulnerability to, 178–80
  mammals, 168–69
  riparian forests, 133, 135, 144
  spartina in Willapa Bay, 96–99
Exposed coast habitats, 164–65
Exxon Valdez oil spill, 210
Eyak-Athapaskan peoples, 257

Farming, ix
Farr, J. A., 373
Feare, C. J., 167
Fens, 78
Fenton, M. B., 115
Ferguson, T. A., 280
Ferns, 71, 72
Fetherston, K. L., 134
Ficken, R. E., 317
Fire
  aboriginal burning, 76
  cycles, 110
  diversity, species, 84
  perhumid temperate rain forest, 74–75
  seasonal rain forests, 74
First Nations in Canada, 201–4, 247–48, 279, 318, 383
Fir trees, 74–75, 76, 233, 235, 236, 241, 246–47
Fish
  anadromous, 84, 169, 214–15
  endangered species, 207
  exotic species, 179
  fisheries, 17–20, 329, 334–35
  freshwater tidal habitats, 160
  income, personal, 333
  life history transitions, 169, 170–71, 172–74
  ocean-type, 218, 219
  overfishing, 190–91, 221
  sticklebacks, 191
  stream-type, 218, 219
  see also Salmon
Fish and Wildlife, Oregon Department of, 308
Fisher, 113
Fisheries, 17–20, 329, 334–35
Fisheries and Oceans, Canada's Department of, 189, 202
Fjords, 150, 153, 157
Fladmark, K. R., 265
Flats, 100
Flexible manufacturing, 349–51
Flooding, 62, 63, 140–41, 160
Florence R. Kluckhorn Center for the Study of Values, 96
Flounder, 161
Fluharty, D. L., 14
Folke, C., 275
Fonda, R. W., 74
Foods and traditional ecological knowledge, 283
Food web interactions and foraging habitats, 165–69
Ford, R. I., 276
Forest cover, 64, 109
Forest Ecosystem Management Assessment Team (FEMAT), 403
Forest ecosystem networks (FENs), 373–74
Forestry and fisheries initiatives
  assessment, forest, 205
  Clatsop Economic Development Council's Salmon Program, 195–97
  ecoforestry, 204–6
  heritage stocks, 207–9
  Institute for Sustainable Forestry, 192–94
  Kennedy Lake technical working group, 189–92
  Pacific Forest Trust, 197–201
  Prince William Sound Science Center, 209–11
  Skenna Watershed Committee, 201–4
  see also Trees
Forest Service, U.S. (USFS), xv
Forests Forever Fund, 200
Forest Stewardship Network, 194
Forest/vertebrate relations, 109

Forman, R. T. T., 114, 371
Fortmann, L., 386, 389, 390
Fossil records, 250
Fragmentation of forest habitats, 85–86
Francis, R. C., 222
Frank, P. W., 167
Franklin, J. F., 53, 73, 82, 104, 115, 230, 233, 365, 367, 371–73, 376, 403
Franklin, Jerry E., xi
Fransen, B. R., 84, 214
Fraser Glaciation, 236, 238, 239–40
Fraser River, 133, 150, 152
Frederick, S. G., 247
Fredin, R. A., 217
Freeman, M. R., 275
French, D., 288
Fresh, K. L., 169
Freshwater tidal habitats, 155, 157, 160–61
Fujimori, T., 104
Funeral rituals, pre-European, 268–70
Furbish, C. E., 169
Future, past as key to the, 248, 250, 366–69, 386

Gadgil, M., 275
Gard, R., 141
Garreau, J., 313
Garton, E. O., 113
Geller, J. B., 179
Geoghegan, J., 403
Geographic information systems (GIS), 180, 299–301
Geological processes influencing ecological systems
  confronting the issues, 66
  ecotones, terrestrial/marine, 150, 154
  geology/geomorphology and ecology, 44–45
  geomorphic provinces, 50–53, 54
  landscape elements, 58–60, 61, 62, 63, 64
  lithotopo units, 43
  plate margins and regional tectonic setting, 45–46, 47, 48–50
  process domains, general, 64–65
  watersheds and lithotopo units, 53, 55–56
  see also Latitudinal patterns of biological diversity
Geomorphic provinces, 50–53, 54
Gill, I., 189, 192
GIS, *see* Geographic information systems
Gisday Wa, 280
Gitksan communities, 201, 202–3
Glaciation, 229
  freshwater tidal habitats, 161
  history, 51–53
  primary responsibility for physical features, 150
  refugia, 81, 241
  unglaciated Oregon coast, 52, 53
  see also Preglacial forests
Glacier Bay, 81
Glantz, M. H., 13, 20, 21
Global and extraregional effects, 6
Global positioning system (GPS), 300
Global warming, 250
Goering, J. J., 20
Gold rush, California, 319
Good, D. A., 53
Gore, J. A., 140
Gorges, 50
Gottesfeld, A. J., 248
Gottesfeld, L. M. J., 248
Gough, B. M., 315
Government and conservation-based development, 390–91, 401–3
Gradients, habitats and ecological, 157, 158–59, 160–61, 162, 163–65, 167
  see also Latitudinal patterns of biological diversity
Granite, 50
Grant, G. E., 56, 60, 62, 174
Grasses and sedges, maritime, 78
Graumlich, L. J., 230
Grays Harbor, 150, 152
Greenhouse gases, 16
Gregory, S. V., 110, 142, 145, 220
Grizzly bears, 89–91, 108
Groot, C., 134, 169, 216
Grossman, P., 353
Grove, L. E., 142
Grumbine, R. E., 364, 365
Guetter, P. J., 230, 236
Gulf of Alaska, 8, 11, 12, 17–20
Gulls, 167
Gunderson, D. R., 174
Gustafsson, L., 86
Gwaganad, 279
Gwaii Haanas reserve, 293

Habitats
  bears, 113
  bird species, 120–27
  estuarine, 161, 162, 163–64, 169
  exposed coast, 164–65
  food web interactions and foraging, 165–69
  forest cover and ubiquitous water, 109

fragmentation of forest, 85–86
freshwater tidal, 157, 160–61
geological processes, 64
gradients and, ecological, 157, 158–59, 160–61, 162, 163–65, 167
high-elevation, 79–80
hyporheic zone, 62, 135, 142, 144
mammals, 128–30
maritime, 77–78
micro, 84
older forests needed for, 113
salmon, 218–19, 220
Hack, J. T., 44
Hagenstein, R. H., 145
Hagenstein, Randall, xv
Haida Gwaii, 81, 241, 257, 293
Haida peoples, 257, 258, 289, 293
Haisla peoples, xv, 2, 89–91
Haiyupis, R., 287
Halibut, 18, 19
Hall, J. D., 220
Hallingback, T., 86
Halupka, K. C., 374
Hamilton, E., 80
Hamilton, K., 18
Hamilton, P., 155, 177
Hansen, A. J., 367, 368
Hardt, M. M., 113
Harestad, A. S., 113
Harmon, M., 367
Harper, P. P., 142, 144
Harr, R. D., 174
Harris, C. C., 340
Harrison, S. P., 229, 230, 236
Harry, E., 282
Hartman, D. L., 11
Harvey, R. B., 155, 177
Hastings, L. W., 318
Hausman, J. A., 340
Hauth, N., 357
Hauth, Nancy, xv
Hawken, P., 399
Hawkins, C. P., 142
Hay, D. E., 161, 167
Hayward, T. L., 17
Healey, M. C., 143, 169, 218
Healey, M. J., 167
Health and social characteristics of the coastal population, 335, 337
Heaths and heathers, 71–72, 79–80
Hebda, R. J., 74, 79, 231, 236, 238, 241, 243, 244, 245–48, 250, 269
Hedgpeth, J. W., 179
Heiltsuk, peoples, 262
Hemlock trees, 81, 233, 235, 236
Hemstrom, M. A., 230
Henderson, J. A., 74
Henderson Lake, 28
Herbicides, 98
Herbs, 71–72, 76, 243
Heritage stocks idea, 207–9
Herring, 18, 173–74
Hesquiaht peoples, 290
Heusser, C. J., 233, 235, 236, 238, 241, 243, 245, 246
Heusser, L. E., 235
Hicks, B. J., 220
Hicock, S. R., 235, 238
Hidden Northwest, The (Cantwell), 318
Higley, D. L., 163
Hill, J., 197
Hillslopes, 57–58, 62, 100
Hill-Tout, C., 279
Hinrichsen, D., 156
Hirano, T., 10
History, environmental, 227
change, why do ecosystems, 229–30
coastal temperate rain forest area, 233, 234, 235
ecosystems and time, 228–29
knowledge of historical dynamics needed, 369–71
past as key to the future, 248, 250, 366–69, 386
people and the forest, 247–48, 249
studying forest history, 230, 231, 232–33
see also Preglacial forests
History, pre-European, 255
biomass estimates, 383
culture and environment, relationship between, 263–64
early period (11,000–5500 B.P.), 265
Kennedy Lake system, 189–90
linguistics, 256–59
Pacific period (5500 B.P. to Contact), 266–72
population statistics, 262–63
social systems, 259, 261–62, 263, 264
Hockey, P. A. R., 167
Holland, M. M., 149
Holland, S. M., 344
Holling, C. S., 377
Hollows, 58, 59, 62, 100
Holm, B., 261
Holocene era, warm and dry early, 240, 242, 243, 244, 245
Holtby, L. B., 143, 178
Hood Canal, 150, 152, 157
Horses, 168–69

Horton, S. P., 113
Houston, D. B., 214
Howard, D. R., 339
Howard, W. R., 229
Hoyt, E., 108
Hrubes, R., 400
Hsieh, W. W., 15, 18
Hudson's Bay Company, 317
Hughes, R. M., 53
Human activities and land-shaping processes, 44, 45
  ecotones, terrestrial/marine, 174–77
  exotic species, 178–79
  fragmented watersheds, 220
  freshwater tidal habitats, 160
  hillslopes, 58
  hollows, 58
  Knowles Creek, 100
  rain-shadow forests, 76–77
  rivers and streams, 144
  shaping activities to permit ecosystem survival, 182
  stream channels, 60
  vegetation from ridgetop to seashore, 85
Hume, Stephen, 90–91
Hunn, E. N., 288
Hunter, M., 114
Hunting, commercial, 89–91
Huryn, A. D., 64
Huston, M. A., 82
Hyatt, K., 189, 192
Hyporheic zone, 62, 135, 142, 144

Ice and aquatic organisms, 144
  see also Glaciation
Iida, T., 57
Imbrie, J., 229
Income, nonwage, 335
Income, personal, 332–35, 336
Indigenous peoples, 275
  see also History, pre-European; Traditional ecological knowledge
Information, local access to, 388
Inselberg, A., 368
Institute for Sustainable Forestry, 192–94
Integration of local/global and environment/people, xiii, 6
Interdisciplinary approach, x, 180
Internal Revenue Service, 199
Interrain Pacific, xv, 300
Invertebrates, subterranean, 114
Investment income, 335
Iris, J., 192–94
Iris, P., 192–94
Irons, D. B., 167
Isaacs, J. D., 17

Jacobsen, T., 204, 206
Jay, D. A., 155, 177
Jenkins, R., 403
Jennings, F., 275
Jewell Praying Wolf James (tse-Sealth), 95
John, D., 402
Johnson, A. C., 134
Johnson, R., 340, 341, 343–45
Jones, D. L., 49
Jones, J. A., 361, 365
Jones, J. L., 113
Jones, K. K., 163
Juday, G., 74, 75, 79
Jungwirth, Lynn, xi

Karr, J. R., 221
Kasischke, E., 378
Kates, R. W., 45
Katz, R. W., 13
Kawasaki, T., 17
Kemmis, D., 397
Kempe, S., 155
Kennedy, D., 277, 279, 282
Kennedy, V. C., 135
Kennedy Lake technical working group, 189–92
Kessler, W. B., 181
King Range National Conservation Area, 93
Kinkade, M. D., 257
Kinship organizations, 261–62
Kirchoff, M. D., 113
Kistritz, R. U., 176
Kitimaat peoples, 257
Kitlope ecosystem, bears in, 89–91
Klamath Mountains, 54
Klenner, L. L., 109, 112–14
Kline, T. C., 20
Knowledge, shared, xi, 277, 288–92
Knowles Creek, 99–102
Knudtson, P., 275
Koch, T., 38
Kodiak Island, 81
Kostopulos, G., 350
Kotok, E. I., 327
Krajina, V. J., 73, 74, 229
Kranz, T. K., 45, 372
Kremsater, L. L., 103, 104, 105, 115
Kroeber, A. L., 256

Kuhnlein, H. V., 281, 282
Kukla, G., 229
Kunz, K., 152
Kuroshio Current, 8, 10
Kusel, J. L., 386, 389, 390
Kutzbach, J. E., 229, 230, 236
Kwagiulh peoples, 282
Kwakiutl peoples, 262

Lajoie, K. R., 150
Lammerz, U., 155
Landownership, concentration of, 389, 399–400
Lands, indigenous peoples identifying with ancestral, 279
Landscape elements, 56
 floodplains, 62, 63
 hillslopes, 57–58
 hollows, 58, 59
 patterns, landscape, 62, 64
 stream channels, 60, 61, 62
Landscape-scale management based on scientific knowledge, 181
Langer, O. E., 176
La Niña, 14–15, 17
 see also El Niño–Southern Oscillation
Lankford, S. V., 339
Large ecosystems, structure and dynamics of, 376
Larson, E., 330
Late-successional ecosystems, 375
Latitudinal patterns of biological diversity
 ecological implications, 142–44
 physical environment, 136, 137, 138, 139–42
 rivers and streams, 131–32
 vegetation from ridgetop to seashore, 81–82
Laurel Mountain (Oregon), 38
Laurentide ice sheet, 229, 236
Lawson, P. W., 221
Leamer, E., 340
Lean, G., 156
Learning and decision making
 Coastal Studies and Technology Center, 309–11
 culturally appropriate, 290–92
 Tillamook Bay Estuary Project, 306–8
 vernacular knowledge, validating, 299–301
 Watershed Health Program, Oregon's, 304–6
 Willapa Bay, local science in, 301–4
Lebovitz, Wendy S., 99
Ledyard, J., 314
Ledyard, John, 314
Lee, K. N., 397
Lemaster, D. C., 385
Leopold, L. B., 60, 138
Lertzman, K. P., 368
Lesher, R. D., 74
Levings, C. D., 161, 168, 176
Levy, D., 152
Lewis, H. T., 280
Lewis, Meriwether, 315–16
Lewis, Randall, xv
Lichatowich, J. A., 133, 207, 217, 218, 220, 222
Lichens, 71, 74
Lienkaemper, G. W., 60
Life history transitions, 169, 170–71, 172–74
Likens, G. E., 45
Lindberg, K., 340, 341, 343–45
Linden, E., 275
Lindsey, C. C., 133, 161
Linguistics and pre-European history, 256–59
Lisle, T. E., 60
Lithology, 43–45
Little Port Walter, 29
Liverworts, 71
Living, The (Dillard), 321
Location of coastal mid-latitude rain forests, 25, 26, 27, 70, 151, 234, 384
Lock, P. A., 134
Logging
 beginning of, 319, 320, 321
 clearcutting, 58, 85, 174
 at an equilibrium, 385
 modernization, ix, x
 old-growth, 93, 95–96
 see also Trees
Long, C., 248
Loons, 110
Lowry, Mike, 98
Lummi Indians, 94–96
Lunar nodal tidal cycle, 18, 19
Lyngbyei's sedge, 161

MacArthur, J. W., 104
MacArthur, R. H., 104
MacDonald, L. H., 132–35, 142, 174, 177
MacKay, D., 319, 324
Mackerel, 14
MacKinnon, A., 71
MacMillan, A., 267

MacMillan Bloedel, 189
Mader, H. J., 115
Magazines fueling speculation/expansion, 323–24
Maine, N., 311
Major, J., 73
Makah peoples, 262
Malone, C. R., 365
Mammals
  distribution of, 107
  endemic to coastal temperate rain forest, 109
  exotic species, 168–69
  forest cover, 110
  habitat use and occurrence, 128–30
  riparian areas, 114
  salmon carcasses, 214
Management of resources
  active management for conservation objectives, 376
  adaptive management, 365, 377
  hatchery practices, 221
  landscape-scale management based on scientific knowledge, 181
  matrix management, 372–73
  rivers and streams, 144–45
  spartina, integrated weed management approach to control, 97, 98
  variability, adapting management to natural, 178–81
  vertebrates, terrestrial, 113–15
  see also Ecosystem dynamics to ecosystem management; Restoring and managing ecosystems
Manis Mastodon archeological site, 265
Manufacturing, flexible, 349–51
Manufacturing-sector employment, 337, 338
Marchak, M. Patricia, xi
Margolis, Ken, 91
Margolis, L., 134, 169, 216
Marine and terrestrial environments linked, *see* Energy transference between ocean/atmosphere and landmasses; Rivers and streams
Markets for nontimber products, 198
Markham, A., 156
Marks, B., 376, 377
Marks, D., 131
Marriages, pre-European, 264
Marsh, C. P., 167
Marshes, 78, 150, 154, 161, 175–77
Marten, 113
Martin, J. H., 21
Maser, C., 84, 106, 110, 134, 163, 164
Mater, J., 400
Mathewes, R. W., 236, 241, 243, 245–48, 269
Mathews, J. T., 45
Mathisen, O. A., 20
Matrilineal kinship organization, 261–62
Matrix management, 372–73
Matsen, B., 215
Matson, R. G., 214, 265
Mattole watershed, 91–94
McAllister, M. K., 377
McCann, R. K., 109, 112–14
McComb, W. C., 367, 369
McCrory, Wayne, 89–90
McFarlane, G. A., 17
McGowan, J. A., 17, 20
McIlwraith, T., 277, 279, 291
McIntire, C. D., 163
McKay, P. J., 164
McKee, A., 72, 142, 367
McLaughlin, W. J., 340
McLoughlin, John, 317
McMinn, B., 206
McMinn, N., 206
McPhail, J. D., 133, 143, 161
Meadows, coastal mountain, 80
Mechanization and decline in wood product jobs, 329
Medicines and materials, gathering of, 282, 283–84
Meehan, W. R., 213, 216
Megahan, W. F., 174
Mehlman, D., 373
Meidinger, D., 80, 104, 106, 112, 233
Meinig, D. W., 317
Mendocino triple junction, 48
Merk, F., 317
Mesozoic accreted terrains, 48
Meyer, W. B., 45
M'Gonigle, M., 390, 400
Microtopographical variations, 134
Miller, J. P., 60
Mink, 168
Minshall, G. W., 138
Mitchell, A. P., 145
Mitchell, Andy, xv
Mitchell, R. C., 340, 345
Mobrand, L. E., 218, 220
Moisture-favoring species, 245–46, 250
Mollusks, 106, 168
Monetizing forest services, 199–201
Monitoring objectives, 206, 369, 395
Montane forests, 74
Montgomery, D. R., 52, 56, 58, 60, 64, 141

Moody, S. P., 319
Moore, Bob, xv
Moore, C., 290
Moore, P. D., 233
Morgan, C. A., 179
Morrison, P., 230, 369
Morrissette, V., 275
Morse, B. A., 155
Mosses, 71, 74, 76, 84
Moulton, G. M., 316
Mountain Grove Demonstration Forest, 206
Mountain hemlock zone, 81, 113
Mountain ranges, 44, 64, 104
Mouse, Sitka, 108
Muellner, Marko, xv
Munger, T., 321
Murphy, M. L., 142
Murphy, P. E., 344
Murray, C. B., 143
Muskeg, 79
Mussels, 165
Mutual Life Insurance Company of New York, 95
Mysak, L. A., 15, 18

Na-Dene Phylum, 257
Naiman, R. J., 132–35, 138, 142, 174, 177
Nakashima, D. J., 275
National Estuary Program, 307–8
National Marine Fisheries Service (NMFS), 310
Natural disturbance type (NDT), 370
Natural-resource-based industries/income, 332, 333, 334
Nature Conservancy, 96
Navarrete, S. A., 168
Navarro, P., 346
Nav'oot'en Band, 201
Neal, V. T., 15
Nehlsen, W., 133, 207, 217, 221
Neilson, R. P., 31, 131
New Bearings compass, xiii
Newell, R., 17
Newsom, D., 323
Newson, M., 56
Nicholls, N., 13
Nicholson, A., 80
Nielsen, J. L., 142, 143
Nielsen, R. P., 144
Nitrogen cycles, 214
Nitrogen input into estuarine habitats, 163
Noble fir, 74
Nonforested vegetation, 77–80
Nooksack River, 94
North, M. E. A., 176
North Equatorial Current, 8
Northern Cascades, 54
Northern lowlands and islands subregion, 136
Northern mainland mountains subregion, 136
North Pacific Current, 8, 10, 11
North Pacific High, 36
North Pacific subtropical anticyclones, 8
Northwest coast, defining the, 256
Northwest Economic Adjustment Initiative, 338
Norton, B. G., 181
Noss, R. F., 82, 397
Novato, A., 17
Nuthatches, 112
Nutrient cycling, 20, 84, 163, 214
Nuu-Chah-Nulth peoples, 262, 276, 278, 287
Nuxalk peoples, 262, 291–92

Oak trees, 246, 247
Obrebski, S., 179
Oceans
  life history, 218–20
  oceanic forcing, 156–57, 158–59
  oscillations, 7
  productivity cycles, 221–22, 223
  see also Eastern North Pacific, oceanography of the
O'Connor, M. D., 132–35, 142, 174, 177
Ogilvie, R. T., 82, 282, 285, 289
Ohmann, J. L., 367, 368
Okunishi, K., 57
Old-growth forests, 93, 95–96, 113, 330
O'Loughlin, C. L., 58, 174
Olson, P. L., 132–35, 142, 174, 177
Olsson, T. I., 142, 144
Olympia Nonglacial Interval, 236
Olympic Mountains, 54, 136, 154
Olympic Peninsula
  draining the western, 155
  exotic species, 135
  financing small business on the, 352–53
  lithotopo units, 55, 56
  process domains of watersheds, 65
  rocky beaches, 150, 153
  soil-forming processes, 243
  transitional forests, 246
  WoodNet, 349–51
Omernik, J. M., 50
Oregon, 52, 53, 107

Oregon Coast, economic/demographic transition on the, 329
  demographic and economic changes, 330, 332–35, 336, 337–38
  tourism's social impacts, 339–46
Oregon Coast Range, 54
Oregon's Department of Fish and Wildlife (ODFW), 195
Oregon Trout, 208
Organizational capacity, 388–89, 397–98
Orographic influences on precipitation, 31
Oswald, E. T., 74, 79
Otters, 168, 169
Our Common Future (Brundtland Report), 275
Overland, J. E., 8, 11
Oxley, D. J., 115
Oysters, 97, 179, 356–57

Pacific Certification Council, 206
Pacific Certified Ecological Forest Products program, 194
Pacific County Spartina Task Force, 97–98
Pacific Fishery Management Council, 334
Pacific Forest Trust (PFT), 197–201
Pacific Monthly, 323
Pacific Northwest Regional Council (PNRC), 195
Pacific Northwest Research Station, xv
Pacific Ocean, *see* Eastern North Pacific, oceanography of the
Pacific period (5500 B.P. to Contact), 266–72
Pacific Rivers Council, 99
Pacific silver fir, 74, 75
Paine, R. T., 165
Painting/sculpture, pre-European, 261
Paleoecological studies, 228
Paleoenvironmental reconstructions, 238
Palmson, R. A., 142
Palynology, 232–33
Parfitt, B., 390, 400
Parker, K. S., 18
Parker, P. L., 20
Parsons, T. R., 15, 18, 168
Past as key to the future, 248, 250, 366–69, 386
Pastor, J., 135
Paul, A., 281
Paul, P., 279
Pearce, A. J., 58
Pearcy, W. G., 10
Pearl, C., 248
Pemberton Lillooet peoples, 278–79
Penutian Phylum, 257
People and the forest, 247–48, 249
  see also Human activities and land-shaping processes
Perhumid temperate rain forest, 74–75
Perry, D., 367
Pess, G., 60
Peteet, D. M., 241, 245, 246
Peter, D. H., 74
Peterman, R. M., 377
Peterson, E. B., 74
Peterson, G., 339
Peterson, N. P., 143
Philander, S. G., 13
Phillips, D. L., 31
Physical energy, transference of, 7
Phytoplankton, 14, 17–20
Pickett, S. T. A., 362
Pierce, A. J., 174
Pinchot, Gifford, 324
Pine trees, 76, 79
  see also Fir trees
Pinkerton, E., 204
Pink salmon, 216
  see also Salmon
Plank houses, pre-European, 259, 260
Plant communities and elevation gradients, 53
  see also Vegetation from ridgetop to seashore
Plate margins and regional tectonic setting, 45–47, 48–50
Pleistocene ice sheet, 133, 150, 161
Poe, P. H., 20
Poff, N. L., 140
Point of Arches, 150, 153
Pojar, J., 71, 74, 79, 80, 104, 106, 112, 233, 245, 246
Policies, weather/climate and effective, 40
Policymakers and conservation-based development, 387
Pollen records, 231–32, 236, 240, 244
Pollock, M. M., 134
Pollutants and estuarine organisms, 177
Poole, G. C., 135
Pope, F., 319
Population issues
  aging, 332, 333
  dynamics involved, 49–50
  growth on the coast, 338
  pre-European demographics, 262–63
Portney, P., 340
Port of Astoria, 195–96

Posey, D. A., 276
Postma, H., 162
Potlatch ceremony, 261
Powell, J., 387, 395
Powell, R. A., 113, 115
Pratt, J., 360
Precipitation and humidity, 27–32, 155
Predation, 165, 191
Prefilters, freshwater tidal habitats as, 160
Preglacial forests
  Fraser Glaciation, 236, 238, 239–40
  Holocene era, warm and dry early, 240, 242, 243, 244, 245
  reforestation after ice, 241, 242, 243
  Wisconsin Glaciation, 235–36, 237
Primack, R. B., 114
Prince of Wales Islands, 81
Prince's-pine, 76
Prince William Sound Science Center, 209–11
PRISM model for measuring precipitation, 31
Private-sector capital and forest stewardship, 198–201, 338
Probst, A., 354
Process domains, general, 64–65
Producer and consumer service sectors, 337–38
Productivity
  biodiversity and ecosystem, 82–83
  of marine and estuarine components,4
  ocean cycles, 221–22, 223
Propagules, 165
Province-level influences on ecosystems, 50–53, 54
Public education, 211, 309–11, 386–87
Puget/Fraser Lowland, 54, 62, 77
Puget Sound, 150, 153
Putnam, R. D., 397

Quality of life, 335
Quammen, M. L., 168
Quaternary alluvium, 48
Queen Charlotte Islands, 46, 54, 241, 257, 293
Quinault River, 155
Quincy Library Group, xi
Quinn, T. P., 220
Quinn, W. H., 15

Raccoons, 168
Radiative energy, 7
Radiocarbon dating, 230
Radner, R., 340
Radtke, H. D., 334, 335, 336
Raedeke, K. J., 113, 114
Rahr, G. R., III, 209
Railroads, 322–23
Rain-shadow forests, 31, 76
Rajala, R., 319, 324
Raphael, M. G., 115
Recruitment (biotic process), 165, 191
Red cedar–hemlock phase, 111
Redmond, K., 38
Redwoods, 198
Reed, J., 134
Reed, R. K., 11, 31
Reeves, G. H., 220
Reforestation after ice, 241, 242, 243
Regallet, G., 275
Regulatory uncertainty, 390, 400–401
Reid, L. M., 143, 174
Reimer, P., 232
Reneau, S. L., 58
Repetto, R., 403
Reproductive failures and toxic compounds, 177
Reptiles, 109–10
Resh, V. H., 140
Resource management/use, 5
  see also Management of resources; Restoring and managing ecosystems
Restoring and managing ecosystems, 182, 392
  bears in Kitlope ecosystem, 89–91
  estuarine habitats, 176
  Knowles Creek, 99–102
  Lummi Indians, 94–96
  Mattole watershed, 91–94
  spartina marsh plant, 96–99
  Watershed Health Program, Oregon's, 304–6
Rexstad, E., 53
Reynolds, M., 326
Richards, J. F., 45
Richardson, H., 168
Ricker, W. E., 133
Ridges, blocking, 15
Riparian forests
  anadromous fish contributing nutrients to, 84
  defining, 75–76
  exotic species, 133, 135, 144
  floodplain areas, 163
  human activities and land-shaping processes, 144
  species richness, 113
  vertebrates, terrestrial, 114
  watershed diversity, 133

Risser, P. G., 149
Rivers and streams
channels, water movements in subsurface, 133
complexity of, reduced, 220
geological and climatic gradients, 131–32
latitudinal gradients in the physical environment, 136, 137, 138, 139–44
management/restoration and conservation, 144–45
Mattole Watershed, 91–94
ocean type of life history, 218–20
overview of the Pacific coastal ecoregion, 133–36
runoff from coastal rain forest rivers, 138, 139, 140, 155–56, 174
size of rivers, 142
Skenna Watershed Committee, 201–4
spawning channels, 202
stream channels, 60, 61, 62
streamflow, 32–33, 34, 138, 139, 140
volcanoes, 49
Watershed Health Program, Oregon's, 304–6
watersheds and lithotopo units, 53, 55–56
see also Ecotones, terrestrial/marine; Estuarine habitats; Salmon; *specific rivers*
Roads and road construction, 115, 174
Robbins, W. G., 318, 324
Rock types, 47, 48, 50–51
Rock uplift, 44
Rocky beaches, 77, 150, 153, 164, 165
Rodents, 113, 168
Roemer, H. L., 81
Roemmich, D., 17, 20
Rogue River valley, 318
Roosevelt, Theodore, 324
Root-diggers, 281
Rose, B., 402
Rosenberg, K. V., 115
Ross, A., 316
Ross, Alexander, 316
Rouse, G. E., 243, 245, 246
Royer, T. C., 11, 18
Rural development initiatives (RDI), 357–60
Russell, D., 308
Russo, Kurt, 96
Ryan, J., 387, 403
Saanich peoples, 282
Salamanders, 53, 108, 110
Salinity, 155, 164, 165
Salishan peoples, 258
Salmon, 11
anadromous, 214–15
Carnation Creek, 178
Clatsop Economic Development Council's Salmon Program, 195–97
crisis facing salmon industry, 334
declining stocks, 330
distribution and adaptation, 142–43
estuaries, 106
fisheries reflecting history of forests, 385
foraging habitats, 166
freshwater tidal habitats, 161
habitat complexity, 220
harpoon used in hunting, 286
heritage stocks idea, 207–9
Kennedy Lake technical working group, 189–92
Knowles Creek, 99–102
life history, 169, 170–71, 172–74, 217–20
Mattole watershed, 91–94
nutrients introduced into terrestrial system by, 20
sockeye, 18
and spirituality, 223–24
status and trends, 220–23
temperature of coastal rain forests, 213–14
traditional ecological knowledge, 280
variable physical environment, 134
water temperature, 143
zoom, 216
Salmon, D. K., 15–18
Salo, E. O., 143, 169
Salt marshes, 150, 154, 161
Salwasser, H., 181
Sampson, T., 282
Samuels, S. R., 271
San Andreas Fault Zone, 46
Sandpipers, Western, 168
Sandstone, Tyee, 100
Sandstones, marine, 50
Sandy beaches, 78, 164–65
San Francisco, 319
Sangamon Interglaciation, 235
Sapir, E., 258
Saprophytes, 72
Saunders, S., 368

Scanlan, R. S., 20
Scarlett, W., 143, 214
Schalk, R. F., 217
Schmidt, K., 60
Schmidt, K. M., 58
Schneider, D., 168
Schoen, J. W., 113
Schofield, W. B., 71
Schoonmaker, P. K., 72, 189, 192
Schowalter, T. D., 368
Schramm, G., 56
Schultes, R. E., 275
Schumacher, J. D., 11
Schuman, H., 340
Science
  Ahousaht GIS project, 299–301
  Coastal Studies and Technology Center, 309–11
  elitism and ethnocentrism in western, 275
  interdisciplinary, 180
  Science of Sound, 211
  technological innovations, 326
  Willapa Bay, local science in, 301–4
Scientific Panel for Sustainable Forest Practices in Clayoquot Sound, xi, 276, 368–69, 377, 402–3
Scoters, 168
Scrivener, J. C., 175, 178
Sculpture/painting, pre-European, 261
Seabirds, 167
Seal predation, 191
Seasonal rain forest, 73–74
Seastars, 165
Sedell, J. R., 84, 106, 110, 134, 138, 142, 163, 164, 220
Sedges and grasses, maritime, 78
Sediment
  ecotones, terrestrial/marine, 154
  erosion, 156
  estuarine habitats, 163
  exposed coast habitats, 164
  floodplains, 62
  hillslopes, 57
  history, studying forest, 230, 231
  hollows, 58
  propagules, 165
  stream channels, 60
Seed Savers Exchange, 208
Seegrist, D. W., 141
Selin, R., 230, 236
Service sector employment, 337–38
Sewid-Smith, D., 282, 288
Shad, American, 179, 180
Shared knowledge, xi, 277, 288–92
Shaw, D. C., 74
Shell middens, 266, 267, 268
Sherwood, C. R., 155, 177
Shi, Y.-B., 174
Shingle beaches, 77–78
Shorebank Corporation, 354
Shorebirds, 167–68
ShoreTrust Trading Group, 354–57
Shrews, 113
Shrubs, 111, 243
Sibley, T. H., 222
Sidle, R. C., 58, 174
Sigman, M., 306, 308
Siltstones, expansive, 50
Silvicultural practices, 367–69
Simberloff, D., 373
Simenstad, C. A., 153, 155, 163, 169, 177, 179
Simpson, George, 317
Sinclair, R. E., 36
Sitka mouse, 108
Sitka spruce, 75, 77, 160
Siuslaw National Forest, 99, 100
Siwallace, M., 280, 290
Skenna Watershed Committee, 201–4
Slide alder, 80
Slugs, 53
Small, L. F., 163
Smelts, 172–73
Smith, J., 400
Smith, R., 60
Snow, 37–38, 155
Snyder, Sid, 98
Social and health characteristics of the coastal population, 335, 337
Social impacts of tourism, 339–46
Social institutions of Northwest Coast peoples, 287–88
Social systems, pre-European, 259, 261–62, 263, 264
Sockeye salmon, 191, 213, 216
  see also Salmon
Soil-forming processes, 56–58, 243
Solow, R., 340
Soscia, M. L., 306
Sound Ecosystem Assessment (SEA), 210
Soutar, A., 17
Southern Cascades, 54
Southern coastal mountains subregion, 136
Southwood, T. R. E., 45
Spartina Coordinating Action Group (SCOAG), 98

Spartina marsh plant, 96–99, 179
Spaulding, W. G., 229, 230, 236
Spawning channels, 202
Species-focused conservation, 373
Species richness in Pacific Northwest, 103–4, 105, 106–7, 113
Speculation/expansion in Northwest Coast (1850–1920), 323
Sphagnum wetlands, 246
Spies, T. A., 115, 367–69, 371, 376
Spirituality and salmon, 214, 223–24
Sporopollenin, 232
Spraying, controlled, 98
Spring transition, 27
Spruce trees, 75, 77, 160, 233, 235
Squamish Nation, xv
Stamp, E., 319
Stanford, J. A., 135, 141–43
Statzner, B., 140
Steam power, 321–22
Stearly, R. F., 50
Steel, E. A., 132–35, 142, 174, 177
Steelhead salmon, 213, 216
  see also Salmon
Steeplands, 80
Stephen, E. J., 286
Stewart, H., 285
Stewart, W. N., 235
Stickleback–sockeye salmon interactions, 191
Stikine River, 175
Stocks, B., 378
Stone, L., 207
Storms
  Aleutian Low pressure system, 8, 11
  Columbus Day storm of 1962, 36
  ecotones, terrestrial/marine, 154
  Gulf of Alaska, 20
  heat transported from the ocean, 12
  Kuroshio Current region, 8
  ridges, blocking, 15
  typical behavior of, 35–36
Stream channels, 60, 61, 62
Streamflow, 32–33, 34, 138, 139, 140
Strickland, R., 167
Stuiver, M., 232
Subalpine forests/species, 74, 76
Subduction zones, 46, 48, 150
Subregions in the coastal temperate rain forest, physiographic, 131, 132, 136
Subsidence, associated large-scale, 27
Substrate type, 164
Succession process, 228
Sullivan, K., 56, 143
Summers, R. W., 167
Supreme Court of Canada, 202
Sustainability
  barriers to, 388, 404
  traditional ecological knowledge and, 277, 280–82, 283–85, 286, 287–88
  WoodNet, 349–51
Sutherland, G., 368
Suttles, W., 247, 256, 264, 287
Suzuki, D., 275
Swanson, F. J., 45, 60, 62, 110, 142, 174, 230, 361, 365, 367–69, 372, 376, 377
Synnott, T., 400

Takelman peoples, 257
Talbot, W., 319
Tanaka, S., 17
Tangborn, W., 33
Taniguchi, A., 17
Tatoosh Island, 37, 156–57
Taub, R. P., 394
Taxes and older forests, 198, 199
Taylor, B., 377
Taylor, E. B., 133, 143, 218, 219
Teaching and learning, culturally appropriate ways of, 290–92
Technological innovations, 326
Tectonic processes, 44–47, 48–50
Teleconnections, 38–39
Temperature, 33–35
  changes in, 16
  river flow, 155–56
  rivers, biotic diversity of, 143–44
  salmon, 213–14
  species distributions, 141–42
Temporal variability, 15–17
Terich, T. A., 164
Terrestrial and marine environments linked, *see* Energy transference between ocean/atmosphere and landmasses
Teversham, J. M., 176
Themes around coastal temperate rain forests, common, 5–6
Thom, R. M., 176
Thomas, D. W., 160
Thomas, G. L., 20
Thomas, J., 113, 282, 285, 289
Thomas, T., 115, 351, 371, 376
Thompson, I. D., 113
Thompson, L. C., 257
Thompson, R. S., 229, 230, 236
Thomson, R. E., 155, 156
Thorson, R. M., 238
Thursby, F. P., 315

Tidal marshes, 78
Tidal ranges, 156–57, 158–59, 160
Tillamook Bay Estuary Project, 306–8
Tilman, D., 83
Timber industry
  beginning of, 319, 320, 321
  clearcutting, 58, 85, 174
  at an equilibrium, 385
  modernization of, ix, x
  old-growth logging, 93, 95–96
  see also Trees
Timberline, 79
Time and ecosystems, 228–29
Tla-o-qui-aht Nation, 189–90, 192
Tlingit peoples, 257
Toba, Y., 17
Tools, woodworking, 247–48, 269, 270–71
Total design method, 340
Tourism, 333, 334, 339–46
Toutle River, 49
Toxic compounds draining from developed landscapes, 177
Trade winds, northeast, 8
Traditional ecological knowledge
  characteristics of, 276, 277
  communication and exchange of, 277, 288–92
  in the future, 292–93
  growing movement toward recognizing, 275–76
  sustainable living, strategies for, 277, 280–82, 283–85, 286, 287–88
  worldview, 277–80
Traffic and tourism, 342, 345
Trajectory of forest over time, focus on, 371
Transitional forests, 245–47
Transitions, life history, 169, 170–71, 172–74
Transverse plate margins, 46, 48
Treaty of Point Elliott in 1855, 95
Trees
  aggregate timber harvest from North Pacific slope (1950–1994), 325
  blowdowns, 110–12
  canopy dominants, 362, 363
  climatic gradients, 233, 235
  corporate domination of timberland, 324, 326
  freshwater tidal habitats, 160
  high-elevation habitats, 79
  Holocene era, warm and dry early, 243
  Institute for Sustainable Forestry, 192–94
  latitudinal patterns of biological diversity, 81
  long-lived, 104
  Lummi Indians, 94–96
  nontimber forest products, 198
  Pacific Forest Trust, 197–201
  perhumid temperate rain forest, 74–75
  productivity, biodiversity and ecosystem, 83
  rain-shadow forests, 76
  reforestation after ice, 241, 242, 243
  riparian forests, 75–76
  seasonal rain forest, 73–74
  silvicultural practices, 367–68
  traditional ecological knowledge and gathering materials from, 282
  trajectory of forest over time, focus on, 371
  transitional forests, 245–47
  warm temperate rain forest, 73
  wetlands, 79
  see also Forestry and fisheries initiatives
Trenberth, K., 16
Triska, F. J., 135
Trout, 161, 213, 216
Trowbridge, R., 74, 79
Tschaplinski, P. J., 166
Tsimshian peoples, 201, 257, 258
Tsukada, M., 241
Turbidity maxima, 163
Turner, B. L., 45
Turner, N. J., 273, 280–82, 284, 285, 288–90
Tyee sandstone, 100
Tyler, M. E., 275

Umhoefer, P. J., 48
Uncertainty, coexisting with, 181–82, 378
Understory plants, 71, 83, 85, 111, 160
Upwelling, coastal, 12, 14, 16, 27
Urbanization, 175, 176
Valleys
  drowned river, 150, 152
  Rogue River, 318
  unchanneled, 58, 59
  u-shaped, 64
  v-shaped, 62
Vancouver, George, 314–15
Vancouver Island, 54, 108, 154, 247, 300
Vannote, R. L., 138
van Zyll de Jong, C. G., 108
Variability/variations
  adapting management to natural, 178–81

Variability/variations (*continued*)
  estuarine habitats, rearing in, 169
  microtopographical, 134
  precipitation, 31–32
  salmon distribution, 217–18
  temporal, 15–17
Vascular plants, 71, 77, 81, 82
Vaughn, Jack H., xv
Veblen, T. T., 85
Vegetation from ridgetop to seashore
  biogeographic patterns, 72–75
  classification/nomenclature of, 288–89
  climatic gradients, 233, 235
  conservation implications, 85–86
  diversity, elements of, 69, 71–72
  estuarine habitats, 161
  exotic species, 86, 96–99, 133, 135, 144, 178–80
  histories of selected regions, 237
  Holocene era, warm and dry early, 243
  latitudinal patterns of biological diversity, 81–82
  moisture, range in, 104
  nonforested vegetation, 77–80
  patterns of biological diversity, 80–84
  recruitment (biotic process), 165
  spartina marsh plant, 96–99
  special forest types, 75–77
  traditional ecological knowledge, 280–81, 282, 283–84
  transition to modern vegetation and climate, 245–46
  understory plants, 71, 83, 85, 111, 160
  variable physical environment, 134
  Wisconsin Glaciation, 235
Venrick, E. L., 17
Verbeek, N. A. M., 168
Vermeer, K., 168
Vermeij, G., 179
Vernacular knowledge, validating, 299–301
Vertebrates, aquatic, 114
Vertebrates, terrestrial
  coastal temperate rain forests, 106–8, 109
  forest/vertebrate relations, 109–13
  management implications, 113–15
  protecting the diversity, 115–16
  species richness, 103–4, 105, 106
Vervier, P., 135
Viereck, L. A., 233
Villages, pre-European, 259
Volcanoes, 49, 50
Vole, Coronation Island, 108
Voyages (Cook), 314
Wainman, N., 245, 246
Waitt, R. B., 238
Wakashan peoples, 257, 258
Wake, D. B., 53
Walker, B., 374
Wallace, J. B., 64
Wallin, D. O., 361, 365, 376, 377
Wallmo, O. C., 113
Walters, C. J., 377
Wantz, J. W., 36
Ward, J. V., 135, 140–43
Ware, D. M., 17, 20, 167
Waring, R. H., 73, 82, 104
Warm temperate rain forest, 73
Washington, 107
Washington Conservation Corps, 98
Watershed Institute, xi
Watersheds, 53, 55–56, 138, 177
  see also Rivers and streams
Water temperatures, 141, 143–44, 155
Watson, A. F., 175
Wave exposure, 164
Wavey, R., 275
Weare, B., 17
Weathering processes, 45
Weather systems connected to terrestrial environments, 19, 20, 21
Weaving styles, pre-European, 261
Webb, J. A., 233
Weed management approach to control spartina, 97, 98
Welsh, H. H., Jr., 115
Wenzlick, K. J., 233
West, S., 134
Western hemlock trees, 81
Western redcedar, 74, 160
West Shore, 323
West Wind Drift, 8, 10
Wetlands, 52, 78–79, 160, 175–77, 246
Whealy, K., 208
Whistler Conference Center, xv
White, P. S., 362
White, R. J., 221
Whitehouse, T. R., 161
Whiting fishery, Pacific, 334
Whitlock, C., 229, 230, 236, 238, 241, 243, 246, 250
Whitlock, P. J., 230
Whittaker, R. H., 81
Whole landscapes, 372–73, 376–77
Wild Iris Forestry, 193–94
Wildwood Forest, 206
Wilimovsky, N. J., 167
Willamette Valley, 54, 62

Willapa Bay
  exotic species, 179
  health of the, key indicators for, 395–96
  science in, local, 301–4
  ShoreTrust Trading Group, 354–57
  spartina, stopping growth of, 96–99
  Willapa Alliance, xi, 98
Willer, C., 389
Williams, J. E., 133, 217
Williams, N. M., 275
Williams, R. G., 104
Willson, M. F., 104, 374
Wilson, C. J., 58
Wilson, J. G., 8, 11
Wilson, J. T., 136
Wilson, W. H., Jr., 168
Winds
  blowdowns, 110–12
  canyon, 37
  disturbance agent, 36
  Eastern North Pacific, overview of, 8, 9–10, 11–12
  ecotones, terrestrial/marine, 155
  storm behavior, typical, 35–36
  surface currents driven by, 27
  trade, northeast, 8
  windthrow, 84
Winter current, 27
Wintergreen family, 72
Wisconsin Glaciation, 235–36, 237
Wissmar, R. C., 134
Wolman, M. G., 60
Woodmansee, R. G., 45, 372
WoodNet (sustainability initiative), 349–51
Woodpeckers, 112
Woodworking tools, 247–48, 269, 270–71
Woody debris, 60, 61, 100, 134, 163–64
Wooster, W. S., 14
Wootton, J. T., 167
Worldview, 277–80
World Wildlife Fund, 206
Worona, M. A., 236, 238, 241, 243, 246
Worster, D., 313, 314

Yanev, K. P., 50
Yellow-cedar, 74, 75
Yoder-Williams, M., 141
Younger Dryas in Europe, 243
Youngs Bay terminal fishery, 196

Zellweger, G. W., 135
Ziemer, R. R., 174
Zipperer, V. T., 179
Zooplankton, 11, 12, 14, 17–19, 163
Zuckerman, S., 194